Designing Crops for Added Value

AGRONOMY

A Series of Monographs

The American Society of Agronomy and Academic Press published the first six books in this series. The General Editor of Monographs 1 to 6 was A. G. Norman. They are available through Academic Press, Inc., 111 Fifth Avenue, New York, NY 10003.

1. C. EDMUND MARSHALL: The Colloid Chemical of the Silicate Minerals. 1949
2. BYRON T. SHAW, *Editor*: Soil Physical Conditions and Plant Growth. 1952
3. K. D. JACOB, *Editor*: Fertilizer Technology and Resources in the United States. 1953
4. W. H. PIERRE and A. G. NORMAN, *Editors*: Soil and Fertilizer Phosphate in Crop Nutrition. 1953.
5. GEORGE F. SPRAGUE, *Editor*: Corn and Corn Improvement. 1955
6. J. LEVITT: The Hardiness of Plants. 1956

The Monographs published since 1957 are available from the American Society of Agronomy, 677 S. Segoe Road, Madison,WI 53711.

7. JAMES N. LUTHIN, *Editor*: Drainage of Agricultural Lands. 1957
8. FRANKLIN A. COFFMAN, Editor: Oats and Oat Improvement. 1961
9. A. KLUTE, *Editor*: Methods of Soil Analysis. 1986
   Part 1—Physical and Mineralogical Methods. Second Edition.
   A. L. PAGE, R.H. MILLER, and D. R. KEENEY, *Editors*: Methods of Soil Analysis. 1982
   Part 2—Chemical and Microbiological Properties. Second Edition.
10. W. V. BARTHOLOMEW and F. E. CLARK, *Editors*: Soil Nitrogen. 1965
    (Out of print; replaced by no. 22)
11. R. M. HAGAN, H. R. HAISE, and T. W. EDMINSTER, *Editors*: Irrigation of Agricultural Lands. 1967
12. FRED ADAMS, *Editor*: Soil Acidity and Liming. Second Edition. 1984
13. E. G. HEYNE, *Editor*: Wheat and Wheat Improvement. Second Edition. 1987
14. A. A. HANSON and F. V. JUSKA, *Editors*: Turfgrass Science. 1969
15. CLARENCE H. HANSON, *Editor*: Alfalfa Science and Technology. 1972
16. J. R. WILCOX, *Editor*: Soybeans: Improvement, Production, and Uses. Second Edition. 1987
17. JAN VAN SCHILFGAARDE, *Editor*: Drainage for Agriculture. 1974
18. G. F. SPRAGUE and J. W. DUDLEY, *Editors*: Corn and Corn Improvement. Third Edition. 1988
19. JACK F. CARTER, *Editor*: Sunflower Science and Technology. 1978
20. ROBERT C. BUCKNER and L. P. BUSH, *Editors*: Tall Fescue. 1979
21. M. T. BEATTY, G. W. PETERSEN, and L. D. SWINDALE, *Editors*: Planning the Uses and Management of Land. 1979
22. F. J. STEVENSON, *Editor*: Nitrogen in Agricultural Soils. 1982
23. H. E. DREGNE and W. O. WILLIS, *Editors*: Dryland Agriculture. 1983
24. R. J. KOHEL and C. F. LEWIS, *Editors*: Cotton. 1984
25. N. L. TAYLOR, *Editor*: Clover Science and Technology. 1985
26. D. C. RASMUSSON, *Editor*: Barley. 1985
27. M. A. TABATABAI, *Editor*: Sulfur in Agriculture. 1986
28. R. A. OLSON and K. J. FREY, *Editors*: Nutritional Quality of Cereal Grains: Genetic and Agronomic Improvement. 1987
29. A. A. HANSON, D. K. BARNES, and R. R. HILL, JR., *Editors*: Alfalfa and Alfalfa Improvement. 1988
30. B. A. STEWART and D. R. NIELSEN, *Editors*: Irrigation of Agricultural Crops. 1990
31. JOHN HANKS and J. T. RITCHIE, *Editors*: Modeling Plant and Soil Systems. 1991
32. D. V. WADDINGTON, R. N. CARROW, and R. C. SHEARMAN, *Editors*: Turfgrass. 1992
33. H. G. MARSHALL and M. E. SORRELLS, *Editors*: Oat Science and Technology. 1992
34. L. E. MOSER, D. R. BUXTON, and M. D. CASLER, *Editors*: Cool-Season Forage Grasses. 1996
35. ALBERT A. SCHNEITER, *Editor*: Sunflower Technology and Production. 1997
36. K. R. BARKER, G. A. PEDERSON, and G. L. WINDHAM, *Editors:* Plant and Nematode Interactions. 1998
37. D. C. ADRIANO, R. C. SIMS, J -M. BOLLAG, and W. T. FRANKENBERGER, JR., *Editors:* Bioremediation of Contaminated Soils. 1999
38. R. W. SKAGGS and JAN VAN SCHILFGAARDE, *Editors:* Agricultural Drainage. 1999
39. R. B. BROWN, J. H. HUDDLESTON, and J. L. ANDERSON, *Editors:* Managing Soils in an Urban Environment. 2000
40. C. F. MURPHY and D. M. PETERSON, *Editors:* Designing Crops for Added Value. 2000
41. R. I. BARNHISEL, W. L. DANIELS, and R. G. DARMODY, *Editors:* Reclamation of Drastically Disturbed Lands. 2000

Designing Crops for Added Value

Editors
C. F. Murphy and D. M. Peterson

Editorial Committee
C. F. Murphy, chair
D. M. Peterson
H. R. Boerma
A. A. Rosielle
H. W. Ohm

Managing Editor: J. M. Bartels

Editor-in-Chief ASA Publications: J. L. Hatfield

Editor-in-Chief CSSA Publications: Jeffrey J. Volenec

Editor-in-Chief SSSA Publications: Jerry M. Bigham

Number 40 in the series
AGRONOMY

American Society of Agronomy, Inc.
Crop Science Society of America, Inc.
Soil Science Society of America, Inc.
Madison, Wisconsin, USA
Publishers
Madison, Wisconsin, USA
2000

American Society of Agronomy, Inc.
Crop Science Society of America, Inc.
Soil Science Society of America, Inc.
677 South Segoe Road, Madison, WI 53711 USA

Library of Congress Catalog Card Number: 00 133615

Printed in the United States of America

CONTENTS

FOREWORD

Adding value to crops that historically have been generic commodities has been a vision of geneticists, breeders, producers, and processors of plant products for decades. The tools of molecular biology have now given the geneticists and breeders the ability to create designer plants with unique properties to enhance their feed, food, fiber and medicinal values or industrial uses.

Adding value is more than adding yield or other agronomic traits of interest primarily to the producer. Adding value is meeting a need, perceived or unperceived, of a customer who is willing to pay more to receive a uniform and distinct cultivar with assurance of more oil, more protein, more malt, better amino acid composition, unique fatty acid composition, longer shelf life, better color, new uses, etc.

Designing Crops for Added Value could just as easily have been entitled "Research and Breeding for the Future." The authors of the various chapters reflect a diversity of expertise outside the crop science and agronomy areas. They reflect a level of cooperation and the multidisciplinary effort perceived by many to be the research mode of the future. This is a forward-looking publication that will be useful to practicing breeders, as well as to students, advanced undergraduates and graduate students. The students will find embedded in the chapters of this publication a guide to many of the career specializations for crop scientists and agronomists in the twenty-first century.

On behalf of the American Society of Agronomy, a thanks is due to the authors and editors for their time and efforts to bring this publication to fruition.

Vernon B. Cardwell, *president*, ASA

PREFACE

Dramatic advances in molecular biology are now allowing plant breeders and geneticists to set their sights on more and more ambitious goals. The primary focus of plant improvement, for most of the past century, has been the necessity of defending plants from biotic and abiotic stresses. Also, specific market demands for some crops, e.g., malting barley, wheat for milling, cotton, and peanuts, forced breeders to meet standardized quality standards. These paradigms have begun to change, however, and it was in anticipation of the rapid acceleration of these changes that this monograph was conceived and developed.

While the need to protect plants and assure optimal production (yield) will continue, we foresee future plant genetic improvement making dramatic strides toward the *design* of crops to meet specific feed, food, and industrial use needs. Among food crops, we anticipate designer combinations of nutritional and health beneficial traits. And we anticipate a rapidly growing need for, and use of, identity preservation. We also believe these opportunities will offer more excitement and satisfaction for plant geneticists, increased profits for producers and processors, and much more desirable products for consumers.

The coverage of the nine chapters is quite diverse, ranging from economic considerations of identity preservation to the biochemistry of starch synthesis. The various topics reflect the expertise and interests of the individual authors, but the common thread is the design, production, and marketing of new cultivars with specific traits. The emphasis is on field crops, although similar considerations apply to vegetable and fruit crops.

This volume is the product of a team effort; dependent upon the cooperation, trust, and dedication of our editorial committee, authors, and reviewers (and with Society support). We express our sincere appreciation to all of them. The senior editor also wishes to express special thanks to the junior editor, who so willingly and ably assumed lead responsibility for the project when necessitated by a health situation.

Co-Editors
Charles F. Murphy
National Program Staff, Agricultural Research Service, USDA, Beltsville, Maryland

David M. Peterson
Cereal Crops Research, Agricultural Research Service, USDA, and Department of Agronomy, University of Wisconsin-Madison

CONTRIBUTORS

P. Stephen Baenziger	Eugene W. Price Professor, Department of Agronomy, 330 Keim Hall, University of Nebraska, Lincoln, NE 68583-0915
William A. Berzonsky	Assistant Professor, Department of Plant Sciences, North Dakota State University, Loftsgard Hall, P.O. Box 5051, Fargo, ND 58105-5051
R.F. Bruns	General Manager, AgriPro Wheat, Berthoud, CO 80513
Gary L. Cromwell	Professor, Department of Animal Sciences, University of Kentucky, Lexington, KY 40546.
Ian B. Edwards	Chief Executive Officer, Grain BioTech Australia Pty Ltd., 10 Whipple St., Balcatta, Western Australia 6021
Kenneth Eskins	Research Leader (deceased), Biomaterials Processsing Research, National Center for Agricultural Utilization Research, Agricultural Research Service, USDA, 1815 N. University Street, Peoria, IL 61604
George F. Fanta	Research Chemist, Biomaterials Processing Research and Plant Polymer Research, National Center for Agricultural Utilization Research, Agricultural Research Service, USDA, 1815 N. University Street, Peoria, IL 61604
Frederick C. Felker	Plant Physiologist, Biomaterials Processing Research, National Center for Agricultural Utilization Research, Agricultural Research Service, USDA, 1815 N. University St., Peoria, IL 61604
Bruce R. Hamaker	Professor, Department of Food Science, Purdue University, W. Lafayette, IN 47907
L. Curtis Hannah	Professor, Horticultural Sciences Department, University of Florida, P.O. Box 110690, Gainesville, FL 32611
William D. Hitz	Senior Research Scientist, Dupont Experimental Station, P.O. Box 80402, Wilmington, DE 19880-0402
Anthony J. Kinney	Senior Research Scientist, Dupont Experimental Station, P.O. Box 80402, Wilmington, DE 19880-0402
Roger A. Kleese	Vice President, Director of Research and Development, Optimum Quality Grains, L.L.C.; current address is 6700 80th Ave. N, Brooklyn Park, MN 55445
April C. Mason	Professor, Department of Foods and Nutrition, Purdue University, West Lafayette, IN 47907
William R. Meredith, Jr.	Research Leader, Jamie Whitten Delta States Research Center, Agricultural Research Service, USDA, P.O. Box. 345, Stoneville, MS 38776
Amit Mitra	Associate Professor, Center for Biotechnology and Department of Plant Pathology, 406 Plant Sciences Hall, University of Nebraska, Lincoln, NE 68583-0722

Charles F. Murphy — Senior National Program Leader, Grain Crops, National Program Staff, Agricultural Research Service, USDA; current address is 7829 Wilton Crescent, University Park, FL 34201

Herbert W. Ohm — Professor, Department of Agronomy, Purdue University, West Lafayette, IN 47907

David M. Peterson — Research Leader, Cereal Crops Research, Agricultural Research Service, USDA, 501 Walnut St., Madison, WI 53705 and Professor, Department of Agronomy, University of Wisconsin-Madison

Richard K. Perrin — Jim Roberts Professor, Department of Agricultural Economics, 102 H.C. Filley Hall, University of Nebraska, Lincoln, NE 68583-0922

Rollin G. Sears — AgriPro Wheat, 12115 Tully Hill Road, Junction City, KS 66441

Mary Ann L. Smith — Professor, Department of Natural Resources and Environmental Sciences, University of Illinois, Urbana, IL 61801

Connie M. Weaver — Department Head and Professor, Department of Foods and Nutrition, Purdue University, 1264 Stone Hall, West Lafayette, IN 47907-1264

Pamela J. White — Professor, Department of Food Science and Human Nutrition, Iowa State University, Ames IA 50011

Conversion Factors for SI and non-SI Units

Conversion Factors for SI and non-SI Units

To convert Column 1 into Column 2, multiply by	Column 1 SI Unit	Column 2 non-SI Units	To convert Column 2 into Column 1, multiply by
		Length	
0.621	kilometer, km (10^3 m)	mile, mi	1.609
1.094	meter, m	yard, yd	0.914
3.28	meter, m	foot, ft	0.304
1.0	micrometer, μm (10^{-6} m)	micron, μ	1.0
3.94×10^{-2}	millimeter, mm (10^{-3} m)	inch, in	25.4
10	nanometer, nm (10^{-9} m)	Angstrom, Å	0.1
		Area	
2.47	hectare, ha	acre	0.405
247	square kilometer, km^2 (10^3 m)^2	acre	4.05×10^{-3}
0.386	square kilometer, km^2 (10^3 m)^2	square mile, mi^2	2.590
2.47×10^{-4}	square meter, m^2	acre	4.05×10^{3}
10.76	square meter, m^2	square foot, ft^2	9.29×10^{-2}
1.55×10^{-3}	square millimeter, mm^2 (10^{-3} m)^2	square inch, in^2	645
		Volume	
9.73×10^{-3}	cubic meter, m^3	acre-inch	102.8
35.3	cubic meter, m^3	cubic foot, ft^3	2.83×10^{-2}
6.10×10^{4}	cubic meter, m^3	cubic inch, in^3	1.64×10^{-5}
2.84×10^{-2}	liter, L (10^{-3} m^3)	bushel, bu	35.24
1.057	liter, L (10^{-3} m^3)	quart (liquid), qt	0.946
3.53×10^{-2}	liter, L (10^{-3} m^3)	cubic foot, ft^3	28.3
0.265	liter, L (10^{-3} m^3)	gallon	3.78
33.78	liter, L (10^{-3} m^3)	ounce (fluid), oz	2.96×10^{-2}
2.11	liter, L (10^{-3} m^3)	pint (fluid), pt	0.473

		Mass	
2.20×10^{-3}	gram, g (10^{-3} kg)	pound, lb	454
3.52×10^{-2}	gram, g (10^{-3} kg)	ounce (avdp), oz	28.4
2.205	kilogram, kg	pound, lb	0.454
0.01	kilogram, kg	quintal (metric), q	100
1.10×10^{-3}	kilogram, kg	ton (2000 lb), ton	907
1.102	megagram, Mg (tonne)	ton (U.S.), ton	0.907
1.102	tonne, t	ton (U.S.), ton	0.907
		Yield and Rate	
0.893	kilogram per hectare, kg ha^{-1}	pound per acre, lb $acre^{-1}$	1.12
7.77×10^{-2}	kilogram per cubic meter, kg m^{-3}	pound per bushel, lb bu^{-1}	12.87
1.49×10^{-2}	kilogram per hectare, kg ha^{-1}	bushel per acre, 60 lb	67.19
1.59×10^{-2}	kilogram per hectare, kg ha^{-1}	bushel per acre, 56 lb	62.71
1.86×10^{-2}	kilogram per hectare, kg ha^{-1}	bushel per acre, 48 lb	53.75
0.107	liter per hectare, L ha^{-1}	gallon per acre	9.35
893	tonne per hectare, t ha^{-1}	pound per acre, lb $acre^{-1}$	1.12×10^{-3}
893	megagram per hectare, Mg ha^{-1}	pound per acre, lb $acre^{-1}$	1.12×10^{-3}
0.446	megagram per hectare, Mg ha^{-1}	ton (2000 lb) per acre, ton $acre^{-1}$	2.24
2.24	meter per second, m s^{-1}	mile per hour	0.447
		Specific Surface	
10	square meter per kilogram, m^2 kg^{-1}	square centimeter per gram, cm^2 g^{-1}	0.1
1000	square meter per kilogram, m^2 kg^{-1}	square millimeter per gram, mm^2 g^{-1}	0.001
		Pressure	
9.90	megapascal, MPa (10^6 Pa)	atmosphere	0.101
10	megapascal, MPa (10^6 Pa)	bar	0.1
1.00	megagram, per cubic meter, Mg m^{-3}	gram per cubic centimeter, g cm^{-3}	1.00
2.09×10^{-2}	pascal, Pa	pound per square foot, lb ft^{-2}	47.9
1.45×10^{-4}	pascal, Pa	pound per square inch, lb in^{-2}	6.90×10^3

(continued on next page)

Conversion Factors for SI and non-SI Units

To convert Column 1 into Column 2, multiply by	Column 1 SI Unit	Column 2 non-SI Units	To convert Column 2 into Column 1, multiply by
		Temperature	
1.00 (K – 273)	kelvin, K	Celsius, °C	1.00 (°C + 273)
(9/5 °C) + 32	Celsius, °C	Fahrenheit, °F	5/9 (°F – 32)
		Energy, Work, Quantity of Heat	
9.52×10^{-4}	joule, J	British thermal unit, Btu	1.05×10^{3}
0.239	joule, J	calorie, cal	4.19
10^{7}	joule, J	erg	10^{-7}
0.735	joule, J	foot-pound	1.36
2.387×10^{-5}	joule per square meter, J m^{-2}	calorie per square centimeter (langley)	4.19×10^{4}
10^{5}	newton, N	dyne	10^{-5}
1.43×10^{-3}	watt per square meter, W m^{-2}	calorie per square centimeter minute (irradiance), cal cm^{-2} min^{-1}	698
		Transpiration and Photosynthesis	
3.60×10^{-2}	milligram per square meter second, mg m^{-2} s^{-1}	gram per square decimeter hour, g dm^{-2} h^{-1}	27.8
5.56×10^{-3}	milligram (H_2O) per square meter second, mg m^{-2} s^{-1}	micromole (H_2O) per square centimeter second, μmol cm^{-2} s^{-1}	180
10^{-4}	milligram per square meter second, mg m^{-2} s^{-1}	milligram per square centimeter second, mg cm^{-2} s^{-1}	10^{4}
35.97	milligram per square meter second, mg m^{-2} s^{-1}	milligram per square decimeter hour, mg dm^{-2} h^{-1}	2.78×10^{-2}
		Plane Angle	
57.3	radian, rad	degrees (angle), °	1.75×10^{-2}

	Electrical Conductivity, Electricity, and Magnetism		
10	siemen per meter, $S\ m^{-1}$	millimho per centimeter, $mmho\ cm^{-1}$	0.1
10^4	tesla, T	gauss, G	10^{-4}
	Water Measurement		
9.73×10^{-3}	cubic meter, m^3	acre-inch, acre-in	102.8
9.81×10^{-3}	cubic meter per hour, $m^3\ h^{-1}$	cubic foot per second, $ft^3\ s^{-1}$	101.9
4.40	cubic meter per hour, $m^3\ h^{-1}$	U.S. gallon per minute, $gal\ min^{-1}$	0.227
8.11	hectare meter, ha m	acre-foot, acre-ft	0.123
97.28	hectare meter, ha m	acre-inch, acre-in	1.03×10^{-2}
8.1×10^{-2}	hectare centimeter, ha cm	acre-foot, acre-ft	12.33
	Concentrations		
1	centimole per kilogram, $cmol\ kg^{-1}$	milliequivalent per 100 grams, $meq\ 100\ g^{-1}$	1
0.1	gram per kilogram, $g\ kg^{-1}$	percent, %	10
1	milligram per kilogram, $mg\ kg^{-1}$	parts per million, ppm	1
	Radioactivity		
2.7×10^{-11}	becquerel, Bq	curie, Ci	3.7×10^{10}
2.7×10^{-2}	becquerel per kilogram, $Bq\ kg^{-1}$	picocurie per gram, $pCi\ g^{-1}$	37
100	gray, Gy (absorbed dose)	rad, rd	0.01
100	sievert, Sv (equivalent dose)	rem (roentgen equivalent man)	0.01
	Plant Nutrient Conversion		
	Elemental	***Oxide***	
2.29	P	P_2O_5	0.437
1.20	K	K_2O	0.830
1.39	Ca	CaO	0.715
1.66	Mg	MgO	0.602

1 Designing Crops for Added Value: A Vision, A Mission

ROGER A. KLEESE

Optimum Quality Grains, L.L.C.
Urbandale, Iowa

New technology and the marketplace have fueled interest in designing crops for added value. Oilseeds with more healthful fatty acid profiles, corn (*Zea mays* L.) with increased energy density and tomato (*Lycopersicon esculentum* L.) with extended shelf life are examples of new hybrids and varieties in the market today. Plant genetic research programs, both public and private, increasingly reflect an interest in improving crops for a particular end-use.

The new biotechnologies of the 1980s stimulated tremendous interest in looking for ways to apply the technologies to improve crop species. Because of the technology's power to manipulate at the molecular level, there was special interest in altering the composition of plants leading to the potential for adding value for end-use. Missing from much of this early interest was an understanding of the market needs and the potential value of a particular product concept.

Clearly the driver for these early efforts to create value was workable technology looking for an application. In addition, the marketplace in the early 1980s was not very interested in the concept of genetically altering the composition of agricultural commodities used as raw materials for food, feed, and industrial purposes.

Food production in the traditional food system occurs in a sequence of separate market segments. It may be thought of as a food chain since it operates as separate links connected in sequence. The first link connects to the second but to no other links (Fig. 1–1.)

Input suppliers for agricultural production, the first link in the chain, develop products for and do business with farmers, the second link in the chain. There is minimal interaction among the input suppliers and grain handlers, feeders, processors and food wholesalers and retailers. Plant breeders have focused on developing products for farmers. In this food system, farmers ARE the customer.

I. A COORDINATED FOOD PRODUCTION SYSTEM

There is little wonder that end-users of grains, meals, and oils had not expressed interest in the use of genetics to add value to agricultural commodities. They were disconnected from the genetic input to the chain. But the marketplace has

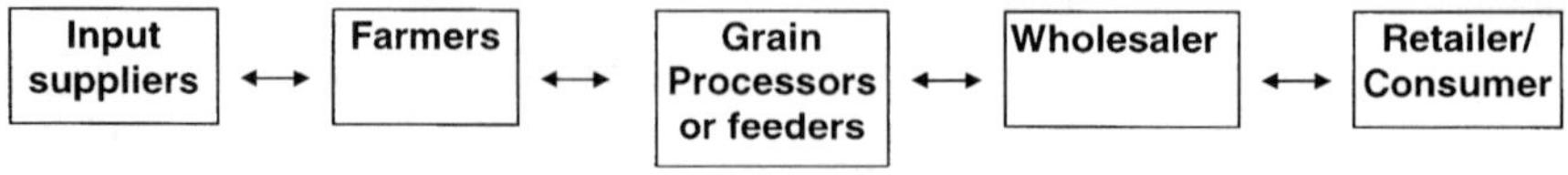

Fig. 1–1. Traditional food production system.

changed and continues to undergo rapid evolution toward a coordinated food production system. This is driven by consumer demands; consumers with substantial disposable income for food who eat half their meals outside the home and who have less and less time to prepare the food that is eaten at home. These are consumers who are concerned about the relationship between choice of food and health and about food safety. The Nutrition Labeling and Education Act of 1990 has enabled the consumer to understand better than ever before the composition of the food they're buying.

Our society is becoming more aware of the need for a food production system in a sustainable environment. Consolidation of livestock and poultry operations is raising concerns about their potential impact on the environment. There is debate about which level of government, i.e., township, county, state, or federal, has jurisdiction as well as the limits of this jurisdiction. This has in some instances pitted crop producers against livestock producers creating a paradox regarding the production and consumption of feed grains. A sustainable environment is a driver with a number of uncertainties for our food production system.

Furthermore, this coordinated food system is driven by economic pressure to produce more with the same or reduced input costs. Major brands are finding stiffer competition from store brands and generic brands. Food companies are being pushed to differentiate their products. Much consolidation has occurred throughout the food system and it continues today. Fewer players with larger operations are working through a variety of relationships or by integrating upstream or downstream. Poultry producers brand their products and market them to retail grocers. Grain handlers process corn grain and add further value by making high fructose corn sugar, vitamins and enzymes. Food corn processors retail their products and also contract for the production of food corn grain. This coordinated food system enables each segment of food production to communicate and do business directly with any other segment (Fig. 1–2).

This coordinated food system is putting the plant breeder in touch with the consumer as well as other segments in the food system. The plant breeder can be-

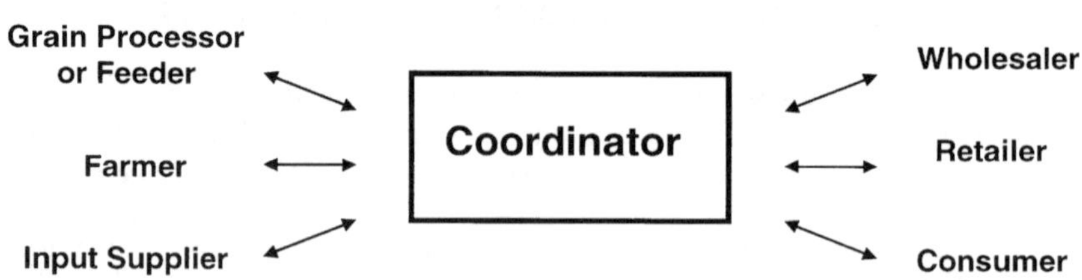

Fig. 1–2. Coordinated food production system.

come informed about the needs and preferences of the consumer also while learning about the operation and constraints of the total food system. Likewise, all segments of the food system have a greater opportunity to understand the potential to use plant genetics as a technology to enhance their segment(s) of the food system.

Plant genetic suppliers are talking with livestock producers, grain handlers, grain and food processors, as well as wholesale and retail food companies to learn about those businesses and to understand their needs. Translating those needs into plant product concepts is a substantial challenge especially when it leads to new product concepts. But the dialogue is taking place, and various players in the food system are beginning to ask for and even expect varieties and hybrids designed to meet their end-use needs.

II. AGRONOMIC CROPS CURRENTLY BRED FOR END-USE

Forages are the group of species in which plant breeders have focused most on end-use. Digestibility, protein and certain plant morphology attributes, e.g., leafiness, are several of the feed value traits that have been studied and selected. This has brought together plant and animal scientists and producers as they realized the value of working together.

Among the cereals, barley (*Hordeum vulgare* L.) and wheat (*Triticum aestivum* L.) stand out as crops bred for an end-user. The malting industry has identified those malting characteristics of value and at times has been very demanding when potential new varieties were considered for commercialization. New malting varieties generally require an endorsement from the malting industry.

Wheat breeders also have found it important to consider milling and baking as significant components of their variety development programs. In fact, it is within wheat that we have seen the divergence of special types for unique end-uses such as bread, cookies, and pasta.

Sugar beet (*Beta vulgaris* L.) is bred to meet the needs of the processors. There is close communication and coordinated evaluation of new varieties between the seed companies and the sugar beet processors. New varieties must perform up to standards set by the processors to be placed on a recommended list.

Canola (*Brassica campestris* L.) represents a new crop bred specifically for the food oil market. Erusic acid was removed and glucosinolates reduced to convert rape (*B. napus* L.) seed into varieties with edible oil. It is interesting to note that added value came not from the addition but rather the removal of constituents. The higher oleic acid content of canola is thought to make it more healthful than most other vegetable oils. This has stimulated research to further alter fatty acid profiles in other oilseeds to improve nutritional and functional properties (Erickson & Frey, 1994).

Low linolenic, low saturate and high oleic soybean (*Glycine max* L.) varieties and high oleic sunflower (*Helianthus annuus* L.) hybrids are available commercially. Each of these oil profiles has a unique combination of functional and/or healthful attributes. It's clear there is significant technology to alter fatty acid profiles. These commercial varieties and hybrids suggest a market opportunity as well.

An additional point to be emphasized here is the transfer of technology across species. Understanding fatty acid biosynthesis creates the opportunity to transfer the technology across species. There will no doubt be species-specific features of fatty acid biosynthesis but the basics hold across species. Many of the opportunities to add value to crops will be based on leveraging the technology developed in one species across additional species.

Food corn hybrids have been developed according to dry milling and masa processors' specifications. But, quality standards can be relaxed to allow for a high-yielding hybrid with borderline quality attributes because of reduced premiums for a yield differential. In addition, waxy and high amylose corn are bred for the specialty starch market. However, these market segments combined represent less than 2% of the total U.S. corn production.

The major end-use market that has not been addressed with varieties or hybrids designed specifically for that market is the feed end-use market that consumes feed grains and oilseed meals. In the USA this market is primarily a corn grain and soybean meal market. Eighty percent of the corn produced in the USA goes into livestock and poultry feed (USDA-ERS, 1997a). Essentially all soybean meal is used in animal and poultry feed (USDA-ERS, 1997b). Combined, corn and soybean represent 50% of the agronomic crop acres in the USA (USDA-NASS, 1997).

Clearly if the concept of designing crops for added value is to take on new meaning, the potential to have a major impact on our food system lies with the development of hybrids and varieties of corn and soybean, respectively, that have improved feed value. Significant efforts in this area are underway (Kuhn, 1996).

III. ADDED VALUE ISSUES

The new technologies and the changing marketplace create demands and raise a number of issues. Several of the more important issues are identified and briefly detailed below.

A. Understanding the End-User's Needs

Identifying end-use needs will be paramount to any product improvement program. Success will vary by end-use market segment. The livestock production industry has been willing to articulate their needs for improved feed value. In contrast, end-users in some of the food and processing market segments have at times been reluctant to share their needs no doubt because of a concern about proprietary information in a competitive marketplace. As a result, understanding needs will at times be a significant challenge.

Nonetheless the needs are many and varied and will require prioritization based on market opportunity, real value of modification, and fit with the technology. Translation of that need into a plant genetic product concept requires qualitative and quantitative definition of the product concept and will no doubt be viewed differently by plant breeders as well as members of an end-use market segment. Settling on the product concept that meets a significant market need and also creates significant value will be a challenge.

B. Substituting Value Versus Creating New Value

The more obvious opportunities to add value will involve an improvement in the grain or seed composition to meet a need currently met by another ingredient. For example, improving essential amino acid content in corn or soybean could be attractive to the livestock producer because of the opportunity to reduce the use of synthetic lysine, methionine or soybean meal. The added-value grain or soybean meal, lysine, methionine and commodity soybean meal are substitution products. The value of the improved grain or soybean meal will be tied closely to the value of the substitution ingredients, lysine, methionine and commodity soybean meal. As the improved grain or meal is used the price of the other substitution ingredients will likely decrease, reducing the value of the improved grain and meal.

In contrast, creating new value where substitution ingredients or technologies don't exist would seem to be a strategy for preserving the value in the improved grain, meal or oil. Improved vegetable oils with unique fatty acid profiles would appear to be a step in this direction.

Another example of creating new value is the reduction in bound P (phytate) in corn. Most of the P in cereals as well as soybean is bound to inositol (phytate) and is not available to monogastric species. As a result, inorganic P is added—at a cost—to the rations of these animals and poultry. The phytate P passes through the animals and is concentrated in the animal waste. Phosphorus in animal waste is becoming a significant environmental concern for high-density livestock and poultry operations.

Corn mutants with a substantial reduction in phytate but with levels of total P comparable to normal corn represent the opportunity to both reduce the cost of adding inorganic P to the ration and reducing P in the animal waste. Estimating the value of increased available P in an animal ration is relatively easy to do given a set of prices for various feed ingredients. However, the greater potential value with this attribute is in the reduction of P in animal waste and the value to the environment. This value varies greatly from situation to situation and is changing rapidly with more demanding regulations on livestock and poultry operations.

These mutants represent the creation of new value. Yet the development of a commercial phytase enzyme that can free phytate-bound P in a feed is a new and competing technology. Only after low-phytate corn hybrids come to market will it be possible to determine the extent to which the phytase represents a substitution product. Creating truly new value will not be easy.

C. Capturing Value

At the heart of creating added value is the question of whether the market will pay for value and how it will be shared. Many opportunities to add value genetically will compete with alternative ingredients or manufacturing processes. The costs associated with those ingredients or processes will respond to new plant genetic ingredients. For example, increasing digestible lysine in corn might be viewed as a target for adding value to corn grain for feed for monogastrics. To understand the potential value of this product requires at least some understanding of the synthetic lysine market, cost of goods for producing lysine, and any intrinsic value to the livestock or poultry producer in being able to add lysine as a separate ingredient to their rations. The market analysis can become very demanding. This simple example suggests the importance of understanding the needs of the livestock producer, the way they think about and operate their business today as well as their views to the future, plus some understanding of the synthetic amino acid industry.

Suddenly one is forced beyond a complex agronomic production system into a livestock and meat production system largely unknown to those engaged in plant genetics. Understanding value and the way in which it may be shared throughout this system will not be easy.

Today, commodity corn is the standard for measuring value in corn grain. It is the standard because of excellent yield and agronomic performance. It is the standard for feed grain in the USA because it is nearly the universal source of energy. High yields have resulted in a low cost for corn grain as this source of energy. The value added from decades of corn breeding has been the low cost per bushel for energy. Characterization of commodity corn hybrids for particular end-uses is a means to identify additional value. Good examples of this already exist within the food corn hybrids. New value-added hybrids will face significant competition from commodity corn.

D. Redefining Value

How will value be determined? For the crop producer, value for most agronomic crops has been yield per acre. It still is today. But if value is to be added for the end-user it will be difficult to maintain yield. Furthermore, even with parity yield to varieties without added value there must be some incentive for the crop producer to choose the variety with added value. Value per acre will have to be redefined from yield acre^{-1} to (yield x added value) acre^{-1}.

E. Will the Technology Work?

There is little question that plant breeding can be successfully used to manipulate the chemical composition of plants. Perhaps the best known examples are the Illinois high and low oil and protein long-term selection experiments that have created corn lines that range in oil from 0.5 to about 20% and another set of lines that range in protein from 4 to about 30% (Dudley & Lambert, 1992). Notwithstanding this tremendous plant breeding accomplishment, corn hybrids with higher levels of oil have only been grown on significant acreages in the past few years.

Corn hybrids with substantially different protein than the majority of yellow dent hybrids are not a significant part of the U.S. seed corn market. Plant breeding can alter chemical composition *but* can it add enough value for end-users to pay for the change?

Molecular and cell biology may represent even more powerful technologies than plant breeding for the manipulation of chemical composition. This is because they are focused on manipulating molecular and cellular systems and because they open the possibility to source germplasm from all of biology. The application of these technologies for adding value is beginning to be realized in the market. Transgenic tomato with extended shelf life is an example of an added-value product. The commercialization of cotton (*Gossypium hirsutum* L.) and corn containing the *Bacillus thuringiensis* endotoxin to provide insect resistance and the commercialization of corn and soybean with herbicide resistance are evidence that transgenic technology is working in agronomic crops.

Transgenic crops pose an additional challenge compared to those developed by plant breeding alone. Various restrictions are being placed on the production and/or the importation of transgenic crops by some countries. Although transgenic plant varieties are being commercialized in the USA, export markets are problematic if the transgenes are banned in other countries. It's not possible today to be certain about the longer-term impact of transgenes on the total market for value-added crops.

Combining the added-value traits with elite germplasm will be necessary to develop the most competitive seed products. This will result in an even greater demand on plant breeding resources since plant breeding will continue to be the technology that ultimately delivers the seed product to the market.

The critical test will be whether total value, that is value to the crop producer and to the end-user, can be increased. High lysine corn developed with the opaque-2 gene did increase feed value for monogastrics (Cromwell et al., 1968). However, yield and other aspects of grain quality were so severely depressed that total value was less than normal yellow dent hybrids. For new added-value traits there is no experience for predicting pleiotropic effects of the trait on yield or other agronomic attributes.

F. Measuring Added Value

It will be necessary to measure added value at points of commerce along the path from the crop producer's field to the end-user. Generally this will require rapid, nondestructive analytical methods to handle many transactions in real time where neither the buyer nor the seller have laboratories. Near infrared spectroscopy is an example of such an analytical procedure that is used by some end-users and is beginning to find its way into grain elevators. It's doubtful that any one technology will fulfill all of these needs for the many and varied added-value attributes. It's not clear where the analytical technologies will come from or who will develop the applications for grains and crop products. This is a need.

G. Preserving Added Value

Agronomic crops are produced by several hundred thousand farmers in the USA. Most are not end-users nor is their production adjacent to the end-user. If commodity grains are produced there is no unique value to be preserved thus the current grain handling system has evolved to operate efficiently with commodities.

Added-value crop products will have to be handled separately to preserve the value. Today there is substantial interest by a variety of players in the grain handling industry to participate in the business of identity preserving value-added grains. Included among these players are farmers with bin capacity, local country elevators, plus regional and multinational grain handlers.

The soybean varieties and sunflower hybrids with unique fatty acid profiles are a new and growing complement of identity-preserved products. All are produced under contract with the farmer. This requires arrangements for storage and delivery to an oilseed crusher. In turn, crushers willing to handle small quantities of soybean or sunflower seeds have to be identified and given incentives to do separate runs. Further downstream oil refiners work under an agreement to refine these unique oils. A significant step in product development is the production of oils extracted and refined under commercial conditions to be tested by the end-users in their various applications. That is to say, laboratory analyses and tests for end-use are predictive but the final evaluation generally has to come from some small-scale, yet commercial production and processing for an end-use.

All of this requires years of development and substantial investments in resources before really knowing the market potential for a new product concept. Today, the up-front costs for the biotechnology and the plant breeding are being carried largely by those commercial entities with the technology. As new product concepts occur in potential commercial varieties or hybrids there is an attempt to distribute the cost and the value across the entire food system.

Understanding the costs and especially the values of a new product concept make it especially challenging to create a system where all the players are comfortable with the way the value is shared. It remains to be seen how large-scale efforts involving millions of acres to produce, move and preserve value will develop.

H. Freedom to Operate

The definition of freedom to operate (FTO) continues to evolve in the seed industry. Owning germplasm through plant variety protection or trade secret has been a part of the seed industry for a number of years. But the more recent use of utility patents is revolutionizing the development of new seed products by controlling the rights to individual genes, components of genes, specific traits and even product concepts. One organization may create the product concept and devote substantial resources to its development but if even a small piece of DNA in that product is covered by another party's utility patent, the developer will not be free to commercialize the product. Intellectual property will be bought, sold, licensed, traded and at times be unavailable. Competitors will do business with each other because of the need to have freedom to operate. The FTO is adding a new dimension to new seed product development. This is not unlike other industries, however, and it can work well.

I. Which Traits to Stack?

We have been stacking traits for years by putting them into a limited number of products that served multiple end-users. An increasing number of agronomic and added-value traits will present a new challenge in offering those packages of traits in a product that has greatest value to an end-use market. Deciding which traits to stack is critical to maximizing the total value of the product line. Too many trait combinations will increase development and inventory costs. Too few products will not address all market segments. Furthermore, too few products are likely to reduce the value of the product since not all customers will be interested in all traits in the product.

Can traits be stacked without reducing agronomic productivity? Plant breeding experience leads one to be cautious about adding traits because of the risk that yields will be reduced. If total value is increased by stacking traits even though yields may be reduced, then we'll have incentive to do so. However, most added-value traits leave little room for a decrease in yield because of the limited value of the trait. Will site-specific transformation or multiple traits in the same genetic construct allow us to use transgenic approaches that don't have a negative impact on yield? This will be a challenge.

We will likely see further specialization in the seed industry since not all players will have access to, nor have the resources to develop all trait combinations. We will see end-use traits for a particular market segment stacked to add greater value to the product. For example, the poultry producer that values increased energy density and amino acid quality will generally prefer and value both of those attributes in the same corn grain rather than having to choose one or the other.

It's apparent the issues are many and varied. Choosing a viable product target will be critical and demanding. But the marketplace and the technology are powerful drivers that will lead to crops designed with added value.

While the analysis in this overview is taken largely from a U.S./Canadian perspective, it is not intended to imply that some of the same phenomena are not operating elsewhere in the world. Certainly the technologies are not limited to the USA and Canada. Furthermore, there are similar marketplace drivers operating in developed food production systems elsewhere in the world. Crops with added value can be expected to impact those food production systems, also.

IV. VISION

Consumer demand will define product needs understood throughout a coordinated food system. This coordination will put plant breeders in touch with consumers more than ever before and will lead to value added plant genetic products that meet those consumer's needs. The market will segment creating new commodities. Raw materials will become value-added ingredients.

Increasingly we will come to understand plants as biochemical factories. Today most of us view plants as sources of food and feed plus certain industrial products made from further processing. In time, we will view plants as factories that can be genetically manipulated to produce complex and valuable chemicals. And the value of crop production will grow substantially.

V. MISSION

Opportunities created via a coordinated food system will challenge those controlling and delivering plant genetic technology to understand the marketplace, to prioritize opportunities and to allocate resources for product development that maximize returns. Implicit within such a product development system will be the need to develop relationships up and down the food system as well as with other input suppliers to source technology and to market products.

REFERENCES

Cromwell, G.L., J.C. Rogler, W.R. Featherston, and T.R. Cline. 1968. A comparison of the nutritive value of opaque-2, floury-2, and normal corn for the chick. Poult. Sci. 47:840–842.

Dudley, J.W., and R.J. Lambert. 1992. Ninety generations of selection for oil and protein in maize. Maydica 37:81–87.

Erickson, M. D., and N. Frey. 1994. Property-enhanced oils in food applications. Food Technol. 48(11):63–67.

Kuhn, W. E. 1996. Designer ingredients for animals, from a plant breeder perspective. J. Appl. Poult. Res. 5:379–385.

Nutrition Labeling and Education Act. 1990. Public Law 101-535.

U.S. Department of Agriculture-Economic Research Service. 1997a. Feed yearbook. ERS, Washington, DC.

U.S. Department of Agriculture-Economic Research Service. 1997b. Oil crops outlook. ERS, Washington, DC.

U.S. Department of Agriculture-National Agriculture Statistics Service. 1997. Crop production. annual summary. Natl. Agric. Stat. Serv., Washington, DC.

2 Economic Opportunities for New Genetic Products

RICHARD K. PERRIN

University of Nebraska
Lincoln, Nebraska

Economic opportunities for developing a specific new crop product may be short-lived and difficult to identify, depending as they do on such changing factors as the availability and cost of substitutes for the product, the potential for future substitutes that may arise to curtail the market life of the product, the cost of current methods available for developing and exploiting it, etc. There are some trends, however, that suggest there will be viable markets for a wider array of differentiated crop products than heretofore. First among these is the trend in scientific capacity to create more differentiated products—a trend manifested in the content of other chapters of this volume. Second is the trend in information technologies—the capacity to measure, monitor, certify and inform at low cost. It is becoming less costly to inform buyers about a myriad of alternative products and to certify their hidden characteristics. It also is becoming less costly to coordinate the breeding, growing and marketing phases of producing a differentiated crop product.

Thus, opportunities for new crop products will surely increase, but the development of any given new product is an entrepreneurial venture, and its success will depend partly on scientific and market knowledge, but also on intuition, inspiration and luck. The purpose of this chapter is to introduce some economic and marketing concepts that provide ways of thinking about the economics of these opportunities. These insights are based on the theory of "improved goods", which would be crop products that have either improved characteristics or a different combination of existing characteristics. "New goods", on the other hand, are those that meet previously unmet (or badly met) objective needs, such as the electric light, the airplane or the television picture tube. They generally require a very long period of adaptation for full adoption and impact, and not much is known about evaluating their economic potential in advance (Bresnahan & Gordon, 1997). New crop products will almost certainly be of the "improved goods" type, so that economic evaluation based on a recombination or improvement of existing characteristics will be useful.

I. CROP PRODUCTS AS BUNDLES OF CHARACTERISTICS

The value of a crop depends upon users of the crop—how much they are willing to pay for it. If a crop is to have added value, it is to the users of the crop, therefore, that we must look to determine as best we can what they would be willing to pay for particular changes. How does the user decide willingness to pay? If the buyer is a consumer, as in the case of a tomato (*Lycopersicon esculentum* L.), there will be a number of tomato characteristics that are relevant, along with such important considerations as the availability and price of substitutes and the size of the budget for food. If the buyer is not a consumer but an intermediate processor, as in the case of soybean (*Glycine max* L.), the story is very much the same—willingness to pay will depend on the characteristics of the soybean and the prices of the products to be produced with the soybean, but also on the availability and price of substitutes for soybean in those uses.

It becomes useful therefore to think of a crop as a bundle of characteristics of interest to the buyer. And it is useful to describe a particular bundle as a *crop product* (since such terms as "hybrid", or "variety" or "cultivar" aren't quite broad enough to encompass all the alternative bundles). Since we are here focusing on value to buyers, it is characteristics such as nutrient content, chemical composition, acidity, smell, etc., that are of primary interest here as opposed to production characteristics such as drought or pest tolerance, even though the production characteristics also will be relevant in designing genetic products. In these terms, the issue posed in this chapter is how many and which of the multitude of feasible genetic products will provide bundles with a net value exceeding that of the existing crop products. (The term "net" is a reminder that any extra production cost and any extra processing cost must be deducted from the users' willingness to pay for a particular change in the bundle of characteristics.)

II. THE ECONOMIC VALUE OF BUNDLES OF CHARACTERISTICS

A. One Characteristic—the Linear City Concept

A simple but useful concept of the market for differentiated product is the "linear city" model first advanced by Hotelling in 1929. Consider a single characteristic, such as the level of an amino acid in corn (*Zea mays* L.), and suppose there are two users of the product—hog producers and chicken producers. Suppose that for hog production, the "ideal" corn product (the product with the characteristics of the highest value) has amino acid content c^h, and that the value of corn in hog production falls linearly as its amino acid content, c, departs from this ideal level. This is depicted in Fig. 2–1, where c^h indicates the ideal level of the amino acid for hogs, and the inverted-V labeled u^h represents the value of corn to hog producers as a function of the level of amino acid. This value function can be expressed algebraically as $u^h = v^h - |c^h - c| * \tau^h$, where v^h represents the value of corn that contains the ideal level of amino acid, the expression $|c^h - c|$ indicates the absolute value of the departure of amino acid level from the ideal level, and τ^h is the slope of the value function.

The value function for corn in chicken production shown in Fig. 2–1 indicates that the ideal amino level is higher for chickens than for hogs, and furthermore that amino acid level is more critical in determining the value of corn for chickens than for hogs. That is, the value of corn drops off more rapidly for chickens than for hogs, as the amino acid content departs from their respective ideal levels. The availability and price of substitutes for the amino acid in corn is what determines the slopes of these value functions. The slope might reflect the cost of reduced gain-to-feed ratios with nonideal levels of the amino acid, or the cost of supplementary amino acid additions to the ration, whichever is the best alternative. The value functions represent "willingness to pay" for a crop with different characteristics because the user would be willing to pay any price up to and including that indicated by the function, but would not buy if the product were priced at a higher level.

The issue addressed now is whether there is an economic opportunity to develop two corn products instead of one. If there were but one product, what level of the amino acid should be produced? If any level were feasible at the same development cost, and if the amount of corn used for hogs and chickens is equal, then the answer would be c^l in Fig. 2–1—that level of amino acid at which the value of the corn is equal for each user at v^l. (None of these assumptions are especially plausible, but relaxation of the assumptions, while quite feasible, obscures the basic principle without adding much insight.) While c^l would be the best single product, if the existing product were other than c^l it would not significantly change the analysis to follow.

How much is to be gained by developing two new crop products, one tailored for each user? Again assuming for simplicity that any level of the characteristic can be developed, the two new products to be developed would contain amino acid levels of c^h for hogs and c^c for chickens. Hog producers gain $v^h - v^l$ per unit of corn and chicken producers gain $v^c - v^l$ per unit. The amount of this gain depends upon the slopes of the willingness to pay functions in Fig. 2–1, which are determined by

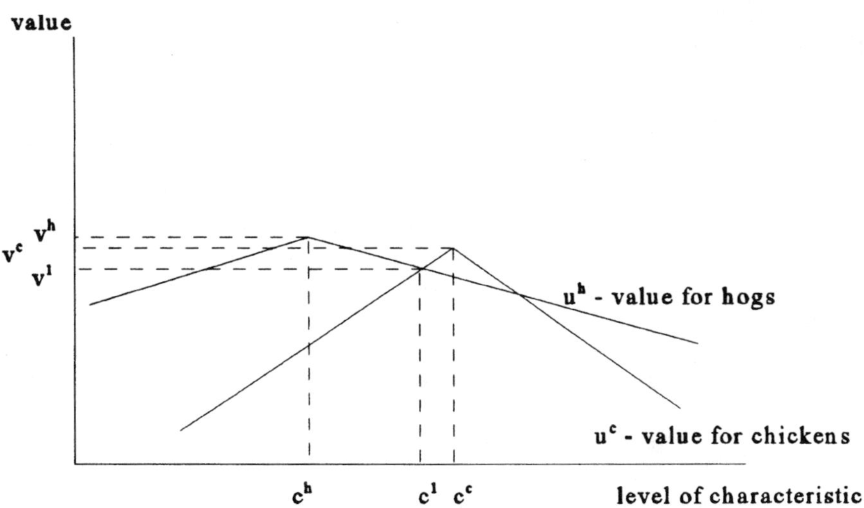

Fig. 2–1. Product differentiation with one characteristic.

the availability and cost of substitutes for the characteristic. (The slope might indicate the cost of supplementing the animal feed with a synthetic amino acid, for example.)

From the viewpoint of the users, multiple products are the obvious thing to do if the price remains the same, since both prefer a crop that meets their ideal specifications. If the cost of producing two different products is the same as for an equivalent amount of just one, then the price *would* normally be the same, and there is definitely an economically viable opportunity for two new products. The costs considered must include the genetic development cost, any changes in farm production costs for the two new products, and increases in identity-preservation costs throughout the production, transportation and storage phases. A useful concept here is the notion of "diseconomies of scope" in producing multiple products. Diseconomies of scope exist if the cost of developing, producing and marketing a million tons each of two products is greater than that for producing two million tons of one product. In general we can assume that diseconomies will exist for multiple crop products. If there were no such diseconomies, then for the case of a single characteristic, it always would be profitable to develop a different ideal product for each group of buyers.

A short digression is in order on the issue of the denominator for the measure of the characteristic. Should it be grams of amino acid per kilogram of product (i.e., percentage), tons of amino acid produced per acre, grams of amino acid per gram of total protein, grams per calorie, or just what (such as relative sweetness on a scale of 1–10)? There is no universal answer, but at this point it is the buyer whose interests are paramount, and the value of the product to the buyer might best be represented by any one of these measures.

The principle illustrated by the linear city model is that factors affecting the economic opportunity for new products include: buyers' willingness to pay for incremental changes in a characteristic, the feasibility and cost of genetic selection for incremental changes in a characteristic, changes in the farm production cost for the new products, and increases in identity-preservation costs for specialized crop products rather than generic commodities.

Application of the linear city model to examine particular opportunities will be much more complicated than the simple illustration of the principle shown here. Instead of two users, there will be many, say *N* users, and there will therefore be the potential for up to *N* "designer" products, some of which could be designed to serve multiple users. And instead of one characteristic, there will be many characteristics of interest to users, and these characteristics are correlated genetically, greatly complicating the "design" task. Principles for product differentiation with multiple characteristics can be illustrated with the two characteristic problem to be described next.

B. Two Characteristics—the Innovation Possibilities Frontier

In Fig. 2–2 three varieties are plotted in characteristics space. For soybean, for example, Characteristic 1 could be protein per ton of bean (*Phaseolus vulgaris* L.) and Characteristic 2, oil per ton. Of the three varieties, Number 1 has the highest ratio of protein to oil, and Number 3 the lowest. Given these three existing ge-

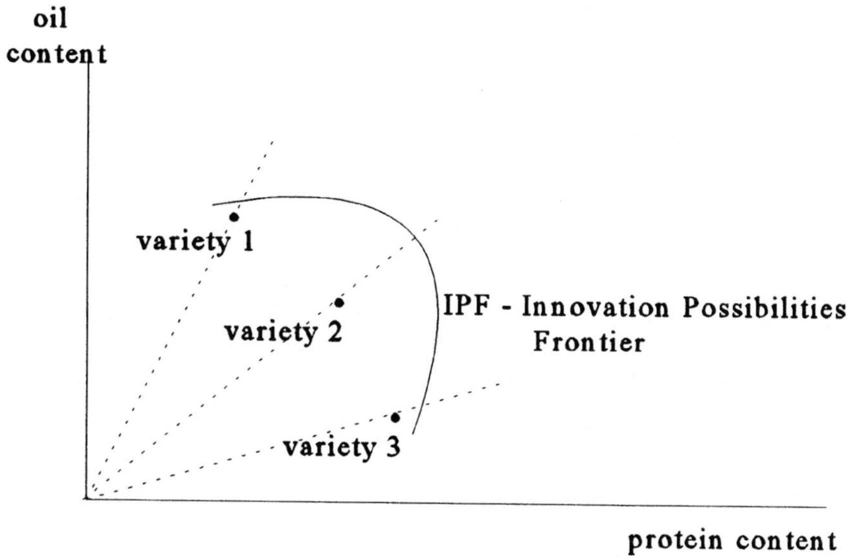

Fig. 2–2. Existing products and the Innovation Possibilities Frontier.

netic lines, it might be possible to develop new varieties that would produce both more oil and more protein, as well as varieties higher in one characteristic, but lower in the other. The *Innovation Possibility Frontier*, labeled IPF in Fig. 2–2, defines that set of possible varieties that could be developed with a given research effort[1]. Clearly, one would not want to develop all of the varieties represented by that IPF. To consider what is desirable, it is necessary to look at the value of alternative bundles from the users' perspectives.

Willingness to pay for a crop product, as argued above, will depend on the nature and price of the buyer's final product, or on personal tastes and income if the buyer is a consumer, and on the cost and availability of substitutes in both cases. Willingness to pay functions for a single characteristic were depicted in Fig. 2–1. For the case of two characteristics, say calories and crude protein content of corn, consider Fig. 2–3 where u^h_o represents an equal-value contour for hog producers. Now, it may be that in some sense the "ideal" protein level for hogs would be p^h, while the "ideal" energy level is e^h. This combination of traits is indicated at point *v,* but the equal-value contours indicate that there are tradeoffs—corn with a little less protein might be just as valuable if it contained enough more energy, and furthermore that if corn were available with higher content of both protein and energy, it would have a higher value than product *v*. Every combination of protein and energy on the line labeled u^h_o has the same value in hog production, value u^h_o. Equal-value contours u^h_1 and u^h_2 represent higher-valued sets of characteristics. (The

[1] The slope of the IPF is determined by genetic potential. As the heritability of the two traits becomes more positively correlated, for example, the IPF will collapse toward a single ray (such as that extending through Variety 2) that defines the fixed ratio of characteristics that is genetically attainable. If heritability were more negatively correlated, the IPF would flatten out toward a downward-sloping straight line, the slope of which would define the genetic tradeoff between the two traits.

dashed lines represent comparable concepts for a different user, say, chicken producers or cornbread makers.)

The shape of the equal-value contours depends on the taste of the user (if a consumer) or the technology of the user (if a producer). An extreme case of an "ideal" combination of traits for the user is represented by an equal-value curve with a right angle at *v*, as indicated by the dotted line. Here, extra levels of energy above e^h do not increase the value of corn at all, and the same is true of levels of protein above p^h . Conversely, if a consumer were to value the corn for its sweetness as a roasting ear, then the consumer is indifferent between a gram of protein and a "gram" of energy, and the equal-value curve would flatten out to a downward-sloping straight line, indicating indifference between the two traits, as long as one or the other is attained.

Now, it is conceivable that from the point of view of the user, one might trace out a set of "ideal" ratios of energy to protein, as the nutrient density of the crop increased, and such a locus of points for hog producers is the line labeled i^h . But the equal-value curves are a reminder that everywhere along this ideal locus there are other combinations of characteristics that have equal value, and users would happily choose a nonideal crop product over an ideal one of equal value if it were cheaper to obtain. In fact, the standard economic analysis is to define the "ideal" combination of characteristics as the one that achieves a given equal-value contour at lowest cost. The loci of these least-cost combinations look just like the rays i^h and i^c in Fig. 2–3, but they are the locus of points at which the equal-value curves have slopes equal to the inverse of the price ratio of the two characteristics. Instead of defining the combinations of characteristics that maximize performance, they define combinations that achieve given levels of value at lowest cost. In what follows,

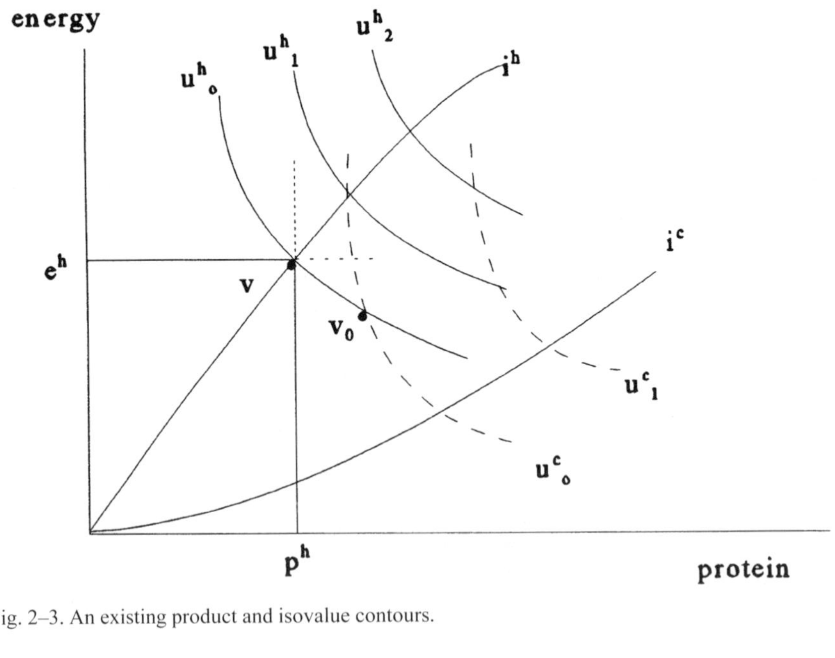

Fig. 2–3. An existing product and isovalue contours.

the ideal product rays will generally be interpreted as least-cost ideals, rather than maximum-performance ideals.

Suppose the current variety v_0 has the protein-energy characteristics as plotted in Fig. 2–3. It is not an ideal product for either user (it is a multipurpose product analogous to c^1 in Fig. 2–1). The economic opportunity for user-specific varieties depends upon the genetic possibilities, as represented by the IPF added in Fig. 2–4. The frontier represented here indicates that considerable genetic progress could be made in developing varieties with more of both energy and protein at the existing ratio (genetic progress along the vector $\mathbf{v_0v_1}$), in addition to progress on either characteristic alone. Three of the possible varieties that could be developed are indicated by v_h, an ideal variety for hogs, v_c, an ideal variety for chickens and v_1, another dual-purpose variety.

A surprising result of the situation depicted is that both buyers will be willing to pay more for the dual purpose variety than for the ideal varieties that could be developed with the same research effort. Variety v_1 is on a higher isovalue curve for each producer. Hog producers in this case for example, find that the extra protein in v_1 has more value than the cost of replacing the foregone energy with another ingredient, while chicken producers find that they can replace the lost protein in v_1 compared to v_c at lower cost than the value to them of the extra energy. This situation illustrates how the IPF interacts with the cost of substitutes to eliminate the economic opportunity for developing user-specific crop products. Suppose on the other hand that the IPF in Fig. 2–4 is more nearly a straight line between v_h and v_c, indicating the genetic opportunity for more progress in increasing either characteristic alone as compared to both simultaneously. In this case, either or both of the users might then be willing to pay more for their respective "ideal" crop prod-

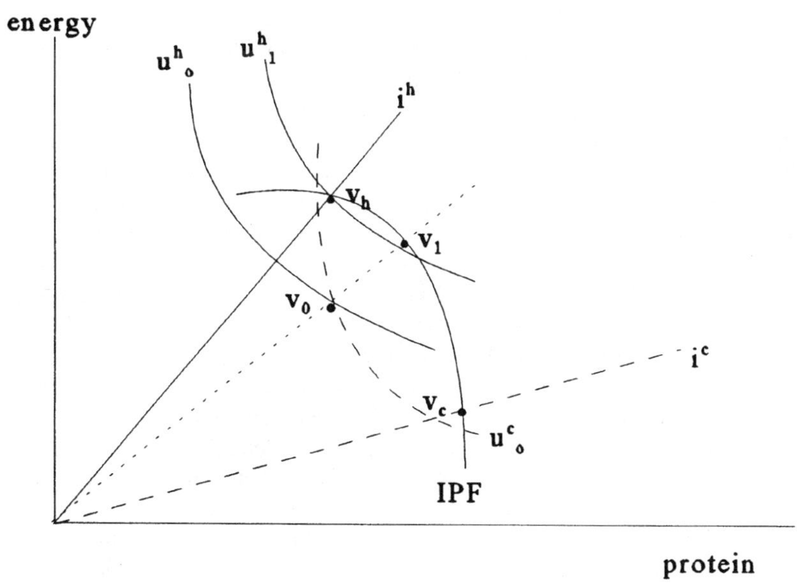

Fig. 2–4. Product differentiation with two characteristics.

ucts than for the equally achievable dual-purpose variety v_1 (that would then lie closer to v_0 than shown in Fig 2–4).

Thus one circumstance that would be most likely to yield economic opportunities for differentiated products is for an IPF that is a straight line passing through v^0. This is the case if it is genetically impossible to increase both protein and energy simultaneously, but possible to give up one in exchange for more of another. In such a case, both buyers will be willing to pay more for the feasible versions of their ideal varieties than for the current variety, and given a linear IPF as implied, these are the only available breeding objectives. As in the case of a single characteristic, efforts to develop these specialized varieties will be economically attractive only if development, production, segregation and marketing costs for the two specialized varieties are sufficiently small relative to the cost of the current variety and the increase in willingness to pay by the two buyers (as represented in Fig. 2–4 by the equal-value contours).

A number of principles are illustrated by the innovation frontier model. Buyers may differ both in their preference for ratios of characteristics and in the cost they suffer from departures from that ideal ratio. These features of the purchasers combine with the innovation possibilities frontier of the developer to determine whether economic opportunities exist for diversifying the crop products available. We've seen here that even if the costs of developing, producing and marketing two crop products are *equal* to those for a single product, it may still be best to continue development of one dual-purpose product if the genetic innovation possibilities permit a faster rate of progress in improving that product as compared to the more specialized products that buyers prefer.

C. The Special Case of Characteristics as Components

Roughly speaking, most of the soybean crop is crushed to separate it into two components, oil and protein meal. The characteristics of the crop can thus be expressed in terms of kilograms of oil and kilograms of 44% protein meal per ton of soybean. This is a simplified abstraction, of course, because soybean plants also are used for other purposes, and oil is not oil—it has many characteristics of its own that are relevant to buyers. But this abstraction is useful in examining the economic opportunities for product diversification when the crop is simply separated into two components, each of which has its own market.

As shown in Fig. 2–5, the equal-value contours become straight lines when the characteristics are separable components. Each point on an equal-value line again represents a combination of meal and oil that will fetch a given dollar value when sold at market prices. (The slope of every equal-value line equals the negative of the ratio of market prices.) Lines u_0 , u_1 , and u_2 represent increasing values. While there might exist two existing varieties v_1 and v_2 with equal value, there is no reason to develop more than one new product, and that one is v_3 in the figure. This new product is at that point on the IPF with slope equal to the equal-value lines. There is no other point on the set of technically feasible product space that achieves a value as high as v_3, and the entrepreneur who produces that product will dominate any competitors.

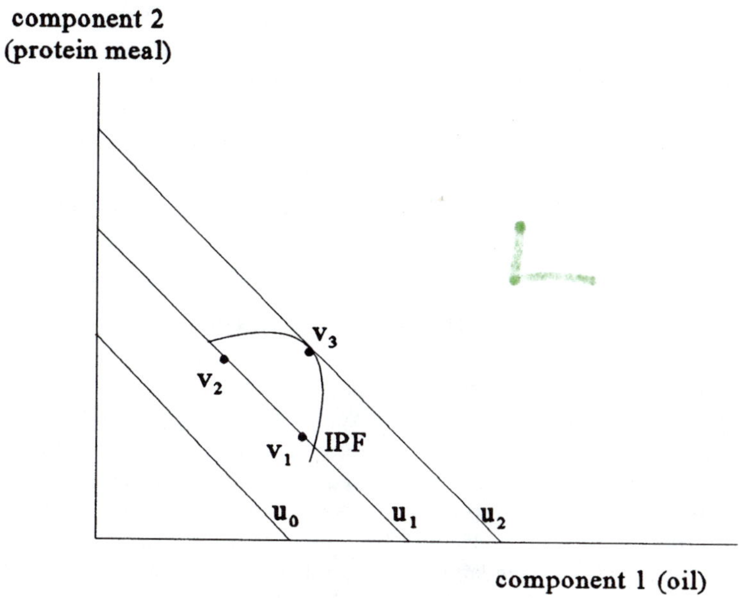

Fig. 2–5. Product differentiation for characteristics as components.

The principle illustrated by the characteristics-as-components concept is that if the buyers care little about departures of the product characteristic from any "ideal" ratio, the equal-value lines become straight lines and there is no economic opportunity for product differentiation.

III. WHAT IF MARKET PRICES DO NOT REFLECT THE SOCIAL VALUE OF CHARACTERISTICS?

In the discussions above, no distinction is made between willingness-to-pay and market price. Yet they need not be the same, and if they are not there may be a difference between market opportunities and social benefits from differentiated products. Consumers may be willing to pay more for a tomato with higher Vitamin A content, but if they cannot observe this characteristic at the point of purchase, the market price may be no higher for the high-Vitamin-A product. Sweet corn eaters might be willing to pay for reduced nitrates in the groundwater, which might be achieved by a N-fixing characteristic in a sweet corn cultivar, but that willingness to pay will not be reflected in the market price for the corn product. An individual wheat (*Triticum aestivum* L.) producer may be willing to pay something to have the countrywide wheat crop genetically diverse to protect from plagues, but unwilling to pay more for a diverse-pedigree variety seed because his individual action in itself has no effect on whether or not plagues occur.

In each of these cases of market failure, customers or the public may be willing to pay for some characteristic, but that willingness is not reflected in market prices actually paid. There is therefore no incentive for private entrepreneurs to provide more of this characteristic because their efforts will not be rewarded, even though the customer or the public will be. Public policy innovations can in some cases help to close this gap, through new kinds of property rights, product labeling requirements, product liability rules, or regulations on minimum or maximum levels of certain characteristics. When the gap between market prices and willingness to pay persists, an important function of public-sector crop researchers is to develop crop products with the socially desired characteristics that cannot be gainfully pursued by the private sector.

IV. THE ECONOMICS OF SELECTION INDEXES

The principles for differentiated products described above have been illustrated with only one or two characteristics. Yet for almost any potential buyer there will be a number of characteristics that determine value. The product differentiation principles above can be extended to the case of many characteristics, although they cannot be conveniently represented graphically. The equal-value levels for buyer h introduced in Fig. 2–3 could be represented as a value function $u^h = u^h(c_1, c_2)$, which is easily extendable to any number of characteristics as $u^h = u^h(c_1, \ldots, c_n)$. In theory, one can determine the ideal product for this buyer by maximizing u^h subject to the constraint imposed by the IPF [that also might be quantified as $I(c_1, \ldots, c_n)$, based on genetic correlations and heritabilities of the characteristics]. This optimization is exactly what the Lush selection index achieves, except that the value function has the very simple linear form $u = a_1c_1 + a_1c_1 + \ldots + a_nc_n$, where the a_j's represent fixed values of incremental changes in each characteristic. In other words, the equal-value curves of the Lush index are linear, as shown in the two-characteristic diagram of Fig. 2–5.

Thus a Lush selection index achieves the highest-valued bundle of characteristics that is genetically feasible from a given research effort. Since value is in the eyes of the buyer, value functions may differ among groups of buyers, and a separate selection index would then be appropriate for each. Although selection indexes are used much less in crop breeding than in animal breeding, they also offer an approach to empirical crop product design corresponding to the approach presented in Figs. 2–4 and 2–5.

REFERENCES

Bresnahan, T., and R. Gordon (ed.) 1997. The economics of new goods. Univ. Chicago Press, Chicago, IL.

Hotelling, H. 1929. Stability in competition. Econ. J. 39:41–57.

Krouse, C.G. 1990. Theory of industrial economics. Basil Blackwell, Oxford, UK.

Shy, O. 1995. Industrial organization: Theory and applications. MIT Press, Cambridge, MA.

3 Food Uses

**CONNIE M. WEAVER, APRIL C. MASON,
AND BRUCE R. HAMAKER**

*Purdue University
West Lafayette, Indiana*

Crops of agronomic importance provide 50% of the world's calories and are the staple of the human diet. In addition to energy, crops provide many nutrients and health protection ingredients important to human health. Nutritional status of an individual greatly influences quality of life, work performance, learning potential, overall health, and protection against disease. Because of the contribution of cereals and grains to good nutrition, the U.S. Department of Agriculture (USDA) placed this group as the foundation of the food guide pyramid. The incorporation of grains and seeds in the diet must satisfy consumer demand for quality and convenience as well as food handling and manufacturing requirements for quality, functional performance, and stability during distribution and storage. This chapter will briefly review the major known relationships between nutrients provided by crops, risk of disease, and the current guidelines for eating for good health. The macronutrients and micronutrients contributed to the American diet by crops will be discussed. Opportunities for increasing nutrients in cereals and grains and areas of research for investigating health protection ingredients in plant foods and their role in health will be identified. The behavior of components in cereals and grains in food systems and the potential for genetic manipulation for food quality will be reviewed. [The latter is covered in more detail in Peterson (2000), Kinney & Hitz (2000), Hannah (2000), White (2000), Smith (2000), see Chapters 7a,b,c,d,f.]

I. NUTRIENT CONTRIBUTION OF SELECTED CROPS

The nutrient composition of selected crops is given in Tables 3–1 and 3–2. Detailed discussions of nutrient quality will be discussed later in this chapter. In this section, an overview of the contribution of important agronomic crops and changing dietary trends will be discussed.

Cereals and grains are the major contributors of complex carbohydrates to the Western diet. Starch is the bulk of this category and contributes energy. The dietary fiber content of cereals and legumes and products containing these crops is given in Tables 3–3 and 3–4. The fiber in cereal grains decreases intestinal transit time and is more effective at increasing fecal bulk than fiber from fruits and veg-

 Designing Crops for Added Value, Agronomy Monograph no. 40.

Table 3–1. Macronutrient composition of selected crops mean amount (kg^{-1}) sample†.

	Wheat, hard red winter	Wheat, soft white	Corn	Rice, brown	Oat	Barley	Sorghum	Millet	Bean, kidney	Bean, navy	Cowpea, common	Soybean
Water (g)	131.0	104.2	103.7	123.7	82.2	94.4	92.0	86.7	117.5	123.6	119.5	85.4
Food energy (kcal)	3270	3400	3650	3700	3890	3540	3390	3780	3330	3350	3360	4160
Protein (g)	126.1	106.9	94.2	75.0	168.9	124.8	113.0	110.2	235.8	223.3	235.2	364.9
Carbohydrate (g)	711.8	753.6	742.6	761.7	662.7	734.8	746.3	728.5	600.1	606.5	600.3	301.6
Total lipid (g)	15.4	19.9	47.4	26.8	69.0	23.0	33.0	42.2	8.3	12.8	12.6	199.4
SFA (g)‡	2.69	3.68	6.67	5.36	12.17	4.82	4.57	7.23	1.20	3.31	3.31	28.84
MFA (g)‡	2.00	2.27	12.51	9.71	21.78	2.95	9.93	7.73	0.64	1.11	1.06	44.04
PUFA (g)‡	6.27	8.37	21.63	9.59	25.35	11.08	13.70	21.34	4.57	5.52	5.42	112.55
Selected fatty acids												
16:0 (g)	2.34	3.46	5.69	4.57	10.34	4.11	4.07	5.28	1.06	3.10	2.54	§
18:0 (g)	0.13	0.18	0.75	0.48	0.65	0.17	0.35	1.54	0.14	0.20	0.53	§
16:1 (g)	0.08	0.10	0.04	0.10	0.13	0.06	0.29	0.14	§	§	0.04	§
18:1 (g)	1.92	2.17	12.47	9.61	21.65	2.41	9.64	7.39	0.64	1.11	0.88	§
18:2 (g)	6.00	8.00	20.97	9.18	24.24	9.99	13.05	20.15	1.78	3.01	3.43	§
18:3 (g)	0.27	0.36	0.65	0.41	1.11	1.10	0.65	1.18	2.79	2.52	1.99	§

† USDA (1989).
‡ SFA = saturated fatty acids, MFA = monounsaturated fatty acids, PUFA = polyunsaturated fatty acids.
§ Blanks mean data unavailable.

Table 3–2. Micronutrient composition of selected crops mean amount (kg^{-1}) sample†.

	Wheat, hard red winter	Wheat, soft white	Corn	Rice, brown	Oat	Barley	Sorghum	Millet	Bean, kidney	Bean, navy	Cowpea, common	Soybean
Ascorbic acid (mg)	0	0	0	0	0	0	0	0	45	30	15	60
Thiamin (mg)	5.04	4.10	3.85	4.13	7.63	6.46	2.37	4.21	5.29	6.45	8.53	8.74
Riboflavin (mg)	1.10	1.07	3.85	0.43	1.39	2.85	1.42	2.90	2.19	2.32	2.26	8.70
Niacin (mg)	57.10	47.66	36.27	43.08	9.61		29.27	47.20	20.60	20.63	20.75	16.23
Vitamin B_6 (mg)	3.36	3.78	6.22	5.09	1.19	3.18		3.84	3.97	4.37	3.57	3.77
Folacin (µg)	430			200	560	190			3941	3697	6326	3751
Fe (mg)	36.0	53.7	27.1	18.0	47.2	36.0	44.0	30.1	82.0	64.4	82.7	157.0
Mg (g)	1.24	0.90	1.27	1.43	1.77	1.33		1.14	1.40	1.73	1.84	2.80
P (g)	3.32	4.02	2.10	2.64	5.23	2.64	2.87	2.85	4.07	4.43	4.24.0	7.04
K (g)	3.40	4.35	2.87	2.68	4.29	4.52	3.50	1.95	14.06	11.40	11.12	17.97
Zn (mg)	27.8	34.6	22.1	20.2	39.7	27.7		16.8	27.9	25.4	33.7	48.9
Cu (mg)	4.10	4.26	3.14	2.77	6.26	4.20		7.50	9.58	8.79	8.45	16.58

† USDA (1989).

Table 3–3. Dietary fiber content of selected cereals and legumes and their fractions

Product	Total dietary fiber, g (100 g^{-1}), dry wt. basis
Cereal	
Wheat flour white	3.6†
Whole wheat flour	10.9†
Wheat bran	41.6†
Whole oat, cv. Sang	31.2†
Oat bran, Sang	28.3†
Rolled oat	15.3‡
Barley meal, cv. Alva	16.4†
Barley bran, Alva	28.5†
Corn bran	62.8†
Rye meal	13.5†
Rye bran	22.7†
Bean	
Soy polysaccharide	78.2†
Pinto, raw	27.0‡
Lentil, raw	21.4‡

† Marlett, 1993.
‡ Chen and Anderson, 1981.

etables (Wisher et al., 1985). Cereals and grains also contribute substantially as a source of protein, vitamins, and minerals. These nutrients are generally found in the highest concentrations in the germ and aleurone layer. Thus, refining cereals substantially reduces the contribution of these nutrients as well as fiber. Cereal grains contribute approximately 50% of the world's calories and 45% of the world's protein (FAO, 1980), but only about 20 to 25% of the energy in North America. Ap-

Table 3–4. Dietary fiber content of selected cereal and legume products

Product	Moisture	Total dietary fiber†
	%	g (100 g^{-1}), fresh wt. basis
Cereal		
Total†	4.0	3.1
Cornflakes‡	10.9	4.3
Cream of wheat, quick§	87.9	0.7
Honey Smacks‡	25.5	2.3
Oatmeal, old fashion¶	84.2	1.9
Rice Krispies‡	7.8	1.9
Special K‡	12.7	2.7
40% Bran Flakes#	3.2	19.5
All Bran‡	5.7	30.1
Frosted Miniwheats‡	15.0	8.2
Product 19‡	3.4	5.5
Legumes		
Kidney beans, canned	77.1	5.2
Pork and beans, canned	75.0	4.4
Peas, green, frozen	82.3	3.5

† General Mills, Minneapolis, MN.
‡ Kelloggs, Battle Creek, MI.
§ Nabisco, East Hanover, NJ.
¶ Quaker Oats, Chicago, IL.
# Post/Kraft, Tarrytown, NY.
†† Marlett, 1992.

proximately three-fourths of this contribution is from wheat-based foods. Hard wheats have higher protein contents than soft wheats. Cereals and grains are a good source of some vitamins and minerals and a poor source of others. Whole grains contribute substantially to the Fe, Mg, Zn, Cu, and Mn content of the diet. Whole grains are good sources of Se, but the concentration is dependent on soil levels of the mineral. Calcium content is low in cereal grains, but because they are consumed in large quantities, cereals and grains contribute about 12% of the Ca in the American diet (Fleming & Heimbach, 1994). Legumes, nuts, and seeds contribute another 1%. Corn tortillas treated with lime have substantial quantities of Ca. Bioavailability of minerals from cereal grains is generally low with the exception of Ca. Cereal grains and products can be modified to improve bioavailability of micronutrients by reduction of phytic acid content. Cereals and grains contribute greatly to the diet in B vitamins, but do not contribute any Vitamin C and are generally poor sources of fat-soluble vitamins. An exception is the rich concentration of Vitamin E in wheat (*Triticum aestivum* L.) germ, oat (*Avena sativa* L.) meal, and brown rice (*Oryza sativa* L.).

Plant crops as a food group are generally notable for their low lipid content with the exception of oil seeds as represented by soybean (*Glycine max* L.) in Table 3–1. Cereals that have the germ removed during refining are particularly low in lipid content. However, because of the volume consumed, white bread, rolls, and crackers contribute more total lipids to the U.S. diet than do whole wheat, rye (*Secale cereale* L.), and other dark breads (4.88% from refined wheat products vs. 0.47% from whole grain breads) (Block et al., 1988). The ratio of polyunsaturated to saturated fats are well above the recommendation of 1.0. The ratios of polyunsaturated to saturated fats exceeds 2.0 for the cereal crops and 1.5 for legumes.

Legumes have a higher protein content than do cereals and grains. Their contribution to carbohydrates is smaller than for cereals, but fiber content is much higher than most processed cereals. The major carbohydrates in soybean are sugars rather than starch as in cereals. Legumes in general, and soybean in particular, are higher in minerals and thiamin and folacin than cereals.

Partly because of the inadequate consumption of whole grain cereals and legumes, the American diet is typified by high caloric content, high total fat, high saturated fat and cholesterol, high salt intakes, and low fiber-containing foods. Total lipid intake has recently decreased somewhat from an average of 40% of calories from 1977 to 1978 to 34% of calories, 12% of which is saturated fats in 1989 to 1990 (Frazão, 1995). Because calorie consumption has increased, this means sugar consumption has increased. Only approximately one-fifth of American men and one-fourth of the women meet the recommendations of 30% for total lipid and 10% of saturated fat (Tippett & Goldman, 1994).

Annual per capita use of grains increased by 3.6 kg between 1968 and 1984 from 66.4 to 70 kg. Increased use of wheat flour and rice further increased the per capita consumption by another 8.2 kg by 1988 (Raper, 1991). This greater use of grain products led to an increase of complex carbohydrates from 22 to 23% of total calories and an increase in thiamin, riboflavin, Fe, and niacin (Table 3–5).

Although consumption of grain products increased between 1977 to 1978 and 1989 to 1990, intakes of Vitamin B-6, Ca, and Zn were still below the recommended levels across all economic groups, and were lowest for poor adults. Americans fell

Table 3–5. Changes in nutrient consumption in average U.S. diet†.

Nutrient	1968	1988
	person^{-1} d^{-1}	
Food energy (kcal)	3300	3600
Carbohydrate (g)	379	428
Protein (g)	98	105
Fat (g)	158	168
Saturated fatty acids (g)	63	60
Monounsaturated fatty acids (g)	64	67
Polyunsaturated fatty acids (g)	24	34
Cholesterol (mg)	500	440
Thiamin (mg)	2.0	2.2
Riboflavin (mg)	2.3	2.4
Niacin (mg)	22	26
Vitamin B_6 (mg)	2.0	2.2
Folate (μg)	270	284
Vitamin B_{12} (μg)	10.2	9.1
Ca (mg)	850	890
P (mg)	1470	1540
Mg (mg)	320	330
Fe (mg)	14.7	17.1
Zn (mg)	12.5	12.7
Cu (mg)	1.6	1.7

† Raper, 1991.

quite short of the National Cancer Institute recommended fiber intakes of 20 to 30 g d^{-1}, at only 13 g d^{-1} from 1989 to 1990.

II. DIET, NUTRITION AND DISEASE

Six of the ten leading causes of death in the USA have diet as a factor in their etiology. These disorders also represent reduced quality of life and productivity. Approximately 11 to 12% of the population are more than 65 yr of age and this group represents more than 35% of health-care expenditure. The diseases and disorders for which crop consumption appears to be relevant include obesity, coronary heart disease, cancer, diabetes, immune function, and neural tube defects.

Obesity is associated with a variety of health problems including diabetes, hypertension, hyperlipidemia, cardiovascular disease, and some types of cancer. More than one-third of Americans are overweight or obese. It may not be surprising, therefore, that the major causes of death are chronic diseases including cardiovascular disease and cancer.

Saturated fatty acids influence coronary heart disease risk by increasing serum cholesterol levels, whereas, unsaturated fatty acids and monounsaturated fatty acids lower serum cholesterol (Kris-Etherton et al., 1988). For every 1% decline in total cholesterol or the highly correlated LDL-cholesterol, the reduction in coronary heart disease is approximately 2 to 3% (NIH, 1993).

Compared to an average American diet, diets low in fat can decrease serum cholesterol by 7%, and addition of 40 to 50 g dietary fiber d^{-1} can reduce serum cholesterol an additional 5% or more (Van Horn et al., 1986; Hunninghake et al.,

1994). Some soluble fibers such as those found in oat bran and bean (*Phaseolus vulgaris* L.) reduce serum cholesterol by as much as 15% (Anderson et al., 1994). Soybean protein also is hypocholesterolemic.

Another potential role of diet in modulating development of atherosclerosis is the presence of antioxidant nutrients. Oxidation of LDL-cholesterol is a key event in cholesterol accumulation. Plants contribute several key antioxidant nutrients including Vitamin E, Se, β-carotene and other flavonoids (Ascherio & Willett, 1995).

Because there are many different kinds of cancer, usually classified by their tissue site, there are different diet etiologies. In the USA in 1993, of the 1 170 000 new cases of cancer, 45% were cancers of the breast, colon, rectum, prostate, or pancreas (Am. Cancer Soc., 1993). Fat intake is associated with prostate and colorectal cancers, and to a lesser extent, lung cancer (Austoker, 1994). The relationship of fat intake and breast cancer in epidemiological studies, where the range of fat intake is relatively small, is equivocal (Graham et al., 1992). Diets high in fiber are associated with a reduced risk of colorectal cancer (Austoker, 1994). Modifications in diet have been estimated to reduce the risk of cancer by one-third to two-thirds (Austoker, 1994). Antioxidant nutrients also may play a role in cancer prevention. A combined supplementation treatment of β-carotene, Vitamin E, and Se for 5 yr reduced gastric cancer death by 11% in rural China (Blot et al., 1993). However, 6 yr of treatment with 20 mg β-carotene and/or 50 mg Vitamin E did not benefit Finnish smokers (The Alpha-Tocopherol, Beta Carotene Prevention Study, 1994). Diets high in fruits, vegetables, and whole grains are associated with reduced risks of colon cancer. A partial explanation may be the contribution of folate from these foods since red blood cell folate was inversely related to polyp risk in men (Bird et al., 1994). Crops may contain other phytochemicals that are chemopreventive. Identifying and exploiting phytochemicals in crops is perhaps the most promising area of research.

A diet rich in complex carbohydrates and low in fat, especially saturated fatty acids, is recommended for noninsulin-dependent diabetes mellitus. A diet high in monounsaturated fatty acids was equally effective in maintaining glucose homeostasis without worsening plasma lipid concentrations in these patients (Bonanome et al., 1991).

A relatively newly appreciated diet-health relationship is the awareness of the importance of folic acid for women in early stages of pregnancy in preventing birth defects, e.g., neural tube defects. Good sources of folic acid include dark green leafy vegetables, citrus fruits, legumes, and fortified breakfast cereals. It is currently difficult to achieve the recommended 400 μg folic acid each day through the diet. Attempts to increase the folic acid content of crops could greatly assist this dilemma, especially if the current attitude against fortification persists.

III. DIETARY GUIDELINES FOR GOOD HEALTH

The relationship between diet and risk of chronic diseases has led many groups to make dietary recommendations for health promotion and disease prevention. The National Cholesterol Education Program (NCEP) recommends that all Americans older than 2 yr of age follow the NCEP/American Heart Association Step-One diet, which specifies a total fat intake of ≤30% of calories, <10% of energy as saturated fatty acids and <300 mg d^{-1} cholesterol, and to achieve and maintain ideal body weight. Both this diet and a diet low in saturated fatty acids and high in monounsaturated fatty acids lead to a reduction in plasma total cholesterol and LDL-cholesterol. However, only about 20% of Americans are consuming a diet that meets the fat guidelines of the Step-One diet.

The low fat content and contributions of plant crops to dietary fiber and important vitamins and minerals have led to the prominent positioning of these foods in the various dietary guidelines. In addition, the gums in legumes and water-soluble fiber in oat bran have cholesterol-lowering properties.

The USDA Food Guide Pyramid emphasizes the Bread, Cereal, Rice and Pasta Group as the foundation. The recommendation is to consume 6 to 11 servings from this group daily. Beans are included in the protein-rich group. These can be marketed as a low fat, high fiber choice in this food group.

The 1995 Dietary Guidelines for Americans published by the USDA and Department of Health and Human Services specifically recommended choosing a diet with plenty of grain products (USDA, 1995). In contrast to the 1990 Dietary Guidelines, grain products were listed before fruits and vegetables to be more consistent with the Food Guide Pyramid. Incorporation of crops into the diet facilitates meeting the other dietary guidelines of eating a variety of foods, maintaining healthy weight and choosing a diet low in fat, saturated fat, and cholesterol. Although awareness of the diet-health link has increased, Americans still fall short of the dietary guidelines, possibly because they do not believe that changing their diet will influence health, they are not motivated to change their diet, or do not understand how to make better food choices (Frazão, 1995).

The USDA evaluates the American diet in meeting the Dietary Guidelines and Food Guide Pyramid by the Health Eating Index. Using the 1989 and 1990 Continuing Survey of Food Intake on 3997 and 3466 individuals in the respective years, consumption of grains received an average score of 6.1 out of the maximum score of 10.0 (USDA, In Press). Only 9.8% of the population received a score of 10. Eating cereals and grains is so basic, this group is often ignored. It may require designing plants that incorporate beneficial constituents to capture marketing attention.

IV. NUTRITIONAL QUALITY—MACRONUTRIENTS

A. Proteins

1. Protein Quality

Protein in the diet is digested to its constituent amino acids, and amino groups are utilized for protein synthesis for proper growth, development, and function in humans. Evaluating protein quality and protein requirements is important, especially for plant foods. The quality of protein is dependent on the amino acid composition of that protein, its digestibility, and the physiological, nutritional, and health status of the individual consuming the protein.

Rassin and Bhatia (1992) described protein nutrient sufficiency as the supply of protein in the range such that normal physiological response results. Young and Pellett (1991) describe the function of proteins as a supply of indispensable (essential) amino acids and N to meet physiological needs. Many different methods have been employed to determine normal physiological needs, and review articles describe protein status determination, including blood urea N, plasma proteins, plasma amino acids, and stable isotope methodology (Rassin & Bhatia, 1992). This chapter will describe the determination of protein quality of plant foods focusing on protein digestibility. The protein quality of specific plant foods will be compared.

Evaluation of protein traditionally was determined by the Protein Efficiency Ratio (PER) method. This measurement of grams of weight gain of an animal per gram of protein intake compared to an accepted standard protein of casein was the protein evaluation method of choice for decades. The limitations of this technique, however, have led to its dismissal as the accepted method of protein quality determination. The PER determines the ability of a protein to support growth in young rapidly growing rats. This use of the rapidly growing rat model, however, leads to the overestimation of protein value of some animal proteins and the underestimation of the value of plant proteins, because during rapid growth there is an increased need for essential amino acids that are limited in plant proteins. Other concerns about the PER technique are that weight gain does not necessarily mean protein synthesis; results of the method vary with food intake and proteins that do not produce growth cannot be evaluated (Satterlee et al., 1979). When plant foods contribute a large percentage of dietary calories, use of the PER method of protein quality determination can be very deceptive in measuring the quality of the diet. In a time of health concern and emphasis on decreased fat and saturated fat in the diet, overemphasis of animal food consumption for its protein value can be a contradiction to recommendations to decrease fat and cholesterol intake (Henley & Kuster, 1994).

Another method, the Net Protein Retention (NPR) method of nutritional evaluation of proteins, has the advantage that maintenance use of protein is taken into account as well as protein used for growth by comparing the weight gain of animals on a 10% protein test diet compared to animals maintained on a no-protein diet. This method is not without drawbacks in that it does not take into account endogenous turnover of protein in the animal. Endogenous protein losses in pancreatic secretions, mucosal cells, endogenous urea ammonia, and free amino acids can have a profound effect on calculated protein use (Sauer & Lange, 1992).

2. Protein Digestibility: In Vivo

The FAO/WHO Expert Consultation report on Protein Quality Evaluation (FAO and WHO, 1990) has carefully considered the limitations of the PER method and other quality determination methods and reported the method of protein quality determination called protein digestibility—corrected amino acid score (PDCAAS) to take its place. In this technique, the true protein digestibility of a test protein is determined in vivo using the fecal index method. The true digestibility calculation shown below is used

$$TD = [I - (F - Fk)/I] \times 100$$

where TD = true digestibility, I = N intake, F = fecal N, and Fk = endogenous fecal N.

Table 3–6. Suggested pattern of amino acid requirements†.

Amino acid	Suggested pattern of requirement preschool child (2–5 yr)
	mg g^{-1} crude protein
Histidine	19
Isoleucine	28
Leucine	66
Lysine	58
Methionine + cystine	25
Phenylalanine + tyrosine	63
Threonine	34
Tryptophan	11
Valine	35
Total	
+ histidine	339
– histidine	320

† FAO and WHO, 1990.

The PD-CAAS determination depends on the essential amino acid ratios of the test protein and a reference pattern. This is the milligrams of an essential amino acid in 1.0 g of test protein mg^{-1} of the same amino acid in 1.0 g of a reference pattern for nine essential amino acids plus tyrosine and cystine. The FAO/WHO suggested reference pattern of amino acid requirements is that for the preschool child (Table 3–6). The essential amino acid requirement pattern of the 2- to 5-yr old is the most demanding pattern other than that of the infant, and will lead to some underestimation of protein value, but not excessively. With the true digestibility known, the PD-CAAS is calculated by multiplying the lowest amino acid ratio in the test protein by the true protein digestibility. A comparison of values from the N balance method and PD-CAAS for several sources is given in Table 3–7.

The Food and Drug Administration has called for the replacement of PER with the PD-CAAS method for nutritional labeling regulation. The FDA states the advantages of the PD-CAAS as using the human amino acid requirements and taking into account the food proteins essential amino acid profile, its digestibility, and its ability to supply essential amino acids in amounts required by humans (Henley & Kuster, 1994).

The nutritional quality of protein has generally been accepted as the amino acid composition of that protein assuming complete hydrolysis through digestion and that resulting free amino acids are absorbed. This assumption does not take into account many interfering factors that can prevent complete hydrolysis of proteins and the absorption of amino acids. These interfering factors include the conformation of the protein, tannins, phytic acid, proteolytic enzyme inhibitors, hemagglutinins, complexing of protein with starch, hemicellulose, minerals, other proteins, glycoproteins, Maillard browning reaction products, or fiber in the diet (Nielsen, 1991; Swaisgood & Catignani, 1991). With plant proteins these factors can be significant in decreasing the nutritional quality of the protein.

In *Phaseolus vulgaris* L., trypsin inhibitors, condensed tannins, and dietary fiber can reduce the digestibility of protein (Barampana & Simard, 1994). Godinez et al. (1992) reported an increased protein digestibility of common black bean protein by nearly 100% after elimination of the seed coat and cooking water. Cooking

Table 3–7. Values for the percentage digestibility of protein in various food products as determined by N balance method and biological value PD-CAAS.

Protein source	N balance†	PD-CAAS‡
Casein	100	100
Pea flour	88	69
Pea, century (autoclaved)	83	
Pinto bean (canned)	79	68
Lentil (autoclaved)	85	52
Fava bean (autoclaved)	86	
Soybean	90	
Soybean protein isolate	98	100
Rapeseed protein concentrate	95	
Peanut	96	
Peanut meal	91	52
Peanut butter	98	
Sunflower meal	90	
Wheat	93	40
Rolled oat	94	57
Rice-wheat-gluten	93	
Wheat flour-casein	95	
Rice-soybean	90	
Corn-pea	83	
Corn-soybean	93	

† FAO and WHO, 1990.
‡ Henley and Kuster, 1994.

and drying significantly increased protein digestibility in chickpea (*Cicer arietinum* L.) and lentil (*Lens culinaris* Medikus) (Estevez et al., 1991). Reported levels of phytic acid in the chickpea dropped from 2.99 to 1.64 mg $(100\ g)^{-1}$ after soaking, and trypsin inhibitors were decreased by 50% after heat treatment. Fermentation of protein decreases inhibiting factors from plants. Barley (*Hordeum vulgare* L.) and wheat fermented with buttermilk (Rabadi) had reduced phytic acid levels from those in the nonfermented product (Gupta et al., 1992; Gupta & Khetarpaul, 1993). Fermentation of soaked-cooked bean also led to a decrease in the phytic acid levels by 32% (Barampama & Simard, 1994).

3. Protein Digestibility: In Vitro

In the food industry, food product protein composition may not change, but processing conditions such as thermal processing, water activity, pH, and component concentration may vary markedly. A rapid method of determining the effect of processing conditions on protein digestibility is necessary. In vitro assays for protein digestibility are fast, economic, and commercially acceptable for protein quality monitoring of large numbers of samples, or as a prescreening procedure in advance of the use of an in vivo method.

In vitro digestibility methods rely on proteolytic enzymes correlated to protein digestion in vivo. Four closely related procedures are the pepsin digest residue (PDR), pepsin digest dialyzate (PDD), pepsin pancreatin digest (PPD), and pepsin pancreatin digest dialyzate (PPDD) methods. As the method's names imply, pepsin or pepsin and pancreatin are used to digest a test protein. The resulting amino acids and small peptides are precipitated and analyzed by automated amino acid analysis. In the PDD and PPDD methods, the resulting digestion products are dialyzed through small exclusion limit dialysis membranes before amino acid analyses (Swaisgood & Catignani, 1991).

A limitation of the PDR and PDD methods is that both methods depend on a single enzyme digestion protocol. Proteins resistant to pepsin would not be digested. Another limitation of these methods is that automated amino acid analysis equipment is expensive and not readily available in all laboratories.

A simpler method, easily standardized in most laboratories, is that of Satterlee et al. (1979). This technique uses a combination of trypsin, chymotrypsin and intestinal peptidase as the digestive enzymes and measures the pH drop following incubation. The technique of Pederson and Eggum (1983) takes this technique one step further. In this method the initial rate of alkali consumption to maintain the initial pH is used to calculate peptide bond hydrolysis. With any of these in vitro methods, special care must be taken to control experimental parameters of protein substrate concentration, enzyme concentration, protein enzyme ratio, pH, ionic strength and incubation time.

Advancements in in vitro protein digestibility methods continue to be made. Swaisgood and coworkers have immobilized the digestive enzymes on a large pore diameter resin material (Chang et al., 1992). This technique, called IDEA for immobilized digestive enzyme assay, has the advantages that the immobilized enzymes do not autolyze and the same immobilized enzyme can be used for a number of assays, conserving valuable digestive enzymes. A novel use of sodium dodecyl sulfate polyacrylamide gel electrophoresis (SDS-PAGE) was introduced by Kim and Barbeau (1991). After the standard pepsin, pancreatin digestion, the resulting peptides were analyzed on a modified SDS-PAGE system allowing for the molecular weight resolution of 1 300 to 100 000 dalton peptides. This technique allows estimation of molecular weight of the polypeptides and peptides remaining after protein hydrolysis.

Whether an in vivo or in vitro method is chosen, protein digestibility is essential to the determination of food quality. Proteins cannot be considered alone when determining the nutrient quality of a plant food.

B. Carbohydrates

1. Simple Carbohydrates and Digestible Starch

The carbohydrates present in plant foods are many and diverse (Table 3–8). Each carbohydrate is digested differently or not at all, depending on its structure and chemical composition. The nutritional impact of a plant food is dependent on the type of carbohydrate present and other factors including: physical structure of food, how long the food is chewed during eating, food processing, intestinal transit time, amylase inhibitors present in the diet, the rest of the meal eaten with the plant food, and the lipids and proteins present with the carbohydrate in the plant food (Tagliabue et al., 1995).

Present in plant foods are simple sugars (mono- and disaccharides), oligosaccharides, and polysaccharides including starch, fiber, and gums. Some of these, mono- and disaccharides and most starch, are highly digestible. Others are resis-

Table 3–8. Main food carbohydrates†.

Monosaccharides
Glucose
Fructose
Disaccharides
Sucrose
Lactose
Oligosaccharides
α-galactosides
Raffinose, stachyose
Fructans
β-glucans
Fructooligosaccharides
Starch‡
Amylose
Amylopectin
Modified food starches
Nonstarch polysaccharides
Cellulose
Hemicellulose
Pectins
Fructans
Gum
Mucilages
Algal polysaccharides

† Adapted from Asp, 1995.
‡ Degree of polymerization > 10 to 20.

tant to digestive enzymes and travel through the small intestine undigested. Digestion of carbohydrates requires polysaccharides to be hydrolyzed to monosaccharides and disaccharides that are absorbed across the small intestine. Nonhydrolyzable polysaccharides pass through the small intestine to the large intestine where they are fermented by intestinal microflora. The end-products of fermentation—predominantly, but not exclusively, short chain fatty acids—can be absorbed in the large intestine and contribute to the nutrition of the animal consuming the food.

The major enzyme involved in starch digestion is amylase. Amylase breaks amylose down to maltodrextrins and glucose. Glucose is absorbed and is transported to tissues. Some food starch is more easily digested than other food starch. Yokoyama et al. (1994) surveyed in vitro starch digestibility of 26 common cereal products and found much variability even from the same grain-derived products. Differences in digestibility were attributed to product particle size, grain, processing used, and the amount of reducing sugars present in the product.

For many years starch was considered a highly digestible carbohydrate source. Today, starch digestibility is understood to be more variable. Englyst et al. (1992) characterized starches into three different categories: rapidly digestible starch (RDS), slowly digestible starch (SDS), and resistant starch (RS). Resistant starch is resistant for a number of different reasons: the starch is physically inaccessible, it is in resistant starch granules, or it is retrograded starch. To compare starch digestibility among food products a starch digestion index (SDI) is determined as the (RDS total starch^{-1}) × 100.

2. Resistant Starch

Starch digestion can be slowed or totally inhibited by the physical form of the food. Starch can be held in rigid plant structures that inhibit access of digestive enzymes to the starch. Chung et al. (1995, 1996) have shown that polymerization of bean cell walls during heating results in individual cells that resist rupture and starch release. The amount of starch gelatinization is inhibited with even 10 min of heat treatment. The starch granules themselves, where starch is stored in the plant cell, can be inhibitory to starch digestion. Starch granules are of three types: A, B, and C, as distinguished by their x-ray diffraction patterns. Cereal starches contain predominantly A type granules and are the most highly digestible. Starch granules of the B and C type are found predominantly in tubers and legumes respectively, and are more resistant to amylase digestion. Starch that is easily removed from granules and is fully gelatinized can still be rendered indigestible if allowed to retrograde. High-amylose starch will retrograde forming a crystalline, tightly bound network that will resist amylase digestion (Eggum et al., 1993).

The rate of glucose release from starch and the absorption of starch into the blood stream determines the glycemic response to a particular starch source. The high levels of resistant starch seen in some food products [green banana (*Musa* × *paradosiaca* L.) and uncooked potato (*Solanum tuberosum* L.)] can be used to the advantage of moderating postprandial glycemic index for individuals with diabetes (Raben et al., 1994).

Processing of plant foods can affect starch digestibility. Extrusion of milled whole grain cereals [wheat, rye, corn (*Zea mays* L.), millet (*Panicum miliaceum* L.) and sorghum (*Sorghum bicolor* L.)] increased the in vitro digestibility especially at extrusion temperatures of 150°C (Dahlin & Lorenz, 1993). A different type of processing or pretreatment can affect starch digestibility. Ghazali and Cheng (1991) showed that germination of dry black gram bean caused an increase in simple sugar content of the bean from enzymatic digestion of starch in the bean.

3. Dietary Fiber

The benefits of dietary fiber (nondigestible cell wall components: cellulose, hemicellulose, pectic substances, mucilages, algal oligosaccharides, and lignin) to health are many. Klurfeld (1987) reviewed the health benefits of increased levels of dietary fiber as including decreased severity of diverticular disease, Crohns disease, and hemorrhoids; decreased gallstone formation and relief of constipation. Controversy surrounds the potential benefit of dietary fiber in the reduction of risk of cancer. Epidemiologic evidence does support broad-scale recommendation of increased levels of dietary fiber to help prevent colon cancer. The direct relationship between high dietary fiber diets preventing certain cancers is clouded by the fact that these same diets are low in fat and usually lower in calories than their comparison low dietary fiber diets (Malville-Shipan & Fleming, 1992).

Evidence of the health benefits of increased levels of complex carbohydrates and dietary fiber in diets is more convincing for disease states like diabetes and heart disease. The work of Anderson and colleagues shows the benefits of high carbohydrate, low fat diets on the severity of diabetes complications. Patients on diets containing 55 to 60% energy as carbohydrate (at least 66% of that from complex carbohydrates) show an improvement in glycemic index, serum atherogenic lipids, and blood pressure and a reduction in body weight in the obese (Anderson & Akanji, 1991). Mechanisms for these improvements are a delay of gastric emptying and/or reduced rate of digestion and absorption of carbohydrates in the small intestine (Liljeberg et al., 1992). Anderson and many others recommend increases in dietary fiber intake for all healthy adults to 20 to 35 g d^{-1} (Anderson et al., 1994). This increase, accomplished over a period of time, is reasonable.

Dietary fiber has a second benefit of a hypocholesterolemic effect in monogastrics. A decrease in serum cholesterol is beneficial in the prevention of heart disease. The dietary fiber contained in barley and oat grains is distinctive from that of wheat and is particularly effective in lowering serum cholesterol levels. Barley and oat contain β-glucans, cell wall polysaccharides of glucopyranosyl units with 1-3 and 1-4 linkages. These β-glucans absorb water and form a viscous solution in the gastrointestinal tract. Monogastrics have very low levels of β-glucanase to digest this carbohydrate in the small intestine. There is poor absorption of many food nutrients from the viscous solution. Poorly reabsorbed are high cholesterol components like bile acids, a potential benefit of β-glucans (Illman et al., 1993), but also, minerals and vitamins from the food source leading to poor growth in animal models (Wang et al., 1992).

The study of dietary fiber and its effect on health has been hampered by methodology. Determination of fiber from complex plant matrixes is difficult. The heterogeneity of components of dietary fiber leads to one method of measurement working for one type of fiber and not for others. The main methods of dietary fiber analysis fall into the categories of gravimetric or chemical methods. Each method has its unique advantages and has led to some confusion in the literature when comparisons are made (Marlett, 1993).

C. Fats

If the health benefits of dietary fiber are controversial, the health benefits of plant fats are even more so. Dietary advice stresses the decreased consumption of fat in American diets to reduce the risk of many chronic diseases, including heart disease, obesity and cancer. Current health recommendation stress decreased total fat, saturated fat and cholesterol consumption as well as increasing the proportion of monounsaturated and polyunsaturaed fats in the diet. Plant lipids have a high proportion of mono- and polyunsaturated fats. A slow decrease in consumption of total fat is seen in American diets, with a slow movement from consumption of highly saturated animal fats to unsaturated plant fats. Putler and Frazao (1991) report USDA dietary intake surveys showing a decrease in total fat intake from 41.8 to 37.3% of calories over the years 1977 to 1985. Consumption of less fat from red meats and higher levels from grains and legumes are reported in these same surveys. Yet, the average American diet still contains more than the recommended 30% of calories from fat. Certainly the consumption of more fruits, vegetables, grains and legumes, with their lower level of total fat, will assist Americans in making this transition to a lower fat diet.

Although lipids are a minor constituent of cereal grains, they are a good source of essential fatty acids (Sridhar & Lakshminarayana, 1994). Table 3–9 summarizes

Table 3–9. Total fat and fatty acid percentage in selected crops.

	Rice bran oil†	Sorghum‡	Soybean§	Common bean¶	Oat#	Wheat††	Barley‡‡
Total fat, %		6	18–20	1.3	8– 10	1.9	2.4
Palmitic (C 16:0)	23.7	11.1	13.6	8.3	15.6	14.2	16.7
Stearic (C 18:0)		2.1	1.8	1.2	1.4	0.79	1.3
Oleic (C 18:1)	36.5	38.2	28.0	9.5	42.3	11.5	16.2
Linoleic (C 18:2)	39.0	36.9	48.2	20.5	38.1	38.8	50.1
Linolenic (C 18:3)		2.5	8.2	40.5	1.3	2.0	5.9

† El-Massry et al., 1992.
‡ Sood et al., 1991.
§ Bhagya and Srinivas, 1992.
¶ Hsieh et al., 1992.
# Molteberg et al., 1995
†† Agriculture Handbook 8-20 USDA, HNIS, 1989.
‡‡ Anjum and Ali, 1991; McIntosh et al., 1993.

the fatty acid profiles of a number of common grains and legumes. The total fat content of each grain is low, but provides a large proportion of essential fatty acids (linoleic, linolenic, arachidonic acids). Soybean, used prodomiantly as an oilseed, is nearly 50% linoleic acid.

V. NUTRITIONAL QUALITY—MICRONUTRIENTS

A. Minerals

Micronutrient deficiencies affect over 2 billion people worldwide. Cereals and legumes are important sources of several minerals and vitamins, but their bioavailability is typically lower than food from animal sources. Mineral bioavailability is typically reduced by constituents found in cereals and legumes including phytic acid, tannins, saponins, and oxalic acid. Fibers found in plant foods also have been implicated as adversely affecting mineral bioavailability. In cereals, phytic acid is located in fiber-rich components of seeds (Weaver et al., 1991). Mounting evidence suggests that phytic acid and perhaps other chemical constituents are more responsible for reducing mineral bioavailability than the fiber, per se. For example, phytate-rich extruded wheat bran cereal has an adverse effect on Ca absorption (Weaver et al., 1991), but purified cellulose, the major fiber in wheat bran, has no impact on Ca absorption (Heaney & Weaver, 1995). Similarly, wheat bran and phytate, but not wheat bran fiber, reduced Zn bioavailability in rats (Davies et al., 1977).

Phytic acid is the hexaphosphate of myoinositol. It is the storage form of P in seeds and grains. Phytic acid exists as crystalline globoids inside protein bodies in the cotyledon of legumes or oilseeds and in the aleurone layer of cereals. The negatively charged phosphate groups complex divalent cations with differential affinity as $Cu^{2+} > Zn^{2+} > Co^{2+} > Fe^{2+} > Ca^{2+}$. Complexes formed between phytic acid and minerals may include amino acids or proteins as peptide-M^{2+}(divalent mineral cation)-phytate. Humans lack sufficient intestinal phytase to degrade the complexes. Thus, mineral bioavailability including P, is frequently reduced in phytate-rich foods. The extent of the effect of phytic acid on Zn bioavailability can be

Table 3–10. Fermentation increases mineral utilization.

Mineral	Test species	Product	Bioavailability (%)	Reference
Ca	Human	Wheat bread	70.3 ± 10.8†	Weaver et al., 1991
		Wheat cookies	64.8 ± 8.7	
		Pinto bean, untreated	23.1 ± 5.3	Weaver et al., 1993
		Pinto bean, phytase treated	31.8 ± 7.1†	

† Significantly greater than unfermented group at $P<0.05$.

predicted by Ca/Zn molar ratios. A dietary phytic acid × Ca/Zn molar ratio > 3.5 moles kg^{-1} in rats or >100 mmoles $(1000\ kcal)^{-1}$ in humans is considered to have an adverse effect on Zn status (Fordyce et al., 1987; Morris & Ellis, 1989).

Tannins, a complex group of phenolic compounds, are another class of chemicals present in pigmented legumes that bind both proteins and minerals, especially Fe (Rao & Prabhavati, 1982). In rats, utilization of proteins is more affected than absorption (Butler, 1989). Tannins do not bind all proteins or minerals equally well. For example, Ca absorption was not influenced by tannin content of three varieties of legumes (Weaver et al., 1993). Similarly, saponins, triterpenes linked to one or more sugars that are found in legumes, can bind Zn and reduce absorption of nutrients (West & Greger, 1978).

Oxalic acid is the most potent inhibitor to Ca absorption known (Heaney & Weaver, 1989). Legumes have high concentrations of both oxalic acid and phytic acid, and therefore, Ca bioavailability from kidney bean, navy bean, and red bean is half that from milk (Weaver et al., 1993). Ironically, although soybean contains a higher concentration of oxalic acid than dried bean, Ca absorption is superior from soybean (Heaney et al., 1991).

Processing and milling procedures that reduce the level of phytate and other constituents can increase mineral bioavailability. Removal of the bran and aleurone layer in milling of cereals increases mineral utilization (Erdman, 1981). Unfortunately, this also decreases mineral and vitamin content, which has led to enrichment of cereal products with Fe, thiamin, niacin, and riboflavin. In contrast, soybean hulls are a rich source of trace elements, and the minerals are well absorbed due to the absence of phytate in the hulls (Levine et al., 1982). Using soybean hulls as a source of Fe for bread results in Fe utilization that is as good as enriching with bakery-grade ferrous sulfate (Johnson et al., 1985).

Fermentation of seeds or products with yeast allows yeast and wheat phytases to efficiently hydrolyze phytic acid (Ranhotra et al., 1974). Fermentation or treatment with phytase has been shown to increase utilization of Zn (Ranhotra et al., 1978) and Ca (Weaver et al., 1991, 1993) as shown in Table 3–10. When phytate is removed from seeds or their products following isoelectric precipitation of proteins, Zn utilization also is improved.

Final pH of processed products also can influence mineral utilization. For example, Zn absorption from neutralized isolated soy protein was less than for the non-neutralized acid precipitated product, where complexation of Zn with amino acids and phytic acid is less likely (Khan et al., 1990).

The literature has been confusing on factors affecting bioavailability of nutrients from cereals and legumes because of poor methodological approaches. For example, use of purified phytic acid, which is highly charged, to demonstrate its effect on mineral absorption likely produces different qualitative results than phytate endogenous to the plant seeds, which may already be chelated to various minerals. When plants are grown with varying levels of P, the seeds have varying levels of phytic acid, and the impact of phytic acid on mineral bioavailability can be clearly demonstrated. This approach has shown the negative impact of phytic acid on Zn, Fe, wheat and Ca in soybean (Beard et al., 1988; Saha et al., 1994; Heaney et al., 1991; Mason et al., 1993). Another controversial method is the use of extrinsically added radiolabels of minerals to plant products for studying mineral bioavailability. Extrinsic labeling has been validated for some minerals in some plant foods, but overestimates absorption for others (Weaver, 1985). Therefore, this discussion was based on studies using intrinsically labeled foods or the slope ratio assay in animals, which does not require isotopic labeling, and foods where endogenous phytate levels were manipulated where possible.

Fortification of plant foods with minerals or processing procedures that enhance bioavailability can improve mineral nutriture. Alternatively, genetic breeding can enhance mineral content or reduce the levels of inhibitors. For example, some plant breeders have identified soybean and corn lines with low levels of phytic acid (Raboy & Gerbasi, 1996). Another approach would be to increase the content of certain amino acids (methionine, cysteine, and lysine) in staple crops that enhance Fe and Zn bioavailability (Graham & Welch, 1996) unless they increase the formation of insoluble complexes with phytic acid.

B. Vitamins

Crops that are low in lipids, i.e., cereals, tend to be low in fat-soluble vitamins. Oil-rich crops or fractions such as oil seeds and corn and wheat germ are sources of Vitamin E. Cereal grains are good sources of tocols (tocopherols and tocotrienols). Tocopherols are the main forms of Vitamin E, but grains contain more tocotrienols than do other food products. Tocols have Vitamin E activity and antioxidants functions. Tocotrienols inhibit cholesterol synthesis (Qureshi et al., 1986). Because tocol concentrations vary with genotype and environment, manipulation by breeding is promising. Brewer's spent grains and the by-product of pearled barley are richer sources of tocols (Peterson, 1994). Cereals contain some carotenoids. These nutrients are relatively stable to acid and heat, but are vulnerable to oxidation upon exposure to ultraviolet light and pro-oxidants.

Whole cereal grains are important sources of B vitamins, particularly thiamin, riboflavin, niacin, and pyridoxine (B_6). Enrichment of flour in the USA provides 0.44 to 0.55 mg thiamin, 0.26 to 0.32 mg riboflavin, and 3.56 to 4.45 mg niacin $(100\ g)^{-1}$ flour (NAS, 1950). These nutrients also are added to processed food cereals. Processing has a variable effect on loss of B vitamins. The most labile of these nutrients is thiamin that can be lost in air at higher pH or through autoclaving. Thus, more than 70% of the thiamin can be lost during baking of cookies made with baking soda (Bock, 1991). Thiamin also can be destroyed by potassium bromate (an oxidizing agent used in bread), by sodium bicarbonate (a leavening agent), and sulfur dioxide (a food preservative). Riboflavin is vulnerable to light. Niacin is stable except to leaching.

Table 3–11. Niacin content and proportion of bound niacin in selected cereals and beans.†

Product	Total niacin (μg g^{-1})	% bound
Corn	25.7	98
Boiled corn	23.8	84
Tortillas	12.6	7
Boiled sorghum	45.5	98
Boiled rice	70.7	76
Boiled wheat	57.3	100
Baked bean	24.0	20

† Wall and Carpenter, 1988.

Cereals are moderate sources of Vitamin B_6. Although enrichment of flour with Vitamin B_6 at a rate of 4.4 mg kg^{-1} (2.0 mg lb^{-1}) has been proposed (NAS, 1974), it has never been implemented. As much as 50 to 70% of Vitamin B_6 is lost during roasting of cereal products (Gregory & Kirk, 1978).

Bioavailability is of concern for some of the vitamins in crops, but not for Vitamin E. Of particular concern is niacin, which is found in free and bound forms (Table 3–11). The bound form is poorly utilized by humans and constitutes a major portion of some cereals prior to enrichment. The complex is poorly characterized and involves carbohydrates, peptides, and phenols and is heterogeneous in nature. The bound forms in cereal grains are associated with nicotinic acid. Such complexes exhibit no niacin activity unless hydrolyzed. Heating releases free nicotinamide from NAD and NADP during boiling of corn (Wall & Carpenter, 1988). In most cereal grains, Vitamin B_6 also is partially present in the glycosylated form as pyridoxine-5-β-D-glucoside. The existence of this form was first proven in rice bran (Yasumoto et al., 1977). The effect of glycosylation on bioavailability is demonstrated in Table 3–12.

Wheat bran is a good source of folic acid at 90 to 300 μg 100 g^{-1}, but most cereals are moderate sources of folic acid (30–90 μg 100 g^{-1}). Legumes are much higher sources of folic acid. Folic acid also is present in different forms, much of it is present as conjugated pteroyl polyglutamates. The polyglutamyl chain of folates must be hydrolyzed by pteroylpolyglutamate hydrolase prior to absorption. Bioavailability of folates averages 50% or less from the whole diet (Sauberlich et al., 1987) and 50 to 60% from cereals (Colman et al., 1975). Soybean contains 40% monoglutamate analogues, which are absorbable. Enrichment of cereal-grain products with folic acid has been mandated by FDA since February 1996 in an attempt to benefit women who become pregnant, with the goal of reducing birth defects such as spina bifida and anencephaly. A concern is that overingestion of total folate masks Vitamin B_{12} deficiency symptoms, which could affect 13 000 Americans each

Table 3–12. Effect of glycosylation on vitamin B_6 bioavailability.†

Foods tested	Total Vitamin B_6 (μg 100 g^{-1})	% glycosylated Vitamin B_6	Average bioavailability (%)
Tuna	316	0	100
Whole wheat bread	169	17	75
Peanut butter	302	18	63

† Kabir et al., 1983.

year compared to the potential reduction of 1000 neural tube defects per year in the USA (Pérez-Escamilla, 1995). Therefore, the range of fortification with folic acid is between 0.946 to 3.08 mg kg^{-1} (0.43–1.4 mg lb^{-1}) product to keep daily intake of folic acid below 1 mg. Folates are vulnerable to oxidative degradation. Analytical methods for folate lag behind those for the other B vitamins used in the enrichment program. Based on microbiological assay of folate in serum of humans fed maize and rice fortified with folate, these foods would be suitable vehicles for fortification (Colman et al., 1975). The vitamin was stable to boiling but underwent some loss upon baking.

Cereals are a moderate source of biotin (100–1000 μg kg^{-1}). The relative content and bioavailability to chicks varies considerably among cereals and is shown in Table 3–13.

C. Strategies to Enhance Micronutrient Quality

Several strategies can be used to enhance the micronutrient content of cereals and legumes including fortification and enhancing the natural levels of those nutrients in seeds. Fortification with vitamins and minerals requires less effort and is relatively inexpensive compared to genetic manipulation. The effectiveness of fortification programs is evidenced by the elimination of most overt micronutrient deficiencies in the USA. However, concern over excessive accumulation of some nutrients, especially Fe, has led to much resistance in fortifying the food supply in the USA. Thus far, manufacturers of breakfast cereals have liberally added vitamins and mineral salts to their products, but many other products are not fortified. The marginal Fe, Ca, Zn, Mg, and antioxidant nutrient status of many people in the USA indicates a need for increased consideration of meeting micronutrient needs in the food supply. Micronutrient intervention strategies may prevent the consequences of marginal malnutrition on cognitive development and growth observed in developing countries (Allen, 1993).

An agronomic strategy to increase nutrient content of crops is to breed or use biotechnology to produce varieties that efficiently accumulate or synthesize the nutrient of interest in their seeds. Mineral efficient plants are more tolerant to mineral-deficient soils. Efficient genotypes extract more of certain minerals from deficient soils. Production of nutrient dense seeds is a low-cost strategy to improving nutrition in developing countries while simultaneously improving seedling vigor and yield since these seeds require less fertilizer. The primary selection criterion is the micronutrient content of the seeds. Prime nutrient candidates for this strategy are Fe, Zn, Mn, Vitamin A and compounds that promote bioavailability of nutri-

Table 3–13. Biotin content and bioavailability to chicks in selected cereals.†

Grain	Biotin (μg kg^{-1})	Bioavailability (%)
Corn	45	100
Wheat	104	0
Barley	144	20
Oat	208	32
Milo	290	20

† Frigg (1976).

ents. As much as one-half of all soils in developing countries are deficient in minerals, which makes this strategy especially appealing (Graham & Welch, 1996). However, it must be remembered that no nutritional advantage is achieved if the extra micronutrients are accumulated in the pericarp and seeds are processed to remove the pericarp prior to consumption.

VI. HEALTH PROTECTANT SUBSTANCES IN GRAINS AND LEGUMES

Increasing attention is being given to bioactive constituents found in small amounts in common foods that can promote health and protect against disease [see Kitts (1994) and Thompson (1994) for reviews of bioactive substances]. A new concept in industry that involves enriching or fortifying foods with these substances is emerging. Such foods are termed functional foods or designer foods. Identification of such constituents in foods and efforts to increase their concentration is one of the potentially most exciting areas of investigation in the next decade. The concentration of a bioactive substance can be increased genetically through conventional plant breeding or biotechnology or through addition during processing.

Food groups considered to have the most diverse and greatest quantity of health protectants include fruits, vegetables, and grains. Functional foods have tremendous implication for cereals with a $4 billion market in the USA. The healthful components are in greater concentrations in the bran and germ; thus whole grains are likely to be more important to health than refined grains. Because of the potential role of health protectants in these foods to prevent cancer, the National Cancer Institutes (NCI) Diet and Cancer Branch has instituted the Experimental Food Program to study how phytochemicals (bioactive plant compounds) may prevent cancer and to develop designer foods. The general role of phytochemicals in cancer prevention is indicated in Fig. 3–1.

Fourteen classes of phytochemicals have been identified in common foods. Cereal grain and soybean contain phytochemicals in eight of these classes as shown in Table 3–14. Therefore, soybean, whole wheat, brown rice, flax (*Linum usitatissimum* L), and oat have been targeted for study by the NCI Experimental Food Program. Soybean is among the six foods identified as the most promising, and cereals are in the second tier of important foods to study.

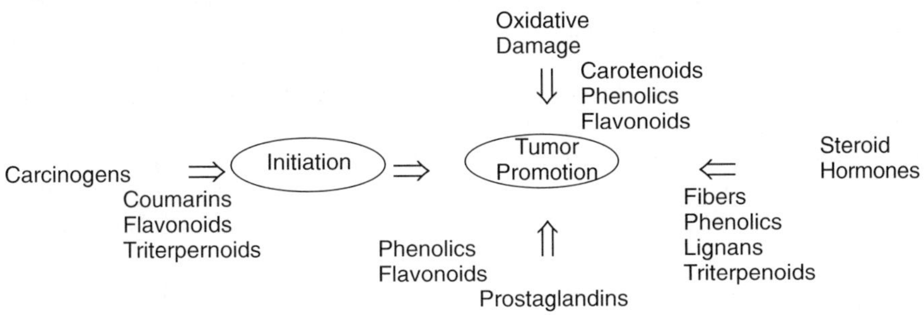

Fig. 3–1. Scheme showing role of phytochemicals in cancer prevention. The phytochemicals are listed by their supposed site of action with the scheme within double lines across arrows denoting inhibition of that step.

Table 3–14. Classes of food phytochemicals contained in soybean and cereals (indicated by X)†.

	Soybean	Cereal grain
Sulfides		
Phytates	X	X
Flavonoids	X	X
Glucarates		X
Carotenoids	X	X
Coumarins	X	X
Mono-terpenes		
Tri-terpenes	X	X
Lignans	X	
Phenolic acids	X	X
Indoles		
Isothiocyanates		
Phthalides		
Polyacetylenes		

† Taken from Caragay, 1992.

Isoflavonoids in legumes are estrogenic compounds that include the glycosides of diadzidein and genistein. Soybean plants are good sources of these estrogenic compounds, which may be metabolized by humans to less or nonestrogenic compounds. Each of these phytoestrogens exhibit varying degrees of affinity for the estrogen receptor, and therefore, exert varying potencies of biological activity. For example, genistein is 50 times less potent than estradiol in exerting estrogen activities, whereas, the binding affinity of daidzidin is much weaker (Bickoff et al., 1962). Because phytoestrogens may bind to mammalian estrogen receptors, they are proposed to modulate estrogen-dependent cancers such as breast cancer. In fact, soybean intake and urinary phytoestrogen have been highly correlated with reduced risk of cancer (Aldercreutz et al., 1990). The mechanism(s) of chemoprevention by phytoestrogens may involve modulation of steroid metabolizing enzymes, inhibition of tyrosine-specific kinases, inhibition of DNA topoisomerases, inhibition of angiogenesis, and action as antioxidants (reviewed in Morton et al., 1997).

While phytoestrogens have been shown to inhibit estrogen-dependent diseases through a receptor-mediated mechanism, soybean also contains other chemopreventive compounds known as protease inhibitors. Protease inhibitors such as the Bowman-Birk protease inhibitor is chemopreventive against chemically induced colon, liver and cheek pouch carcinogenesis in experimental animals (reviewed in Kennedy, 1993). Soy-derived protease inhibitors offer a nonpharmacologic alternative for the prevention of cancer. Recent studies have shown that oral gavage of effective chemopreventive doses of the soy-derived fraction containing the 8-kDa Bowman-Birk inhibitor protein in mice did not stimulate an allergic response (Maki et al., 1994). However, it has been shown that diets containing pure trypsin inhibitors enhanced pancreatic carcinogenesis in experimental animals (Roebuck, 1987).

Flax is the most potent source of plant lignans known. However, whole-grain cereals, legumes, vegetables, and fruit contribute total lignans to the diet because of their volume of consumption. Lignans are a group of diphenolic compounds with characteristics similar to phytoestrogens. Intestinal microflora convert both lignan and phytoestrogens to compounds with estrogenic activity. Flax is being studied for breast and colon cancer risk reduction (Wrick, 1993). Flax incorporated into the diets of rats reduced epithelial cell proliferation in the colon (Serraino & Thompson, 1991b) and mammary gland (Serraiano & Thompson, 1991a) and reduced tumor formation (Serraino & Thompson, 1992).

Fibers in cereals, bran and legumes may be protective against cancer through reduction in serum estrone and estradiol levels as has been shown for wheat bran but not oat or corn bran (Rose et al., 1991). Dietary fiber can bind deconjugated estrogens and increase their fecal output (Whitten & Shultz, 1988). Alternatively, fiber may inhibit carcinogenesis by binding bile acids. In the case of colon cancer, fiber may work by increasing rapid intestinal clearance (especially insoluble fiber, e.g., cellulose, in wheat bran) or fostering bacterial fermentation, which increases volatile fatty acid production that has anticarcinogenic benefits. Phytic acid found in legumes and cereal bran may reduce cancer risk by binding Fe, a potent pro-oxidant (Graf & Eaton, 1993).

Antioxidant nutrients act by scavenging reactive oxygen species such as superoxide anion and hydroxyl radicals. Free radicals are thought to be involved in the early processes of both cancer and cardiovascular disease. Free radicals can denature proteins, oxidize lipids, and damage nucleic acids. Increasing evidence suggests that oxidized low-density lipoproteins may be involved in atherogenesis. The three most studied antioxidant components are Vitamin C, Vitamin E, and carotenoids. Vitamin E is especially important as an antioxidant within lipid membranes. However, not all intervention trials have shown that a particular supplement decreases the incidence of disease. The Alpha-Tocopherol, Beta Carotene Cancer Prevention Study (1994) showed no benefit of either β-carotene or α-tocopherol alone or combined in reducing in 29 000 middle-aged Finnish male smokers. In fact, the group randomized to β-carotene supplementation had significantly more new cases of lung cancer, although prostate cancer was reduced among those receiving Vitamin E. It is likely that whole foods including fruits, vegetables, and cereals show more positive health promoting effects because we do not fully understand the constituents that are important or the necessary concentrations and in what combinations. Phenolic acids in cereals as well as Vitamins E and C may inhibit formation of N-nitroso compounds. Selenium in grains may be an anticarcinogen by regulating glutathione peroxidase, which scavenges free radicals. Copper, Zn, and Mn may perform a similar role since they are components of enzymes performing antioxidant functions.

Consumption of flavonoids including quercetin, kaempferol, myricetin, apigenin, and luteolin has been associated with reduced mortality from coronary heart disease (Rahman et al., 1991). However, those consuming the highest third of flavanoids also consumed more dietary energy, Vitamin C, and beta-carotene. Epidemiological studies can uncover relationships, but cannot be used to determine cause and effect, partially because of such confounding factors.

Fibers also are effective in reducing risk of coronary heart disease. Gums found mainly in legumes and oat are especially effective dietary fibers in reducing serum cholesterol. The β-glucans from oat bran increase bile acid excretion, and thus, lower serum lipids (Lia et al., 1995). Increasing the use of fiber-rich cereal fractions takes advantage of their health promoting attributes. Saponins in legumes also are hypocholesterolemic because they bind cholesterol and bile acids (Thompson, 1993).

A barrier to use of foods designated to be enriched with health protectants is how they are to be regulated and labeled for health claims. New regulation will be required. Currently, each one must be approved according to the regulatory category it best fits. Health claims for non-nutritional food constituents would require classification and regulation as a drug requiring premarket demonstration of effectiveness (Hathcock, 1993).

VII. GENETIC MANIPULATION TO INFLUENCE FOOD QUALITY

A. Proteins

Differences in cereal, legume, and oilseed grain protein amount and composition can have a profound effect on the quality of the cooked grain or further processed product. For example, protein quantity has a large effect on how wheats are utilized. Lower protein soft wheats are used for making cakes, cookies, and biscuits and higher protein hard wheats are appropriate for making breads. Protein content alone, however, is far from the only important criterion when considering wheat varietal quality differences, as wheats of similar protein content can have dramatically different bread-making potentials. In this respect, it has long been known that wheat varieties that differ in specific proteins or groups of proteins produce bread of varying quality. Thus, much research has been devoted to elucidating the individual proteins, or combinations of proteins, that must be present to obtain optimum quality.

In wheat, the gluten proteins, composed of gliadins and glutenins, are responsible for the formation of the viscoelastic complexes of doughs that trap carbon dioxide produced by yeasts, thereby forming the spongy and chewy texture of bread. The glutenin fraction, in a qualitative sense, likely has the largest influence on flour quality (Weegels et al., 1996) and, more specifically, a group of high molecular weight (HMW) protein subunits may be most responsible for gluten elasticity that confers dough strength through formation of HMW polymers linked by disulfide cross links (Shewry et al., 1995). Gliadins also have been shown to influence bread-making properties of wheat (Van Lonkhuijsen et al., 1992). In terms of quantity and composition, the HMW wheat proteins appear to have a large influence on the quality of breads (Shewry et al., 1995). Genetic variability exists among the 3 to 5 HMW subunits that coincides with differences in bread-making quality (Shewry et al., 1992).

With the advent of molecular methods of transforming plants by transfer of genes, the identification of specific quality-related proteins in wheat has led to the cloning of their encoding genes and successful transformation of plants derived from tissue cultures from immature scutella (Shewry et al., 1995). Genetic manipulation of glutenins was achieved by transforming near-isogenic wheat lines deficient in the HMW subunits with genes encoding subunits 1 to 5. Endosperm specific expression was attained.

In other cereals the role of specific proteins in the textural quality of processed foods has received little attention compared to that of wheat proteins and their breadmaking potential. This is partly because relationships between specific proteins, type and amount, and whole grain or flour functionalities have not been thoroughly investigated, but also because other cereal proteins do not behave like gluten.

In rice there have been a few reports that suggest that proteins may influence cooked rice quality or further processed products containing rice flour. Starch, both amylose and amylopectin, has been most studied and is a major determinant of textural attributes of cooked rice, such as stickiness and firmness. However, potential relationships have been postulated between specific endosperm proteins and cooked rice quality (Hamaker et al., 1991; Hamaker & Griffin, 1993; Chrastil, 1990). Granule-bound starch synthase (60 kDa) that is responsible for amylose synthesis may affect the gelatinization behavior of starch granules, and may, in turn, influence cooked rice texture (Hamaker et al., 1991). Rice varieties that had high amounts of this starch granule-bound protein had a relatively low degree of starch gelatinization and when cooked produced firm, nonsticky grains. However, these relationships are not well established and at this point could not be used to genetic manipulate rice to achieve specific qualities useful to processors or desirable for consumers.

The relationship between proteins of corn, or its closely related relative sorghum, and quality of products made from them also are not well established. The majority of corn proteins are zeins (storage proteins) and are found encapsulated in rather rigid protein bodies. There has been scant work, and therefore little evidence, that these bodies break apart and release zein during processing, though such a release could have a large influence on textural properties of corn doughs and end-products made from corn. Lawton (1992) showed that isolated zein added to starch and heated above its glass transition temperature (temperature at which the protein moves from a glassy to a rubbery state) changed so that a cohesive and elastic dough could be made. This was dependent on the protein component, and suggests that zein could influence product textural qualities if released from protein bodies.

Quality protein maize (QPM), the modified hard endosperm/high-lysine corn, has somewhat improved processing characteristics that are presumably related to the higher amount of γ-zein (containing high cysteine content) present in the QPM protein bodies compared to normal corn. In Brazil, investigators showed that QPM flour could be added to wheat flour in higher amounts than normal corn flour without negatively affecting breadmaking quality (Magnavaca, 1992).

Another recent topic of interest is that of flavor enhancement by certain cereal proteins. If, in the future, specific proteins are identified with a flavor role in cereal-based foods, then there may be a way to subtly manipulate flavor in cereals by changing protein composition.

Of the noncereal grains, soybean proteins have been studied the most in their relation to food uses. Unlike cereal proteins, soybean proteins are widely used as ingredients in foods, or as textured protein food products. Following oil extraction, soybean meal may be further processed to a protein concentrate (>70% protein) or protein isolate (>90% protein) or spun into texturized protein fibers (Snyder & Kwon, 1987). Glycinin (11S) and ß-conglycinin (7S) are the two major storage proteins of soybean. Soybean proteins offer many desirable functional properties to foods such as ability to absorb and bind water, increase viscosity of soups and gravies, form gels, and form emulsions. Genetic modification of glycinin has been approached using molecular biology techniques involving alteration of the gene nucleotide sequence to produce proteins with changed functionalities (Utsumi, 1992). An example is the modification of glycinin in the C-terminal region to increase emulsifying activity (Kim et al., 1990). Use of molecular techniques promises not only the potential to design functionally improved proteins, but to provide basic information as to how protein sequence and structure determine its functionality.

Efforts have been made through classical breeding and genetic engineering to improve the nutritional quality of corn and soybean proteins. Cereals are generally most limiting in lysine and oilseeds in methionine, hence, researchers have focused on increasing the levels of these two amino acids in the respective grains. Quality Protein Maize contains the *opaque-2* mutation (Mertz et al., 1964) that elevates synthesis of the high lysine/tryptophan-containing nonzein proteins and reduces synthesis of the essentially lysine-free zein fraction (Habben et al., 1993). Recent development of an ELISA (enzyme-linked immunosorbent assay)-based assay that quantifies the corn cytoplasmic protein, elongation factor 1-α, has been used to screen segregants and advanced lines for lysine content (Habben et al., 1995). In soybean molecular biologists have successfully developed transgenic plants that express a high methionine brazil nut (*Bertholletia excelsa* Humb. & Bonpl.) protein (Nordless et al., 1996). However, it was found that this protein was allergenic and had retained its allergenic properties when expressed in the soybean (Nordlee et al., 1994), making this genetic transformation a good example of what is achievable, but in this case not a practical solution to improving nutritional quality. Advances in the plant molecular biology area in combination with classical breeding approaches promise more rapid development of improved genotypes that have enhanced food quality or nutritional characteristics.

B. Carbohydrates

Starch is the major component of all cereals and some legumes. Accordingly, behavior of flours in further processed products or cooked whole grains are largely affected by its properties. Starch is composed of two types of polymers, amylose and amylopectin. Amylose is a linear chain of α-1,4-linked D-glucopyranosyl units, that contain some α-1,6-linked branches. Amylopectin also is a glucose polymer, although it is much larger and contains many short glucan chains linked through α-1,6 linkages. Its highly branched structure contains about 4 to 5% α-1,6 branch-point linkages. Starch is synthesized in amyloplasts forming granules that differ in size and shape depending on the genus. The granule has alternating rings containing crystalline and noncrystalline regions.

Interest in developing different starch types for food use stems from the unique properties of gelatinized starches. The gelatinization process involves disruption of crystalline regions of the starch granule in the presence of water and heat, followed by swelling, leaching of amylose out of the swollen granular structure, and finally disruption of the swollen granule when a shear force is applied (Morris, 1994). Gelatinized normal starch forms a gel on cooling through a process called retrogradation, which initially entails reassociation of linear portions of amylose, and later amylopectin, by H bonding. Over time further H bonding among the glucans causes removal of water from the gel, a process called syneresis. Waxy starches contain little to no amylose and have the desirable property of forming a gelatinous paste that does not form a gel. Therefore, they can be used in gravies and sauces to maintain viscosity regardless of temperature changes that the food undergoes. High amylose starches, on the other hand, form strong gels on cooling.

Corn starch, the main commercial starch in the USA, is often chemically modified to achieve properties that normal or waxy starches do not have. For example, waxy corn starch has a tendency to undergo substantial shear thinning or loss of paste viscosity when shear force is applied. Hence, waxy starches are often chemically cross-linked to stabilize the swollen granule to make it resist breakdown. Examples of other chemical modifications of starch include hydroxypropylation to add a bulky group that retards retrogradation and gives the starch improved freeze-thaw stability, and acetylation of starch to lower gelatinization temperature and improve paste clarity (BeMiller & Whistler, 1996).

Due to a desire to reduce the content of chemically modifed ingredients in processed foods, much effort has recently been put on finding natural starch mutants that behave similarly to chemically modified starches. Numerous starch mutants have been identified that have unique properties useful as ingredients in further processed foods. These starch mutants, some of which are commercially available in the USA, are visible examples of a successful endeavor to provide new food ingredients and products through genetic manipulation. The most common examples of commercially successful mutant starches in the USA are waxy (*wx*) corn starch and amylose-extender (*ae*) starches; however, many other starch mutants, and combinations of mutants, have been found (Nelson & Pan, 1995). Starch mutants may result from simply changing the amylose/amylopectin ratio, or by altering the fine structure of starch molecules to affect gelatinization and pasting behavior.

In addition to the single starch mutants, classical breeding approaches have produced a number of useful double and triple starch mutants, some of which have functionalities similar to chemically treated starches. Single corn starch mutants include *wx*, *ae*, dull (*du*), horny (*h*), shrunken (*sh*), brittle (*bt*), floury (*fl*), and sugary (*su*). More recent efforts have focused on developing double and triple mutants for starches that have properties like chemically modified starches. Patents were issued in the late 1980s and early 1990s on *wxsh2*, *aedu*, *dush2*, *duwx*, *wxsh1*, and *aewx* with claims of having similar properties to such processed starches as chemically modified thick-thin and cross-linked waxy starches. Patented corn starch mutant combinations and claims on functionalities are listed in Table 3–15. The triple mutant *aedush1* was found to be unique because it contains high amylose content and has low gelatinization temperature that is comparable to chemically cross-linked starches. Pearlstein (1995) reported that there is a dose effect in the triploid endosperm of maize where one gene copy of the dull gene in ae*Dududu*wx resulted in pasting properties most like a cross-linked starch.

The use of genetic engineering to alter starch properties has been successful in the potato. Transgenic potato plants have been developed by modulating the activity of the starch synthesizing enzyme, ADP-glucose pyrophosphorylase, to either increase or decrease starch content (Müller-Röber & Koßmann, 1994). Decrease in starch content was achieved by down-regulating the ADP-glucose pyrophosphorylase using the antisense RNA technique. Increase in starch content was accomplished by a group at Monsanto (Stark et al., 1992) using a gene encoding AG-

Table 3–15. Patents existing on corn starch mutant combinations and their claimed functionalities.†

Genotype	Claim
wx su2	Freeze-thaw stability
wx sh1	Freeze-thaw stability
ae du sh1	High amylose content without high gelatinization temperature, behaves like a cross-linked starch
du h	Suitable for canning of acidic foods
wx fl1	Thickens like a suitably cross-linked waxy maize starch
du wx	Provides high-amylose characteristics without amylose
ae du	More than 50% amylose without high gelatinization temperature, good thickener
du su2	High amylose content without high gelatinization temperature
wx sh2	High amylose content with lowered gelatinization temperature
ae wx	Stability in acidic foodstuffs comparable to a cross-linked ex maize starch
su2	Thickener for acidic foods, freeze-thaw stability
ae su2	High amylose content without high gelatinization temperature

† With permission from BeMiller, 1997.

Pase originating from a mutant *Escherichia coli* K-12 strain that produced higher glycogen content than its wild-type parent. In the latter case, transgenic potato plants were produced that averaged 35% more starch than controls. This has practical application for the potato chip industry.

New knowledge about the starch synthesizing enzymes and their isoforms, and the ways that they are turned on and off (Preiss, 1991), has led to speculations that synthesis of new starch types may be a reality in the foreseeable future. Current limitations, however, are not only from the genetic perspective, but exist from the lack of knowledge of what molecular structure is necessary in starch to achieve a particular functionality. The ability of geneticists and molecular biologists to change starch structure should greatly expand our understanding of structure-function relationships in starch.

In another areas of genetic modification of carbohydrates for improved food quality, there have been efforts to reduce oligosaccharide content in legumes, particularly in soybean (Hartwig, 1996), and to increase β-glucan content in barley (*Hordeum vulgare* L.) (Ullrich et al., 1986). Raffinosaccharides in legumes cause flatulence problems in humans and may reduce seed meal digestibility in monogastric animals. Decreased raffinosaccharide levels in soybean has been achieved by selection through breeding (Hartwig, 1996); however, it is not yet known whether elimination of this oligosaccharide is possible while still retaining seed viability. Much is now known regarding the pathways of synthesis of the oligosaccharides of legumes, suggesting that future control of the levels of these compounds may someday be accomplished through transgenic manipulation. The β-glucans have been shown to have cholesterol-lowering effects in laboratory animals and humans (McIntosh, 1995). Waxy barley has higher β-glucan content compared to normal genotypes.

C. Oil

Oil content of cereals is principally determined by the proportion of germ, which has a relatively high amount of oil, to the whole kernel. Thus, corn, which proportionally has the largest germ of the major cereal crops, has the highest oil content of the cereals. Wheat, barley, oat, and rye have higher endosperm oil contents than endosperm of the coarse cereal grains, with oat containing the highest endosperm oil content. Grain with high oil content is of value as a high energy feed grain, as well as for commercial extraction of oil. Breeders at the University of Illinois have developed a high oil corn that has an even larger germ than normal (Mievic et al., 1989).

In the commercial oilseeds, which include soybean, cotton seed (*Gossypium hirsutum* L.), safflower (*Carthamus tinctorius* L.), and canola, much of the present research emphasis is on altering fatty acid profiles. These seed oils, plus peanut (*Arachis hypogaea* L.) and corn oils, in varying proportions contain mostly oleic (18:1), linoleic (18:2), and linolenic (18:3) acids. The polyunsaturated fatty acids, linoleic and linolenic, undergo oxidative breakdown resulting in formation of off-flavors and decrease the shelf stability of an oil. Linolenic acid is many times more susceptible to oxidative degradation than linoleic acid, and for this reason there have been efforts to genetically reduce its level. In commercial operations, soybean oil, which contains about 8% linolenic acid, is routinely hydrogenated to reduce the level of this fatty acid. However, ideally this would be better done through genetic modification, as there are increasing concerns about the formation of trans-fatty acids from this process.

Mutants have been found in soybean (Wilcox & Cavins, 1987) and canola (Zhegong et al., 1992) that have reduced linolenic acid levels with better oxidative stability compared to normal, untreated oils. High oleic acid mutants, which result in lower polyunsaturated oils and improved stability, have been identified in many crops including soybean, canola, sunflower (*Helianthus anuus* L.), safflower, and corn (Kinney, 1994).

There also has been some success in altering fatty acid profiles of oils using molecular biology techniques. Transgenic lines have been produced in soybean and canola with suppressed levels of desaturase gene expression that has resulted in seed oils with lower levels of linoleic and linolenic acids and higher levels of oleic acid (Hitz et al., 1994). Further testing will show the stability of these traits. In a non-food oil application, transgenic rapeseed containing the enzyme, acyl-ACP-thioesterase, specific for synthesis of lauric acid (12:0) has been produced (Voelker et al., 1992). The seed contains 40% lauric acid in the total oil that can be used by the soap and surfactant industries.

In a related area soybean plants have been developed that have low lipoxygenase levels (Davies & Nielsen, 1987). Lipoxygenase is the enzyme that catalyzes the oxidation of polyunsaturated fatty acids (free and esterfied forms). Soybean plants have three isoforms of lipoxygenase, and mutants now exist that eliminate all three, thereby reducing off-flavors and aromas (Furuta et al., 1996).

REFERENCES

Adlercreutz, H. 1990. Western diet and Western disease: Some hormonal and biochemical mechanisms and associations. Scand. J. Clin. Lab. Invest. 50:3–23.

Allen, L.H. 1993. The nutrition CRSP. What is marginal, malnutrition, and does it affect human function? Nutr. Rev. 51:255–267.

American Cancer Society. 1993. Cancer facts and figures—1993. Am. Cancer Soc., Atlanta.

Anderson, J.W., and A.O. Akanji. 1991. Dietary fiber—an overview. Diabet. Care 12:1126–1131.

Anderson, J.W., B.M. Smith, and J. Gustafson. 1994. Health benefits and practical aspects of high-fiber diets. Am. J. Clin. Nutr. 59(Suppl.):1242S–1247S.

Anjum, F.M., and A. Ali. 1991. Fatty acids, mineral composition and functional (Bread and Chapati) properties of high protein and high lysine barley lines. J. Sci. Food Agric. 55:511–519.

Ascherio, A., and W.C. Willett. 1995. New directions in dietary studies of coronary heart disease. J. Nutr. 125:647S–655S.

Asp, N.G.L. 1995. Classification and methodology of food carbohydrates as related to nutritional effects. Am. J. Clin. Nutr. 61:930S–937S.

Austoker, J. 1994. Diet and cancer. Br. Med. J. 308:1610–1614.

Barampama, Z., and R.E. Simard. 1994. Oligosaccharides, antinutritional factors, and protein digestibility of dry beans as affected by processing. J. Food Sci. 59:833–838.

Beard, J.L., C.M. Weaver, S. Lynch, S.D. Johnson, S. Dassenko, and J.D. Cook. 1988. The effect of soybean phosphate and phytate content in iron bioavailability. Nutr. Res. 8:345–352.

BeMiller, J.N. 1997. Starch modification: Challenges and prospects. Starch/Starke 4S:127–131.

BeMiller, J.N., and R.L. Whistler. 1996. Carbohydrates. *In* O.R. Fennema (ed.) Food chemistry. 3rd ed. Marcel Dekker, Inc., New York.

Bhagya, S., and H. Srinivas. 1992. Extraction of soybean (*Glycine max*) with hexane-acetic acid: Effect on oil quality. Food Chem. 44:123–125.

Bickoff, E.M., A.L. Livingstone, A.P. Hendrickson, and A.N. Booth. 1962. Relative potencies of several oestrogen-like compounds found in forages. Agric Food Chem. 10:410–412.

Bird, C.L., M.E. Swendseid, J.S. Witte, G.H McIntosh, G.H. Newman, and C.W. Newman. 1994. Red cell and plasma folate, folate consumption, and the risk of colorectal adenomatores polyps. Cancer Epidemiol., Biomarkers and Prevent. 4:709–714.

Block, G., C.M. Dressner, A.M. Hartman, and M.D. Carroll. 1988. Nutrient sources in the American Diet: Quantitative data from the NHANES II survey. II. Macronutrients and fats. Am. J. Epidemiol. 122:27–40.

Blot, W.J., J-Y. Li, P.R. Taylor, W. Guo, S. Dawsey, G.Q. Wang, C.S. Yang, S.F. Zheng, M. Gail, G.Y. Li, Y. Yu, B.-Q. Lui, J. Tangrea, Y.-H. Sun, F. Liu, J. Fraumeni, Jr., Y.-H. Zheng, and B. Li. 1993. Nutrition intervention trials in Linxian, China: Supplementation with specific vitamin/mineral combinations, cancer incidence, and disease-specific mortality in the general population. J. Natl. Cancer Inst. 85:1483–1492.

Bock, M.A. 1991. Minor constituents of cereals. p. 555–594. *In* K.J. Lorenz and K. Kulp (ed.) Handbook of cereal science and technology. Vol. 11. Marcel Dekker, Inc., New York.

Bonanome, A., A. Visonã, L. Lusiani, G. Beltramello, L. Confortin, S. Biffanti, F. Sorgato, F. Costa, and A. Pagnan. 1991. Carbohydrate and lipid metabolism in patients with non-insulin-dependent diabetes mellitus; effects of a low-fat, high carbohydrate diet vs. a diet high in monounsaturated fatty acids. Am. J. Clin. Nutr. 54:586–590.

Butler, L.G. 1989. New perspectives on the antinutritive effects of tannins. p. 402–409. *In* J.E. Kinsella and W.G. Soucie (ed.) Food proteins. AOCS, Champaign, IL.

Caragay, A.B. 1992. Cancer-preventive foods and ingredients. Food Technol. 46:65–68.

Chang, H.I., G.L. Catignani, and H.E. Swaisgood. 1992. Effects of sucrose, starch and oil on the in vitro determination of protein digestibility using the immobilized digestive enzyme assay (IDEA) system. J. Food Biochem. 16:133–140.

Chen, W.-J.L., and J.W. Anderson. 1981. Soluble and insoluble plant fiber in selected cereals in vegetables. Am. J. Clin. Nutr. 34:1077–1082.

Chrastil, J. 1990. Protein-starch interactions in rice grains. Influence of storage on oryzenin and starch. J. Agric. Food Chem. 38:1804–1809.

Chung, Y.S., M.A. Uebersax, M.R. Bennink, and G.L. Hosfield. 1995. Indigestible starch components in leguminous seed. p. 85. *In* Book of abstracts IFT Annu. Meet., Anaheim, CA.

Chung, Y.S., M.A. Uebersax, M.R. Bennink, and G.L. Hosfield. 1996. Heat induced changes in cell wall structures reduce bioavailabiolity of bean starch. p. 90. *In* Book of abstracts. IFT Annu. Meet., New Orleans, LA.

Colman, N., R. Green, and J. Metz. 1975. Prevention of folate deficiency by food fortification. II. Absorption of folic acid from fortified stable foods. Am. J. Clin. Nutr. 28:459–464.

Dahlin, K.M., and K.J. Lorenz. 1993. Carbohydrate digestibility of laboratory-extruded cereal grains. Cereal Chem. 70:329–333.

Davies, C.S., and N.C. Nielsen. 1987. Registration of soybean germplasm that lacks lipoxygenase isoenzymes. Crop Sci. 27:370–371.

Davies, N.T., V. Hristic, and A.A. Flett. 1977. Phytate rather than fiber in bran as the major determinant of zinc availability to rats. Nutr. Rep. Int. 15:207–214.

Eggum, B.O., B.O. Juliano, C.M. Perez, and E.F. Acedo. 1993. The resistant starch, undigestible energy and undigestible protein contents of raw and cooked milled rice. J. Cer. Sci. 18:159–170.

El-Massry, R.A., S.A. Taha, and M.M. Afifi. 1992. Fatty acids and unsaponifiable matter composition of rice oil and rice germ oil as well as their dewaxed residue. Acta Agron. Hungar. 41:69–73.

Englyst, H.N., S.M. Kingman, and J.H. Cummings. 1992. Classification and measurement of nutritionally important starch fractions. Eur. J. Clin. Nutr. 46:S33–S50.

Estevez, A.M., E. Castillo, F. Figuerola, and E. Yanez. 1991. Effect of processing on some chemical and nutritional characteristics of pre-cooked and dehydrated legumes. Plant Foods Human Nutr. 41:193–201.

Food and Agriculture Organization. 1980. Food balance sheets, 1975–1977. U.N. FAO, Rome.

Food and Agriculture Organization and World Health Organization. 1990. Protein quality evaluation. Joint FAO/WHO Rep. Expert Consult. U.N. FAO, Rome, Italy.

Erdman, J.W., Jr. 1981. Bioavailability of trace elements from cereals legumes. Cer. Chem. 58:21–26.

Fleming, K.H., and J.T. Heimbach. 1994. Consumption of calcium in the U.S.: Food sources and intake levels. J. Nutr. 124:1426S–1430S.

Fordyce, E.J., R.M. Forbes, R.R. Robbins, and J.W. Erdman, Jr. 1987. Phytate × calcium/zinc molar ratios: Are they predictive of zinc bioavailability? J. Food Sci. 52:440–444.

Frazão, E. 1995. The American diet: Health and economic consequences. USDA ERS Agric. Inform. Bull. 711. U.S. Gov. Print. Office, Washington, DC.

Frigg, M. 1976. Bio-availability of biotin in cereals. Poult. Sci. 55:2310–2318.

Furuta, S., Y. Nishiba, and M. Hajika. 1996. DETBA value and hexanal production with the combination of unsaturated fatty acids and extracts prepared from soybean seeds lacking two or three lipoxygenase isoenzymes. J. Agric. Food Chem. 44:236–239.

Ghazali, H.M., and S.C. Cheng. 1991. The effect of germination on the physico-chemical properties of black gram (*Vigna mungo* L.). Food Chem. 41:99–106.

Godinez, C.M., R. Bressani, and M. Melgar. 1992. Apparent digestibility of bean protein evaluated in humans, rats and in vitro assays. Nutr. Res. 12:235–246.

Graf, E., and J.W. Eaton. 1993. Suppression of colonic cancer by dietary phytic acid. Nutr. Cancer 19:11–19.

Graham, R.D., and R.M. Welch. 1996. Breeding for staple food crops with high micronutrient density. Int. Food Policy Res. Inst., Washington, DC.

Graham, S., M. Zielezny, J. Marshall, X. Priore, J. Frendenheim, J. Brasure, B. Haughey, P. Nasca, and M. Zdeb. 1992. Diet in the epidemiology of postmenopausal breast cancer in the New York state cohort. Am. J. Epidemiol. 136:1327–1337.

Gregory, J.F., and J.R. Kirk. 1978. Assessment of roasting effects on vitamin B_6 stability and bioavailability in dehydrated food systems. J. Food Sci. 43:1585–1589.

Gupta, M., and N. Khetarpaul. 1993. Effect of rabadi fermentation on phytic acid and in vitro digestibility of barley. Die Nahrung 37:141–146.

Gupta, M., N. Khetarpaul, and B.M. Chauhan. 1992. Rabadi fermentation of wheat: Changes in phytic acid content and in vitro digestibility. Plant Foods Human Nutr. 42:109–116.

Habben, J.E., A.W. Kirleis, and B.A. Larkins. 1993. The origin of lysine-containing proteins in opaque-2 maize endosperm. Plant Mol. Biol. 23:825–838.

Habben, J.E., G.L. Moro, B.G. Hunter, B.R. Hamaker, and B.A. Larkins. 1995. Elongation factor 1α concentration is highly correlated with lysine content of maize endosperm. Proc. Natl. Acad. Sci. USA 92:8640–8644.

Hamaker, B.R., and V.K. Griffin. 1993. Effect of a disulfide bond-containing protein on rice starch gelatinization and pasting. Cer. Chem. 70:377–380.

Hamaker, B.R., V.K. Griffin, and K.A.K. Moldenhauer. 1991. Potential influence of a starch granule-associated protein on cooked rice stickiness. J. Food Sci. 56:1327–1329.

Hannah, L.C. 2000. Starch biosynthesis and genetic potential. p. 181–199. *In* C.F. Murphy and D.M. Peterson (ed.) Designing crops for added value. Agron. Monogr. 40. ASA, CSSA, and SSSA, Madison, WI.

Hartwig, E.E. 1996. Registration of soybean germplasm ine D90-7256 having high seed protein and low oligosaccharides. Crop Sci. 36:212.

Hathcock, J.N. 1993. Safety and regulatory issues for phytochemical sources: "Designer foods". Nutr. Today 28:23–25.

Heaney, R.P., and C.M. Weaver. 1989. Oxalate: Effect on calcium absorbability. Am. J. Clin. Nutr. 50:830–832.

Heaney, R.P., and C.M. Weaver. 1995. Effect of psyllium and absorption of co-ingested calcium. J. Am. Ger. Soc. 43:1–3.

Heaney, R.P., C.M. Weaver, and M.L. Fitzsimmons. 1991. The effect of soybean phytate content on calcium absorption. Am. J. Clin. Nutr. 53:745–747.

Henley E.C., and J.M. Kuster. 1994. Protein quality evaluation by protein digestibility-corrected amino acid scoring. Food Tech. 48:74–77.

Hitz, W.D., T.J. Carlson, J.R. Booth, A.J. Kinney, K.L. Stecca, and N.S. Yadav. 1994. Cloning of a higher plant plastid omega-6 desaturase cDNA and its expression in a Cyanobacterium. Plant Physiol. 105:635–641.

Hsieh, H.M., Y. Pomeranz, and B.G. Swanson. 1992. Composition, cooking time, and maturation of Azuki (*Vigna angularis*) and common beans (*Phaseolus vulgaris*). Cer. Chem. 69:244–248.

Hunninghake, D.B., V.T. Miller, J.C. LaRosa, B. Kinosian, T., Jacobson, V. Brown, J. Howard, D.A. Edelman, and R.R. O'Connor. 1994. Long-term treatment of hypercholesterolemia with dietary fiber. Am. J. Med. 97:504–508.

Illman, R.J., G.B. Storer, and D.L. Topping. 1993. White wheat flour lowers plasma cholesterol and increases cecal steroids relative to whole wheat flour, wheat bran and wheat pollard in rats. J. Nutr. 123:1094–1100.

Johnson, C.D., M.F. Berry, and C.M. Weaver. 1985. Soybean hulls as on iron source for bread enrichment. J. Food Sci. 50:1275–1277, 1305.

Kabir, H., J.E. Leklem, and L.T. Miller, 1983. Relationship of the glycosylated vitamin B_6 content of foods to vitamin B_6 bioavailability in humans. Nutr. Rep. Int. 28:709–715.

Kennedy, A.R. 1993. Overview: Anticarcinogenic activity of protease inhibitors. p. 9–64. *In* W. Troll and A.R. Kennedy (ed.) Protease inhibitors as cancer chemopreventive agents. Plenum Press, New York.

Khan, A., C.M. Weaver, and S. Sathe. 1990. Association of zinc with soy proteins as affected by heat and pH. J. Food Sci. 55:263–264, 266.

Kim, C.S., S. Kamiya, T. Sato, S. Utsumi, and M. Kito. 1990. Improvement of nutritional value and functional properties of soybean glycinin by protein engineering. Protein Eng. 3:725–731.

Kim, Y.A., and W.E. Barbeau. 1991. Evaluation of SDS-PAGE method for estimating protein digestibility. J. Food Sci. 56:1082–1086.

Kinney, A.J. 1994. Genetic modification of the storage lipids of plants. Curr. Opin. Biotechnol. 5:144–151.

Kinney, A.J., and W.D. Hitz. 2000. Oils. p. 167–180. *In* C.F. Murphy and D.M. Peterson (ed.) Designing crops for added value. Agron. Monogr. 40. ASA, CSSA, and SSSA, Madison, WI.

Kitts, D.D. 1994. Bioactive substances in food: Identification and potential uses. Can. J. Physiol. Pharmacol. 72:423–434.

Klurfeld, D.M. 1987. The role of dietary fiber in gastrointestinal disease. J. Am. Diet. Assoc. 9:1172–1177.

Kris-Etherton, P.M., D. Krummel, M.E. Russell, D. Dreon, S. Mackey, J. Borchers, and P.D. Wood. 1988. The effect of diet on plasma lipids, lipoproteins, and coronary heart disease. J. Am. Diet. Assoc. 88:1373–1400.

Lawton, J.W. 1992. Viscoelasticity of zein-starch doughs. Cer. Chem. 69:351–355.

Levine, S.E., C.M. Weaver, and A.W. Kirleis. 1982. Accumulation of selected trace elements in hydroponically grown soybeans and distribution of the elements in processed soybean fractions. J. Food Sci. 47:1283–1287.

Lia, A., G. Hallmans, A.-S. Sandberg, B. Sundberg, P. Åman, and H. Anderson, 1995. Oat β-glucan increases bile acid excretion and barley fraction increases cholesterol excretion in ileostomy subjects. Am. J. Clin. Nutr. 62:1245–1251.

Liljeberg, H., Y. Granfeldt, and I. Bjorck. 1992. Metabolic responses to starch in bread containing intact kernels versus milled flour. Eur. J. Clin. Nutr. 46:561–575.

Magnavaca, R. 1992. Quality protein maize development in Brazil. *In* E.T. Mertz (ed.) Quality protein maize. Am. Assoc. Cer. Chem., St. Paul, MN.

Maki, P.A., Y. Paterson, and A.R. Kennedy. 1994. Studies related to the potential antigenicity of the Bowman-Birk inhibitor, an anticariogenic protease inhibitor isolated from soybeans. Nutr. Cancer 22:185–193.

Malville-Shipan, K., and S.E. Fleming. 1992. Wheat bran and corn oil do not influence proliferation in the colon of healthy rats when energy intakes are equivalent. J. Nutr. 122:37–45.

Marlett, J.A. 1992. Content and composition of dietary fiber in 117 frequently consumed foods. J. Am. Diet. Assoc. 92:175–186.

Marlett, J.A. 1993. Comparison of dietary fiber and selected nutrient compositions of oat and other grain fractions. p. 49–92. *In* P. Wood (ed.) Oat bran. Am. Assoc. Cer. Chem., Inc., St. Paul, MN.

Mason, A.C., C.M. Weaver, S. Kimmel, and R.K. Brown. 1993. The effect of soybean phytate content on calcium bioavailability in mature and immature rats. J. Agric. Food Chem. 41:246–259.

McIntosh, G.H. 1995. Barley foods and their influence on cholesterol metabolism. World View Nutr. Diet. 77:89–108.

McIntosh, G.H., R.K. LeLeu, A. Kerry, and M. Goldring. 1993. Barley grain for human food use. Food Australia 45:392–394.

Mertz, E.T., L.S. Bates, and O.E. Nelson, 1964. Mutant gene than changes protein composition and increases lysine content of maize endosperm. Science (Washington, DC) 145:279–280.

Mievic, D., A. Maric, and D.E. Alexander. 1989. Population cross diallel among high oil populations of maize. Crop Sci. 29:613–617.

Molteberg, E.L., G. Vogt, A. Nilsson, and W. Frolich. 1995. Effects of storage and heat processing on the content and composition of free fatty acids in oats. Cer. Chem. 72:88–93.

Morris, E.R., and R. Ellis, 1989. Usefulness of the dietary phytic acid/zinc molar ratio as an index of zinc bioavailability to rats and humans. Biol. Trace Elem. Res. 19:107–117.

Morris, V.J. 1994. Starch gelation and retrogradation. Trends Food Sci. Technol. 1:2–6.

Morton, M.S., N. Bundred, D.F. McMichael-Phillipis, C. Harding, R.I. Nicholson, and K. Griffiths. 1997. Phyto-estrogens and breast cancer. Endocr.-Relat. Cancer 4:331–339.

Müller-Röber, B., and J. Koßmann. 1994. Approaches to influence starch quantity and starch quality in transgenic plants. Plant Cell Environ. 17:601–613.

National Academy of Science. 1950. Flour and bread enrichment 1945–1950. NAS, Washington, DC.

National Academy of Science. 1974. Proposed fortification policy for cereal-grain products. NAS, Washington, DC.

National Institutes of Health Consensus Development Panel on Triglyceride, High-Density Lipoprotein, and Coronary Heart Disease. 1993. Triglyceride, high-density lipoprotein, and coronary heart disease. JAMA 269:505–510.

Nelson, O., and D. Pan. 1995. Starch synthesis in maize endosperms. Annu. Rev. Plant Physiol. Plant Mol. Biol. 46:475–496.

Nielsen, S.S. 1991. Digestibility of legume proteins. Food Technol. 45:112–118.

Nordlee, J.A., S.L. Taylor, J.A.Townsend, L.A. Thomas, and R.K. Bush. 1996. Identification ofa Brazil-nut allergen in transgenic soybeans. New Engl. J. Med. 334:668–692.

Nordlee, J.A., S.L. Taylor, J.A. Townsend, L.A. Thomas, N.E. Lincoln, and I.A. Johnston, 1994. High methionine brazil nut protein binds human IgE. J. Allergy Clin. Immunol. 93:209.

Pearlstein, R.W. 1995. New starch functionality from novel genetics. Cer. Foods World 40:672(abstract).

Pederson, B., and B.O. Eggum. 1983. Prediction of protein digestibility by an in vitro enzymatic pH-stat procedure. A. Tierphysiol., Tierernaehr. Futtermittelkde. 49:265–277.

Pérez-Escamilla, R. 1995. Periconceptional folic acid and neural tube defects: Public health issues. Bull. DAHO 29:250–253.

Peterson, D.M. 1994. Barley tocols: Effects of milling, malting, and mashing. Cer. Chem. 71:42–44.

Peterson, D.M. 2000. Protein. p. 151–166. *In* C.F. Murphy and D.M. Peterson (ed.) Designing crops for added value. Agron. Monogr. 40. ASA, CSSA, and SSSA.

Preiss, J. 1991. Biology and molecular biology of starch synthesis and its regulation. p. 59–114. *In* B.J. Miflin and H.F. Miflin (ed.) Oxford surveys of plant molecular and cell biology. Oxford Univ. Press, Oxford, England.

Proulx, W.R., C.M. Weaver, and M.A. Bock. 1993. Trypsin inhibitor activity and tannin content do not affect calcium bioavailability of three commonly consumed legumes. J. Food Sci. 58:382–384.

Putler, D., and E. Frazao. 1991. Diet/Health concerns about fat intake. Food Review. January–March p. 16–20.

Qureshi, A.A., W.C. Burger, D.M. Peterson, and C.E. Elson. 1986. The structure of an inhibitor of cholesterol biosynthesis isolated from barley. J. Biol. Chem. 261:10 544–10 550.

Raben, A., A. Tagliabue, N.J. Christensen, J. Madsen, J.J. Holst, and A. Astrup. 1994. Resistant starch: The effect on postprandial glycemia, hormonal response, and satiety. Am. J. Clin. Nutr. 60:544–51.

Raboy, V., and P. Gerbasi, 1996. Genetics of myo-inositol phosphate synthesis and accumulation. p. 257–285. *In* B.B. Biswas and S. Biswas (ed.) Subcellular biochemistry. Vol. 26. Plenum Press, New York,

Rahman, M.M., M.A. Wahed, and M.A. Ali, 1991. Beta-carotene losses during different methods of cooking green leafy vegetables in Bangladesh. J. Food Comp. Anal. 3:47–53.

Ranhotra, G.S., C. Lee, and J.A. Gelroth, 1978. Bioavailability of zinc in soy-fortified wheat bread. Nutr. Rep. Int. 18:487–494.

Ranhotra, G.S., R.J. Loewe, and L.V. Payat. 1974. Phytic acid in soy and its hydrolysis during bread making. J. Food Sci. 39:1023–1025.

Rao, B.S.N., and J. Prabhavati. 1982. Tannin content of foods commonly consumed in India and its influence on ionizable iron. J. Sci. Food Agric. 33:89–96.

Raper, N. 1991. Nutrient content of the U.S. food supply. Food Review. July–September p. 13–18.

Rassin, D.K., and J. Bhatia. 1992. Evaluation of protein status in humans. p. 145–165. *In* S. Nissen (ed.) Methods in protein nutrition and metabolism. Acad. Press, New York.

Roebuck, B.D. 1987. Trypsin inhibitors: Potential concern for humans. J. Nutr. 117:398–400.

Rose, D.P., M. Goldman, J.M. Connolly, and L.E. Strong. 1991. High-fiber diet reduces serum estrogen concentrations in premenopausal women. Am. J. Clin. Nutr. 54:520–525.

Saha, P.R., C.M. Weaver, and A.C. Mason. 1994. Mineral bioavailability in rats from intrinsically labled whole wheat flour of various phytate levels. J. Agric. Food Chem. 42:2531–2535.

Satterlee, L.D., H.F. Marshall, and J.M. Tennyson. 1979. Measuring protein quality. J. Am. Oil Chem. Soc. 56:103–109.

Sauberlich, H.E., M.J. Kretsch, J.H. Skala, H.L. Johnson, and P.C. Taylor. 1987. Folate requirements and metabolism in nonpregnant women. Am. J. Clin. Nutr. 46:1016–1028.

Sauer, W.C., and K. de Lange. 1992. Novel methods for determining protein and amino acid digestibilities in foodstuffs. p. 87–120. *In* S. Nissen (ed.) Methods in protein nutrition and metabolism. Acad. Press, New York.

Serraino, M., and L.U. Thompson. 1991a. The effect of flaxseed supplementation on early risk markers for mammary carcinogenesis. Cancer Lett. 60:135–142.

Serraino, M., and L.U. Thompson. 1991b. Flaxseed supplementation and early markers of colon carcinogenesis. Cancer Lett. 63:159–165.

Serraino, M., and L.U. Thompson. 1992. The effect of flaxseed supplementation on the initiation and promotional stages of mammary tumorigenesis. Nutr. Cancer 17:153–189.

Shewry, P.R., N.G. Halford, and A.S. Tatham. 1992. The high molecular weight subunits of wheat glutenin. J. Cer. Sci. 15:105–120.

Shewry, P.R., A.S. Tatham, F. Barro, P. Barcelo, and P. Lazzeri. 1995. Biotechnology of breadmaking: Unraveling and manipulating the multi-protein gluten complex. Bio/Technol. 13:1185–1190.

Smith, M.A.L. 2000. Functional food value of designer crops. p. 229–237. *In* C.F. Murphy and D.M. Peterson (ed.) Designing crops for added value. Agron. Monogr. 40. ASA, CSSA, and SSSA, Madison, WI.

Snyder, H., and T.W. Kwon (ed.) 1987. Soybean utilization. Van Nostrand Reinhold, New York.

Sood, N., S.K. Sachdeva, and R.K. Raheja. 1991. Fatty acid and free amino acid profiles in six species of Sorghum (L.) Moench. J. Plant Sci. Res. 7:12–13.

Sridhar, R., and G. Lakshminarayana. 1994. Contents of total lipids and lipid classes and composition of fatty acids in small millets: Foxtail (*Setaria italica*), Proso (*Panicum miliaceum*), and Finger (*Eleusin coracana*). Cer. Chem. 71:355–359.

Stark, D.M., K.P. Timmermann, G.F. Barry, J. Preiss, and G.M. Kishore. 1992. Regulation of the amount of starch in plant tissues by ADP glucose pyrophosphorylase. Science (Washington, DC) 258:287–292.

Swaisgood, H.E., and G.L. Catignani. 1991. Protein digestibility: *In vitro* methods of assessment. p. 185–236. *In* J. E. Kinsella (ed.) Advances in food and nutrition research. Vol. 35. Acad. Press, New York.

Tagliabue, A., A. Raben, M.L. Heijnen, P. Deurenberg, and D. Pasquali. 1995. The effect of raw potato starch on energy expenditure and substrate oxidation. Am. J. Clin. Nutr. 61:1070–1075.

The Alpha-Tocopherol, Beta Carotene Cancer Prevention Study Groups. 1994. The effect of vitamin E and beta carotene on the incidence of lung cancer and other cancers in male smokers. N. Engl. J. Med. 330:1029–1035.

Thompson, H.J., A. Wilson, J. Lu, M. Singh, C. Jiang, P. Upadhyaya, K. El-Bayoumy, and C. Ip. 1994. Comparison of the effects of an organic and an inorganic form of selenium on a mammary carcinoma cell line. Carcinogenesis 15:183–186.

Thompson, L.U. 1993. Potential health benefits and problems associated with antinutrients in foods. Food Res. Int. 26:131–149.

Thompson, L.U. 1994. Antioxidants and hormone-mediated health benefits of whole grains. Crit. Rev. Food Sci. Nutr. 34:413–497.

Tippett, K.S., and J.D. Goldman. 1994. Diets more healthful, but still fall short of dietary guidelines. Food Rev. 17:8–14.

Ullrich, S.E., J.A. Clancy, R.F. Eslick, and R.C.M. Lance. 1986. β-Glucan content and viscosity of extracts from waxy barley. J. Cer. Sci. 4:279–285.

U.S. Department of Agriculture. 1995. Nutrition and youth health: Dietary guidelines for Americans. 4th ed. Home and Garden Bull. no. 232.

U.S. Department of Agriculture. The healthy eating index. Publ. CNPP-I. (Available online at http://www.nal.usda.gov/fnic/hei/hei.html.)

U.S. Department of Agriculture, HNIS. 1989. Agriculture handbook 8-20. U.S. Gov. Print. Office, Washington, DC.

Utsumi, S. 1992. Plant food protein engineering. Adv. Food Nutr. Res. 36:89–208.

Van Horn, L.V., K. Liu, D. Parker, L. Emidy, Y. Liao, W.H. Pan, D. Giumetti, J. Hewitt, and J. Stamler. 1986. Serum lipid response to rat product intake with a fat-modified diet. J. Am. Diet. Assoc. 86:759–764.

Van Lonkhuijsen, H.J., R.J. Hamer, and C. Schreuder, 1992. Influence of specific gliadins on the breadmaking quality of wheat. Cer. Chem. 69:174–177.

Voelker, T.A., A.C. Worrell, L. Anderson, J. Bleibaum, C. Fan, D.J. Hawkins, S.E. Radke, and H. Maelor Davies. 1992. Fatty acid biosynthesis redirected to medium chains in transgenic oilseed plants. Science (Washington, DC) 257:72–74.

Wall, J.S., and K.J. Carpenter. 1988. Variation in availability of niacin in grain products. Changes in chemical composition during grain development and processing affect the nutritional availability of niacin. Food Technol. 42:198–204.

Wang, L., R.K. Newman, C.W. Newman, and P.J. Hofer. 1992. Barley β-glucans alter intestinal viscosoty and reduce plasma cholesterol concentrations in chicks. J. Nutr. 122:2292–2297.

Weaver, C.M. 1985. Intrinsic mineral labeling of edible plants: Methods and uses. CRC Crit. Rev. Food Sci. Nutr. 23:75–101.

Weaver, C.M., R.P. Heaney, B.R. Martin, and M.L. Fitzsimmons. 1991. Human calcium absorption from whole-wheat products. J. Nutr. 121:1769–1775.

Weaver, C.M., R.P. Heaney, W.K. Proulx, S.M. Hinders, and P.T. Packard. 1993. Absorbability of calcium from common beans. J. Food Sci. 58:1401–1403.

Weegels, P.L., R.J. Hamer, and J.D. Schofield. 1996. Functional properties of wheat glutenin. J. Cer. Sci. 23:1–18.

West, L.G., and J.L. Greger. 1978. In-vitro studies of saponin-vitamin complexation. J. Food Sci. 43:1340–1345.

White, P.J. 2000. Dietary fiber. p. 201–213. *In* C.F. Murphy and D.M. Peterson (ed.) Designing crops for added value. Agron. Monogr. 40. ASA, CSSA, and SSSA, Madison, WI.

Whitten, C.G., and T.D. Shultz. 1988. Binding of steroid hormones in vitro by water-insoluble dietary fiber. Nutr. Res. 8:1223.

Wilcox, J.R., and J.F. Cavins. 1987. Gene symbol assigned for linolenic acid mutant in soybean. J. Hered. 66:1622–1624.

Wisher, E., W. Feldheim, Y. Powerange, and F. Meuser. 1985. Dietary fiber in cereals. p. 169–238. *In* Y. Pomeranz (ed.) Advances in cereal science and technology. Vol. 3. Am. Assoc. Cer. Chem., St. Paul, MN.

Wrick, K.L. 1993. Functional foods: Cereal products at the food-drug interface. Cer. Foods World 38:205–214.

Yasumoto, K., H. Tsuji, K. Iwami, and H. Mitsuda. 1977. Isolation from rice bran of a bound form of vitamin B_6 and its identification as 5'-0-(β-D-glucopyranosyl) pyridoxine. Agric. Biol. Chem. 41:1061–1067.

Yokoyama, W.H., C.A. Hudson, M.M .Chiu, and A.A. Betschart. 1994. Effect of formulation and processing on the enzymatic digestion of common cereal products. Cer. Foods World 39:439–443.

Young, V.R., and P.L. Pellett. 1991. Protein Evaluation, amino acid scoring and the Food and Drug Administrations proposed food labeling regulation. J. Nutr. 121:145–150.

Zhegong, F., L. Debonte, and W. Loh. 1992. Brassica napus plants produce seed having reduced linolenic acid and glucosinolate and yield oil with low sulfur, improved sensory characteristics and increased oxidative stability. Int. Patent Appl. WO 93/06714. Year issued: 1993.

4 An Animal Nutritionist's Perspective

GARY L. CROMWELL

University of Kentucky
Lexington, Kentucky

Farm animals are highly dependent upon crops and crop-derived products to produce meat, milk, eggs, and fiber that are used by humans. Ruminant animals (beef and dairy cattle [*Bos* spp.], sheep [*Ovis aries*]) depend heavily on all types of forage, including pasture grasses and legumes, hay, silage, and other forms of roughage in addition to cereal grains, oilseed meals, and various byproducts to supply nutrients for maintenance, growth, reproduction, lactation, and production of wool. Nonruminants [swine (*Sus scrofa*), chickens (*Gallus domesticus*), turkeys (*Meleagris gallopavo*), etc.] rely mainly on cereal grains, oilseed meals, and other forms of concentrate feedstuffs and are less dependent on fibrous feedstuffs to supply nutrients for maintenance, growth, reproduction, lactation, and egg production.

The nutrients supplied by feedstuffs that are consumed by animals are generally grouped into six general classes. These include carbohydrates, fat, protein (amino acids), minerals, vitamins, and water. The primary function of carbohydrates and fat is to produce energy. Protein serves primarily as a source of amino acids that are used by animals to synthesize the proteins needed for muscle growth, fetal tissues, milk, eggs, and other body proteins, as well as to replace body proteins that are lost during turnover (i.e., maintenance). Ruminants utilize dietary protein as a N source for synthesis of bacterial protein in the rumen that, in turn, is digested and the amino acids are used in a similar manner as for the nonruminant. Amino acids in excess of the animal's requirement also are utilized as an energy source. Minerals and vitamins serve many functions in the animal's body, mainly as coenzymes and cofactors. In addition, some of the minerals serve as important components of bone (e.g., Ca, P) and blood (Fe). The final class of nutrients, water, is probably the most important of all nutrients. Water serves to cool the body, gives turgidity to cells, serves as a medium for chemical reactions, transports nutrients, eliminates wastes, and serves many other functions. Along with drinking water, animals utilize water from feeds to meet their needs.

In general, most animals have the ability to utilize nearly all types of crop-derived feedstuffs to some degree. Naturally, the higher the quality of a feedstuff, the better it is utilized by animals. For example, cattle can utilize poor quality forage to meet their maintenance needs, but high-quality forage is needed if the goal is to produce rapid and efficient growth or a high level of lactational performance. Similarly, swine can utilize poor quality cereal grain and/or oilseed meals, but they need high-quality ingredients in order to maximize body weight gains and efficiency of feed utilization. The same is true for poultry.

With respect to crop-derived feedstuffs, the term "quality" is generally associated with environmental factors such as how the crop grew, the stage of maturity when it was harvested, how it was preserved during storage, how free it is from foreign material, and other such factors. For example, a high-quality legume hay is one that was produced with an abundance of leaves and minimal stems, harvested at the proper stage of development, free from weeds, allowed to cure under ideal conditions, and stored under optimal conditions. A high-quality grain is one that was produced under good growing conditions with ample soil nutrients and moisture, harvested at the proper time, dried (if necessary) properly, free from weed seeds and other contaminating material, and free from any molds and/or toxins.

While environmental factors play a major role in determining the quality of crop-derived feedstuffs, genetics also plays a significant role in determining the nutritional quality of a particular feedstuff. However, until recently, attempts to improve the nutritional attributes of crop-derived feedstuffs received little attention. With few exceptions, most of the efforts of plant geneticists and plant breeders were devoted to producing crops with superior yields and a high degree of resistance to diseases, insects, and drought. However, the tide seems to be changing with new emphasis now being placed on breeding crops that have enhanced nutritional properties.

In order for plant breeders to make genetic improvements in the nutritional properties of forages, cereal grains, oilseeds, and other vegetative feedstuffs, they must be aware of those properties that are important from a nutritional standpoint. For this to occur, plant geneticists/breeders and animal nutritionists must communicate with one another. For many years, this communication did not occur on a formal basis. Instead, it only happened at lunches, coffee breaks, and similar chance meetings when the conversation may have turned to topics of common interest.

However, this is changing. Seven years ago—in 1993—I was asked to represent the animal nutrition community and participate at a symposium on "Designing Crops for Specific Uses" at the annual meeting of the Crop Science Society of America (Cromwell, 1993). That same year, an invited paper by Kleese (1993) at the annual meeting of the American Society of Animal Science also addressed the subject of breeding crops with enhanced nutritional value. These two meetings were the first of this type where plant breeders and animal nutritionists were on the same program. Since those two events, numerous other symposia have addressed this important subject.

If plant geneticists and plant breeders are to make genetic improvements, they need to know which factors are the most important from a nutritional perspective and which of these factors are amenable to change so that emphasis can be placed on improving those traits. The identification of these factors will be somewhat dependant upon who is asked the question.

I. WHAT DOES THE RUMINANT NUTRITIONIST WANT IN A FEED?

As a nonruminant nutritionist, I am more familiar with the issues dealing with nutritional quality of feeds for pigs and poultry and less familiar with those relat-

Table 4–1. Protein content of tall fescue treated with a plant growth regulator (Mefluidide).†

Date	Control	Treated
	Crude protein, %	
1 June	10.8	11.4
1 July	8.7	10.0
1 August	7.8	13.4
1 September	7.6	13.5
1 October	9.6	13.9
1 November	10.3	11.0

† Glenn et al. (1981).

ing to cattle and sheep. Therefore, most of the emphasis in this chapter will deal with nonruminant issues. However, a few years ago, a number of ruminant nutritionists in the Animal Sciences Department at the University of Kentucky were asked to identify the major factors that they considered to be important in describing an ideal forage and an ideal feed grain for ruminant animals. The items cited were as follows (not necessarily in order of importance): forages—grasses that remain longer in the vegetative state, grasses to be more like legumes in their properties, increased digestibility of nutrients, increased true protein [not just nonprotein N (NPN)], increased bypass protein (protein that is not degraded in the rumen), decreased fiber and increased soluble carbohydrates, elimination of growth inhibitors (e.g., alkaloids in fescue); grains—increased energy, increased starch, reduced crystallinity of starch, increased fermentability, increased crude protein with improved amino acid profile, and increased protein degradability in the rumen.

Several of these items are interrelated and all are important in terms of improving the nutritional properties of forages and grains. For example, keeping plants in the vegetative state longer decreases the fiber content and increases the soluble carbohydrate content of the forage and increases the protein concentration. Table 4–1 shows the effects of treating fescue (*Festuca arundinacea* Schreb.) with mefluidide, a plant growth regulator, on the protein concentration of the forage during late summer and early fall in Kentucky, a period when fescue normally reaches a more mature state and the protein content drops. Keeping the plant in the vegetative state results in a higher protein content and superior growth in cattle grazing this type of forage.

II. WHAT DOES THE NONRUMINANT NUTRITIONIST WANT IN A FEED?

A similar survey as the one mentioned in the previous section was conducted several years ago at the University of Kentucky. Both swine and poultry nutritionists were asked to identify those factors that were the most important if genetic changes could be made in feedstuffs consumed by pigs and poultry. Items were restricted to those pertaining only to cereal grains and oilseed meals; forages were not considered because of their limited importance and usage in nonruminants. The items listed were as follows (not necessarily in order of importance): grains—increased energy, improved protein quality (especially lysine), improved P utilization by reducing phytic acid, free from mycotoxins; oilseed meals—improved protein quality (especially methionine in soybean meal), improved P utilization by reducing phytic acid, reduction or elimination of growth inhibitors.

It is interesting that improved protein quality and improved P utilization were listed for both cereal grains and oilseed meals. These two factors are especially important with respect to corn (*Zea mays* L.) and soybean [*Glycine max* (L.) Merr.] meal because these two ingredients constitute 80 to 90% of all cereal grains and oilseed meals that are fed to swine and poultry. Improved P utilization is particularly important because of the heightened concern associated with excessive P excretion in animal waste and its impact on the environment.

The following discussion addresses historical changes that have occurred in breeding of crops for improved nutritional properties as well as the more recent advances in plant breeding that have impacted the nutritional quality of feedstuffs.

III. INCREASING THE QUANTITY AND QUALITY OF PROTEIN IN CEREAL GRAINS

Cereal grains are fed to animals primarily as a source of energy that arises mainly from the starch component of the endosperm. While cereal grains are quite low in protein, they contribute significantly to the pool of dietary amino acids available to the animal simply because of the high percentage of cereal grain in the diet. For example, a typical swine finishing diet contains from 80 to 90% corn. This means that over half of the total protein (amino acids) in a typical finisher diet comes from corn. Unfortunately, the overall balance (or profile) of the amino acids in cereal protein is very poor. Figure 4–1 shows the essential amino acid requirements of a growing chick and pig compared with the amounts of those amino acids in corn. Note that most of the amino acids in corn are severely deficient relative to the animal's requirements.

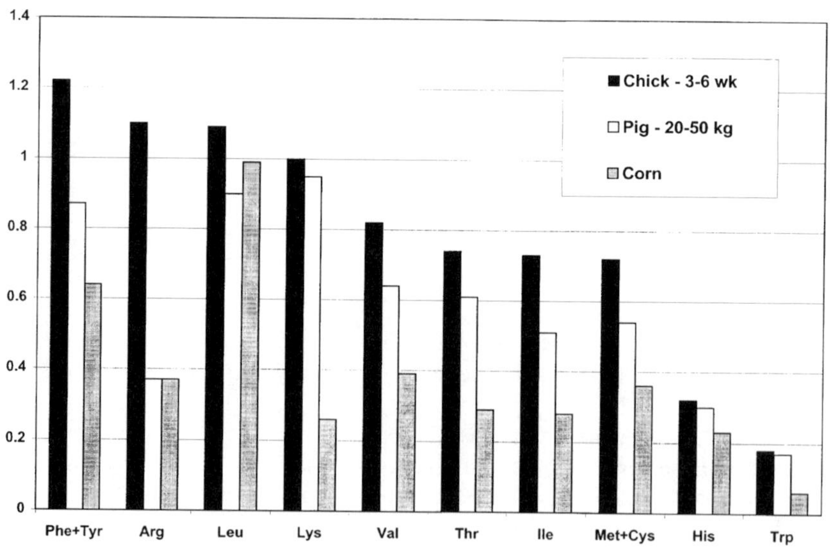

Fig. 4–1. Essential amino acid requirements of growing pigs (20–50 kg body weight) and chicks (3–6 wk age) compared with the levels of essential amino acids in corn (NRC, 1994, 1998).

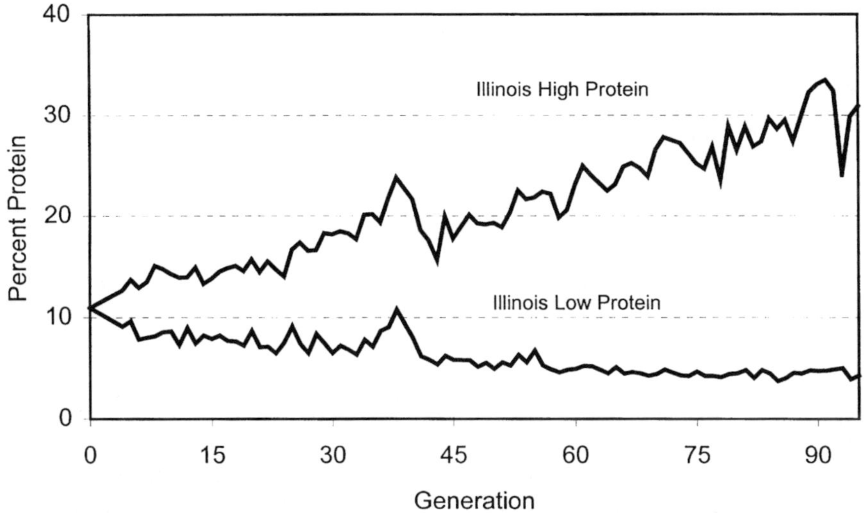

Fig. 4–2. Changes in crude protein content of corn following 95 generations of selection for high or low crude protein content (courtesy of J.W. Dudley, Univ. Illinois).

There have been numerous attempts to increase the protein content of cereal grains. One of the first studies was initiated at the University of Illinois in 1896 by C.G. Hopkins who began with a single open pollinated variety of corn (Burr White). Selection for high and low protein has continued for over a century and the results clearly show that the protein content of corn grain can be altered by selection pressure (Dudley & Lambert, 1992). Figure 4–2 shows the protein content of corn kernels after 95 generations of continuous selection for high protein content in the high protein line and for low protein content in the low protein line. The protein content of the high protein line now exceeds 30% while the protein content of the low protein line is less than 5%. Unfortunately, most of the increase in crude protein in the high protein line has been due to a marked accumulation of a protein called "zein", mainly found in the vitreous portion of the endosperm. Zein is poorly digested and has an extremely poor amino acid profile, being nearly devoid of lysine and tryptophan and having an unusually high proportion of leucine (Table 4–2).

Table 4–2. Partial amino acid composition of zein and glutelin, the two major protein fractions in corn endosperm protein.†

Amino acid	Zein	Glutelin
	% of protein	
Leucine	23.7	11.8
Isoleucine	7.3	4.1
Valine	3.0	6.1
Arginine	1.8	5.5
Cystine	1.0	3.1
Tryptophan	0.1	0.9
Lysine	0.0	3.6

† Mertz (1963).

Table 4–3. Composition of normal and *opaque-2* corn.

	Normal corn	*Opaque-2* corn
Protein fraction in corn†		
Zein, % of total protein	47.1	15.7
Glutelin, % of total protein	23.5	42.3
Lysine in protein		
In endosperm, % of protein†	1.8	4.0
In entire kernel, % of protein‡	2.6	4.2
Amino acids in corn‡		
Lysine, %	0.24	0.49
Tryptophan, %	0.09	0.15

† Mertz (1963).
‡ Cromwell et al. (1966).

As a result, the nutritional value of the protein in high protein corn is poorer than in normal corn.

In general, this same relationship between protein quantity and protein quality (amino acid profile) holds for other cereal grains when the protein content is increased in the kernel. The negative relationship between protein content and lysine as a percentage of the protein and, more importantly, between protein content and yield has been a significant deterrent on the development of high protein cereal grains.

Undoubtedly, one of the most significant breakthroughs in plant breeding occurred in the 1960s with the discovery of "high lysine" corn. This discovery was made by E.T. Mertz, a biochemist at Purdue University, and O.E. Nelson, a plant breeder at the same institution (Mertz, 1963; Mertz et al., 1964). These researchers found that the mutant recessive *opaque-2* gene of corn, when present in the homozygous state, resulted in a favorable shift in the glutelin/zein in corn endosperm protein (Table 4–3). The *opaque-2* gene causes a reduction in the synthesis of zein, which is very low in lysine and tryptophan and has a very poor balance of amino acids, and a concomitant increase in the synthesis of glutelin, which is consider-

Table 4–4. Nutritional value of normal and *opaque-2* corn for rats and pigs.

	Normal corn	*Opaque-2* corn	Normal corn + SBM
Rat study†			
Weight gain, g§	27	97	
Protein efficiency ratio§	1.6	2.8	
Pig study‡			
Daily gain, kg§	0.12	0.43	0.42
Feed/gain§	6.88	3.32	3.23

† Six rats per treatment initially weighing 40 to 57 grams were fed diets that contained 90% corn and were fortified with oil, minerals and vitamins for 28 days. Protein efficiency ratio is weight gain divided by protein consumed. The normal and *opaque-2* corn analyzed 0.30 and 0.49% lysine, respectively, Mertz et al. (1965).

‡ Six pigs per treatment (three pens, two pigs per pen) initially averaging 13.8 kg were fed diets containing normal or *opaque-2* corn with minerals and vitamins (8.8 and 11.2% protein, respectively) for 28 d. The normal corn + SBM diet consisted of normal corn, soybean meal, minerals and vitamins at the same protein level (11.2%) as the *opaque-2* corn diet. The normal and *opaque-2* corns analyzed 0.24 and 0.49% lysine, respectively, Cromwell et al. (1966).

§ Normal vs. *opaque-2* corn ($P < 0.01$).

Table 4–5. Nutritional value of normal and *opaque-2* corn for growing pigs.†

	Daily gain, kg	Feed/gain
Experiment 1‡		
Opaque-2 corn	0.26	3.46
Normal corn	0.06§	9.98§
+ lysine	0.06§	8.52§
+ lysine + tryptophan	0.21	4.16
+ soybean meal	0.21	4.41
Experiment 2¶		
Opaque-2 corn	0.36	3.22
Normal corn	0.08§	8.39§
+ lysine	0.07§	6.19§
+ tryptophan	0.14§	5.62§
+ lysine + tryptophan	0.30	3.34

† Normal corn analyzed 0.24% lysine and 0.09% tryptophan and *opaque-2* corn analyzed 0.49% lysine and 0.15% tryptophan. Supplemental lysine and tryptophan were added to the normal corn to equal the levels in *opaque-2* corn. All diets were supplemented with minerals and vitamins, Cromwell et al. (1967a).

‡ Six individually penned pigs per treatment initially averaging 10.6 kg were fed the diets for 35 d.

§ Differs from other treatments ($P < 0.05$).

¶ Six individually penned pigs per treatment initially averaging 12.6 kg were fed the diets for 28 d.

ably higher in lysine and tryptophan and has a much better amino acid profile (Table 4–2). In the initial investigation, Mertz (1993) found that lysine comprised 4.0% of the endosperm protein, more than twice that of typical hybrid corn (Table 4–3). The first feeding studies with rats (Mertz et al., 1965) and pigs (Cromwell et al., 1966) showed that *opaque-2* corn was clearly superior to normal corn in terms of its protein quality (Table 4–4), and later the superior nutritional value was found to be due to its higher levels of both lysine and tryptophan (Cromwell et al., 1967a; Table 4–5). *Opaque-2* corn also was found to be superior to normal corn for chicks, but only when *opaque-2* corn diets were supplemented with methionine (Cromwell et al., 1967b; Table 4–6).

Table 4–6. Nutritional value of normal, *opaque-2*, and *floury-2* corn for chicks.

	Weight gain, g	Feed/gain
Study 1†		
Normal corn	329	1.85§
Opaque-2 corn	332	1.81
Study 2†		
Normal corn + methionine	286§	1.84§
Opaque-2 corn + methionine	369	1.64
Study 3‡		
Normal corn	347¶	1.80¶
Opaque-2 corn	305	1.84
Floury-2 corn	370	1.74

† Four pens of 10 chicks per treatment in each of the two 21-d experiments, 15% dietary protein. The normal and *opaque-2* corns analyzed 0.24 and 0.49% lysine and 0.15 and 0.16% methionine, respectively, Cromwell et al. (1967b).

‡ Four pens of 10 chicks per treatment, 21-d experiment, 15% dietary protein. The normal, *opaque-2*, and *floury-2* corns analyzed 0.30, 0.54, and 0.45% lysine and 0.15, 0.14, and 0.23% methionine, respectively, Cromwell et al. (1968).

§ Normal vs. *opaque-2* corn ($P < 0.01$).

§ Normal vs. *opaque-2* vs. *floury-2* corn ($P < 0.05$).

Table 4–7. Nutritional value of double mutant corns for pigs.†

	Daily gain, g	Feed/gain
All-corn diets‡		
Normal corn	72§	12.62§
Opaque-2 corn	228	5.90
Opaque-2/floury-2 corn	195	6.33
Low-protein diets (12% soybean meal)¶		
Normal corn	422#	2.87††
Opaque-2/sugary-2 corn	558	2.53
Opaque-2/waxy corn	490	2.78

† The lysine levels in the corns were: normal, 0.26%; *opaque-2*, 0.42%; *opaque-2/floury-2*, 0.38%; *opaque-2/sugary-2*, 0.43%; *opaque-2/waxy*, 0.38%, Cromwell et al. (1984).
‡ Eight individually penned pigs per treatment, 12.4 kg initial weight, 28-d test.
§ Normal vs. double mutant corns ($P < 0.01$).
¶ Three pens of four pigs per treatment, 17.8 kg initial weight, 34-d test.
# Normal vs. *opaque-2/sugary-2* vs. *opaque-2/waxy* corn ($P < 0.01$).
†† *Opaque-2/sugary-2* vs. other corns ($P < 0.05$).

Shortly after the *opaque-2* corn discovery, another recessive mutant gene, *floury-2*, was found to result in an increase in the methionine content of corn protein along with increases in lysine and tryptophan (Nelson et al., 1965). This corn was found to be equivalent in nutritional value to *opaque-2* corn for pigs and superior to *opaque-2* corn for chicks (Table 4–6), due to its increased methionine content (Cromwell et al., 1968).

In the ensuing years, other genetic mutants were found to have a positive impact on the amino acid content of corn protein. *Sugary-2*, *waxy*, and double mutants of *opaque-2/floury-2*, *opaque-2/sugary-2*, and *opaque-2/waxy* were found to be nutritionally superior to normal corn for pigs (Cromwell et al., 1984; Table 4–7). The shift in the amylose/amylopectin in starch from 25:75 in normal corn to 0:100 in waxy corn does not impact the nutritional value of the corn for pigs, but it seems to improve the utilization of the starch by ruminants (McDonald & Mente, 1973).

Unfortunately, *opaque-2* corn had numerous agronomic problems that limited its acceptance in agriculture. Mainly, the lighter kernel weights resulting from the more floury endosperm were associated with a 5 to 10% reduction in yield. In addition, the softer, less vitreous kernels were more susceptible to diseases in the field, to mechanical damage when harvested, and to molds and insect damage in storage. While a few commercial seed companies still produce and market high lysine corn, most of the seed companies have discontinued it. Breeding programs at CIMMYT in Mexico have led to genotypes of corn with the improved amino acid characteristics of *opaque-2* corn, but with improved yields and more vitreous kernels (Villeagas et al., 1992). These genotypes, called quality protein maize (QPM), have made important contributions in a number of tropical and subtropical countries, and are especially important in those regions of the world where corn is a staple for humans.

Probably the most significant impact of the *opaque-2* gene discovery was that it alerted plant breeders to the fact that mutant genes existed in seeds that have a profound influence on the amino acid content of cereal protein. Shortly after the discovery of the *opaque-2* and *floury-2* genes, there were discoveries and development of improved protein cultivars of barley (*Hordeum vulgare* L.: 'Hiproly',

'Notch', 'Risø 1508'), oat (*Avena sativa* L.: 'Dal', 'Spear', 'Hinoat', 'Otee'), wheat (*Triticum aestivum* L.: 'Atlas 66', 'Nap Hal', 'Lancota'), grain sorghum (*Sorghum bicolor* [L.] Moench), and other cereal grains.

Although this section primarily deals with cereal grains, one cannot ignore the importance of protein quality in other protein sources, such as the oilseed meals. Geneticists have attempted to modify soybean to increase their S amino acid content. If this could be accomplished, it would have a major impact on the nutritional value of soybean meal because the S amino acids, methionine and cystine, are quite low in soybean and the first limiting amino acids in soybean protein (NRC, 1994, 1998).

An important advantage in using feedstuffs with improved amino acid profiles is that the animal's amino acid requirements can be met with diets that contain lower amounts of dietary protein. Feeding lower protein diets means that less of the amino acid excesses will have to be degraded and less N (resulting from the excesses of unneeded amino acids) will be excreted (Carter et al., 1996). A reduction in excreted N has important environmental implications because excessive N in animal manure applied as fertilizer to crop land can potentially contribute to groundwater pollution. Excessive N in manure also results in greater amounts of ammonia volatilization, which reduces air quality and contributes to odor.

IV. INCREASING THE ENERGY CONTENT OF CEREAL GRAINS

As previously indicated, the major role that cereal grains play in livestock feeds is as a source of energy. In the USA, energy requirements of animals and energy contributions of feedstuffs are typically measured as digestible energy (DE) or metabolizable energy (ME), and less frequently as net energy (NE). The units of energy are commonly expressed as kilocalories (kcal) or megacalories (Mcal). Corn is an excellent source of energy for swine, poultry, and ruminants due to its high starch content (located mainly in the endosperm), its low fiber content (mainly in the hull) and a small amount of oil (mainly in the germ). Typically, corn is considered to contain approximately 3420 kcal of ME kg^{-1} on an air-dry basis for the pig (NRC, 1998). Wheat is similar to corn in energy content (3305 kcal of ME kg^{-1}), as is grain sorghum (3340), whereas the ME in barley and oat is less (2910 and 2710 kcal of ME kg^{-1}, respectively) due to their higher fiber content (NRC, 1998). The ME values of these same cereal grains are slightly lower for chicks (corn, 3350 kcal kg^{-1}; wheat, 2900 kcal kg^{-1}; grain sorghum, 3290 kcal kg^{-1}; barley, 2640 kcal kg^{-1}; and oat, 2550 kcal kg^{-1}; NRC, 1994).

Studies have demonstrated that the oil content of corn can be altered by genetic selection (Dudley & Lambert, 1992). This was clearly demonstrated by a long-term study initiated in the early 1900s at the University of Illinois in which kernels that differed in oil content were selected from a common parent line. Figure 4–3 shows the oil content of kernels after 95 generations of continuous selection for high oil content in the high oil line and for low oil content in the low oil line. The oil content of the high oil line is now over 20% (Fig. 4–3) while the oil content of the low oil line is less than 0.5% and too low to accurately measure.

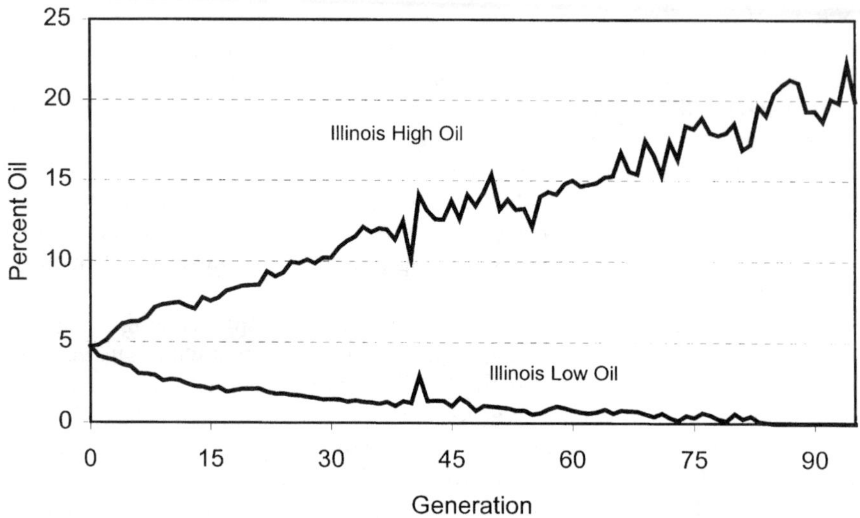

Fig. 4–3. Changes in oil content of corn following 95 generations of selection for high or low oil content (courtesy of J.W. Dudley, Univ. Illinois).

Several years ago, high oil hybrid corn was released for commercial use. Typically, these high oil cultivars are about twice as high in their oil content as normal hybrid corn (6–7% vs. 3.5– 4%; Table 4–8). The 2.25-fold higher energy content

Table 4–8. Composition of normal and high oil corn.†

	Normal corn	High oil corn
Crude fat, %	3.47	6.33
Crude protein, %	7.94	8.49
Crude fiber, %	1.95	2.06
Starch, %	61.4	58.4
Metabolizable energy, kcal/kg	3362	3516
Amino acids, %		
Lysine	0.25	0.29
Threonine	0.28	0.31
Tryptophan	0.07	0.07
Methionine	0.18	0.20
Cystine	0.18	0.19
Fatty acids, % of total		
Palmitic acid	10.5	11.2
Stearic acid	2.1	2.7
Oleic acid	28.0	34.8
Linoleic acid	56.4	48.9
Linolenic acid	1.5	1.0
Minerals, %		
Ca	0.01	0.01
P	0.24	0.26
Vitamin E, mg/kg	22.0	28.0

† Unpublished data from Optimum Quality Grains, L.L.C., Des Moines, Iowa, 1998. Analyses from more than 300 samples over a 3-yr period. Composition is on an 86% dry matter basis.

Table 4–9. Performance of growing-finishing pigs fed corn-soybean meal diets with normal or high oil corn and corn-soybean meal diets with supplemental fat.

	Normal corn-SBM	High oil corn-SBM	Normal corn-SBM + fat†
Experiment 1‡			
Avg. daily gain, kg	0.85	0.85	0.87
Avg. daily feed, kg§	2.31	2.20	2.25
Feed/gain§	2.72	2.58	2.58
Backfat, cm	1.80	1.73	1.88
Experiment 2¶			
Avg. daily gain, kg		0.86	0.84
Avg. daily feed, kg		2.38	2.36
Feed/gain		2.77	2.80
Backfat, cm		2.59	2.49
Carcass lean, %		50.0	50.2

† Choice white grease added such that diet was isocaloric to the high oil corn-soybean meal diet.
‡ Experiment involved 184 pigs (16 pens of four pigs/treatment) from 20 to 115 kg body weight, Risley and Bajjalieh (1996).
§ Normal corn-soybean meal diet vs. other treatments ($P < 0.01$).
¶ Experiment involved 96 pigs (12 pens of 4 pigs/treatment) from 28 to 113 kg body weight, Lohrmann et al. (1998).

of oil compared with starch results in approximately 150 kcal more kilocalories of ME kg^{-1} in high oil corn than typically found in normal corn.

The higher oil content of high oil corn is due to its larger germ. Because the germ also is higher in protein than the endosperm, and the proportion of lysine and certain other amino acids is higher in germ than in endosperm, the larger germ in high oil corn results in slightly higher protein and lysine than found in normal corn (Table 4–8). Additionally, the fatty acid profile of the oil is slightly different, being slightly higher in oleic acid and lower in linoleic acid (Table 4–8).

The feeding of high oil corn to growing and finishing pigs results in similar to slightly faster gains, reduced feed intake (due to the greater energy density), and improved efficiency of feed utilization (Risley & Bajjalieh, 1996; Lohrmann et al., 1998) as shown in Table 4–9. Similar results have been reported with chicks (Araba & Dale, 1998). The improvement in growth performance is essentially the same as when corn oil (or other sources of feed-grade fat) is added to the diet to provide the same level of oil as contributed by the high oil corn.

Probably the greatest benefit associated with high oil corn is the positive effect on the physical properties of feeds. The presence of additional fat in diets reduces feed dust, and it has been clearly documented that reduced feed dust in swine facilities results in healthier pigs with fewer lung lesions and associated respiratory problems (because air-borne microorganisms are carried on dust particles; Gordon, 1963; Curtis et al., 1975). Furthermore, less dust in hog buildings results in a healthier environment for people working in those buildings. The additional oil in corn has further benefits such as: (i) it increases the throughput of hammer mills (more than 10% in one study), thus reducing the amount of electrical energy needed to grind the corn; (ii) it results in better mixing efficiency; (iii) it reduces segregation of particles in mixed feed; and (iv) it results in improved pellet quality when diets are pelleted (Crum & Stilborn, 1997). The "prepackaged oil" in high oil corn has an additional benefit to small producers who mix their own feed, and who may not have sufficient production volume to justify keeping a heated fat tank on the farm.

High oil corn can be grown from either traditional single cross hybrids or using the TopCross (a registered trademark of Optimum Quality Grains, L.L.C., Des Moines, Iowa) grain production system. In the latter, a seed mixture of 90 to 92% male sterile grain parent seeds and 8 to 10% high oil pollinator seeds is used. High oil corn grain harvested from single cross hybrids will typically contain less oil than grain produced with the TopCross production system. When using the TopCross system, a 10% increase in planting rate is recommended when plant populations are less than 30 000 plants per acre. In this production system, high oil corn has been reported to yield as well as normal corn.

V. REDUCING THE PHYTATE CONTENT OF CEREAL GRAINS AND OILSEED MEALS

The majority of the P in cereal grains and oilseed meals is organically bound in a complex called *myo*-inositol 1,2,3,4,5,6, hexakisphosphate, commonly referred to as phytic acid or phytate (Nelson et al., 1968; Fig. 4–4). From 60 to 80% of the P in cereal grains and oilseed meals is in this form (Table 4–10). In order for animals to absorb and utilize P in this form, the phosphate groups must first be enzymatically released from the complex, a process that requires the enzyme, phytase. This enzyme is lacking in the digestive tract of pigs and chickens, so phytic acid is poorly utilized by these and other nonruminant species. Ruminants have no problem utilizing phytate phosphorus because of the abundance of phytase produced by the rumen microorganisms. Studies at the University of Kentucky indicate that the P is only 10 to 15% bioavailable in corn and 20 to 25% bioavailable in soybean meal for nonruminants (Cromwell, 1993). The P in wheat is more bioavailable (50%) due to the presence of endogenous phytase in the seed coat. Table 4–11 gives estimates of the P bioavailability of a number of feedstuffs that are commonly fed to pigs and poultry.

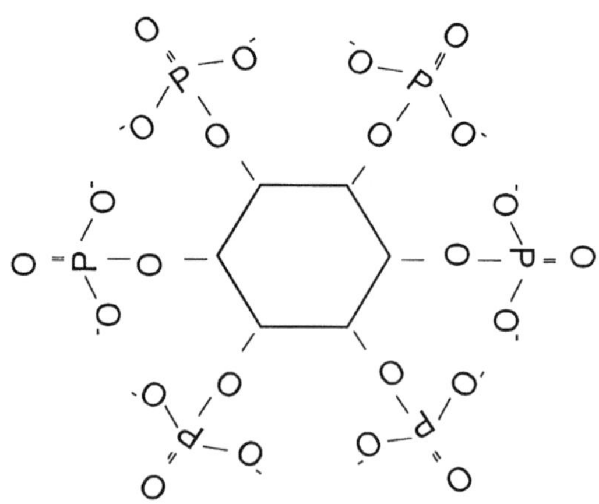

Fig. 4–4. Structure of phytic acid.

Table 4–10. Phytate P content of some common feedstuffs.†

Feedstuff	Total P (%)	Phytate (% of total P)
Corn	0.26	66
Wheat	0.30	67
Grain sorghum	0.31	68
Barley	0.34	56
Oats	0.34	56
Wheat middlings	0.47	74
Soybean meal	0.61	61
Cottonseed meal	1.07	70
Sesame meal	1.27	81
Wheat bran	1.37	70
Alfalfa meal	1.40	0

† Adapted from Nelson et al. (1968).

Because of the poor utilization of phytate P by swine and poultry, diets must be supplemented with relatively large amounts of inorganic P in order to meet their P requirements. The large amounts of dietary P coupled with the poorly utilized phytate P in the diet results in excessive amounts of P excreted in the manure of these

Table 4–11. Bioavailability of P in cereal grains, grain by-products, high protein meals, inorganic phosphates, and phytase supplemented diets for pigs.†

Feedstuff	Bioavailability of P (%)‡	Feedstuff	Bioavailability of P (%)‡
Cereal grains		High protein meals—plant origin	
Corn, pelleted	12	Cottonseed meal	1
Corn	14	Sunflower meal	3
Grain sorghum	20	Palm kernel cake	11
Oat	22	Peanut meal	12
Barley	30	Canola meal	16
Triticale	46	Soybean meal, dehulled	23
Wheat	50	Soybean meal, 44% CP	31
Corn, high moisture	53	Soybean meal, dehulled,	
Corn, low phytate	77	low phytate	49
Grain by-products		High protein meals—animal origin	
Oat groats	14	Feather meal	31
Hominy feed	14	Meat and bone meal	85
Corn gluten meal	15	Dried skim milk	91
Rice bran	25	Blood meal	92
Wheat bran	29	Fish meal	93
Brewers grains	33	Dried whey	96
Cooked cereal fines	40	Inorganic phosphates	
Wheat middlings	41	Steamed bone meal	82
Corn gluten feed	59	Defluorinated phosphate	87
Distillers grains plus solubles	76	Monocalcium phosphate	100
Miscellaneous		Dicalcium phosphate	100
Soybean hulls	78	Phytase addition	
Alfalfa meal	100	Corn-soy diet without phytase	15
		Corn-soy diet with phytase	45

† Research at the University of Kentucky as reviewed by Cromwell (1993). The value for meat and bone meal was modified based on recent research. The value for low phytate corn is from Cromwell et al. (1998a). The value for low phytate soybean meal is from Cromwell et al. (2000a).

‡ Relative to the availability of P in monosodium or monocalcium phosphate, which is given a value of 100.

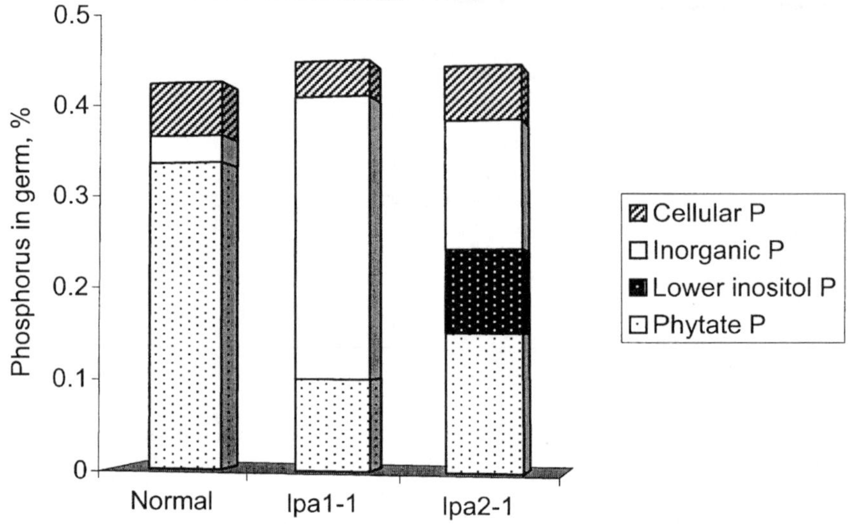

Fig. 4–5. Distribution of P in the germ of normal corn and two low phytate corns containing the mutant *lpa1-1* and *lpa1-2* genes (Raboy et al., 1990).

two species. Excess P in animal manure is a major environmental concern in agriculture. In addition, phytic acid complexes with Zn, Cu, and other trace elements and interferes with absorption of these minerals.

In 1990, Victor Raboy, a USDA plant breeder then at Montana State University, reported the discovery of mutant genes in corn that suppressed the synthesis of phytic acid in the seed without affecting the amount of total seed P (Raboy et al., 1990). As a result, the inorganic P was markedly increased in the seed (Fig. 4–5).

In 1996, the Nutrition and Industry Markets group of Pioneer Hi-Bred International (now part of Optimum Quality Grains, L.L.C, a joint venture of Pioneer Hi-Bred International and DuPont Agricultural Products) produced sufficient quantities of corn containing the mutant low phytic acid gene (*lpa1*) for animal testing. This corn contained half as much phytate P and 3.5 times as much inorganic P as a near-isogenic normal corn (Table 4–12). In initial testing at the University of Kentucky using slope-ratio procedures, we found that the P in low phytate corn was about three to four times more bioavailable than the P in normal corn (Fig. 4–6).

Table 4–12. Composition of normal and low phytate corn.†

	Normal corn	Low phytate corn
Total P, %	0.25	0.28
Phytate P, %	0.20	0.10
Nonphytate (inorganic) P, %	0.05	0.18
Crude protein, %	8.6	8.6
Lysine, %	0.23	0.24

† Source of corn was Optimum Quality Grains, L.L.C., a joint venture of Pioneer Hi-Bred International and DuPont Agricultural Products. The normal and low phytate corns were near-isogenic except for the *lpal* gene.

Specifically, in pig studies, the low phytate gene increased the bioavailability of P from approximately 20% in normal corn to over 75% in low phytate corn (Cromwell et al., 1998a; Fig. 4–6). In chicks, the bioavailability of P was increased from 10% in normal corn to 50% in low phytate corn (Cromwell et al., 1998b; Fig. 4–6). Sim-

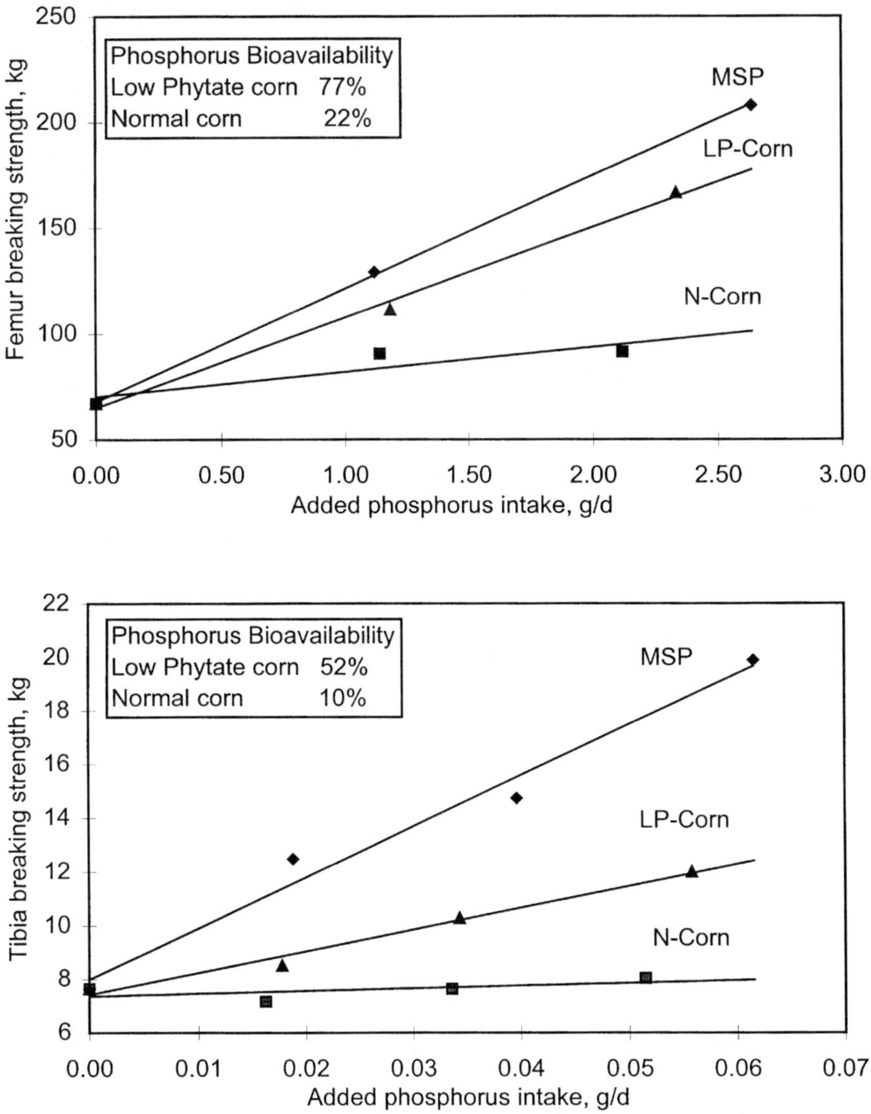

Fig. 4–6. Slope ratio analysis of breaking strength of femus from pigs (top plate) and of tibias from chicks (bottom plate) regressed on daily P intake illustrating the bioavailability of low phytate corn (LP-corn) and a near-isogenic normal corn (N-corn) relative to monosodium phosphate (MSP), considered as 100 bioavailable (Cromwell et al., 1998ab). Pig diets included a low P basal (0.22% P) without and with 0.06 or 0.12% added P from the two corns or MSP. Chick diets included a basal (0.45% P) without and with 0.04, 0.08, or 0.12% added P from the two corns or MSP.

Table 4–13. Performance of growing pigs fed corn-soybean meal diets with normal or low phytate corn and varying amounts of supplemental P.†

	Normal corn				Low phytate corn			
Total P, %‡	0.59	0.50	0.42	0.33	0.59	0.50	0.42	0.33
Bioavailable P, %§	0.35	0.26	0.18	0.09	0.45	0.37	0.28	0.20
Daily gain, kg¶	0.81	0.80	0.74	0.62	0.80	0.78	0.79	0.77
Feed/gain¶	2.37	2.37	2.44	2.79	2.39	2.54	2.38	2.35
Bone strength								
Femur, kg¶	308	301	234	125	332	332	298	208
MM, kg¶#	90	86	66	41	87	89	85	64
MM ash, %¶	50.6	50.6	48.7	45.2	51.0	50.9	50.4	46.9

† Each treatment consisted of five pens of two pigs each from 23 to 51 kg body weight. Dietary lysine was 0.80% and dietary Ca was 0.65%, Pierce et al. (1998a).
‡ The diets with 0.33% total P had no supplemental inorganic P.
§ Based on the following bioavailabilities: normal corn, 20%; low phytate corn, 75%; soybean meal, 25%; dicalcium phosphate, 100%.
¶ Effect of dietary P ($P < 0.01$); dietary P × corn-type interaction ($P < 0.05$).
# MM = average of third and fourth metacarpal and metatarsal bones.

ilar findings were reported by researchers at the University of Missouri (Spencer et al., 1998).

Results of feeding tests with both growing pigs (Table 4–13) and finishing pigs (Table 4–14) indicate that feeding pigs low phytate corn-soybean meal diets with 0.10 to 0.12 of a percentage point less total P than normal results in similar performance and bone mineralization as in pigs fed normal corn-soybean meal diets

Table 4–14. Performance of finishing pigs fed corn-soybean meal diets with normal or low phytate corn and varying amounts of supplemental P.†

	Normal corn			Low phytate corn		
Total P						
Phase 1, %‡	0.45	0.39	0.33	0.34	0.39	0.33
Phase 2, %‡§	0.42	0.36	0.30	0.42	0.36	0.30
Bioavailable P, %	0.24	0.17	0.10	0.34	0.27	0.20
Phase 1						
Daily gain, kg#	0.97	0.96	0.91	0.99	0.98	0.98
Feed/gain††	3.23	3.10	3.31	3.12	3.11	3.07
Entire experiment						
Daily gain, kg	0.98	0.98	0.94	0.97	0.98	0.99
Feed/gain	3.44	3.38	3.48	3.39	3.39	3.35
Bone and carcass traits						
MM strength, kg††,‡‡	181	177	156	181	185	187
Metatarsal ash, %††	53.2	52.9	52.2	53.7	53.5	53.4
Carcass lean, %	53.7	53.9	53.1	54.0	54.0	53.7

† Each treatment consisted of six pens of five pigs each from 52 to 105 kg body weight, Pierce et al. (1998b).
‡ Phase 1 was from 52 to 77 kg and Phase 2 was from 77 to 105 kg body weight. Dietary lysine was 0.85 and 0.65%, and dietary Ca was 0.54 and 0.52% during Phases 1 and 2, respectively.
§ The diets with 0.30% total P had no supplemental P.
¶ Based on the following bioavailabilities: normal corn, 20%; low phytate corn, 75%; soybean meal, 25%; dicalcium phosphate, 100%.
# Effect of dietary P level in normal corn diets ($P < 0.05$).
†† Effect of dietary P level in normal corn diets ($P < 0.01$).
‡‡ MM = average of third and fourth metacarpal and metatarsal bones.

Table 4–15. Performance and P excretion of growing-finishing pigs fed corn-soybean meal diets with normal or low phytate corn.†

	Normal corn	Low phytate corn
Total P, %	0.50–0.45–0.40‡	0.40–0.35–0.30‡
Bioavailable P, %	0.26–0.23–0.20‡	0.26–0.23–0.20‡
Daily gain, kg	0.85	0.84
Feed/gain	2.94	2.88
Bone strength, kg	209	202
P digestibility, %§	33	45
P in feces, % of dry matter§	2.5	1.5
P excreted, g/d§	8.1	5.0
Reduction in P excretion, %	--	38

† Each treatment consisted of five pens of five pigs each from 28 to 113 kg body weight, Pierce and Cromwell (1999a).
‡ These P levels were fed during Phase 1 (28–52 kg), Phase 2 (52–85 kg), and Phase 3 (85–113 kg), respectively.
§ Normal vs. low phytate corn ($P < 0.01$).

(Pierce et al., 1998a, 1998b). This reduction in total dietary P along with the greater bioavailability of P in low phytate corn is associated with a 38 to 43% reduction in excreted P (Pierce & Cromwell, 1999a) as shown in Tables 4–15 and 4–16. When supplemental microbial phytase is included in corn-soybean meal diets containing low phytate corn, reductions in P excretion by pigs of 50% or more have been reported (Pierce & Cromwell, 1999b; Table 4–17).

Preliminary testing by Optimum Quality Grains, L.L.C., indicates that low phytate corn has yields that are comparable to those of normal corn. Also, the protein, lysine, starch, and other important nutritional traits are not affected by the low phytate gene. The gene apparently affects only the type of P in the corn grain. A side benefit of the reduced phytic acid (that is a strong chelating agent) is that bind-

Table 4–16. Phosphorus balance of finishing pigs fed corn-soybean meal diets with normal or low phytate corn.†‡

	Normal corn	Low phytate corn
Total P, %	0.40	0.30
Bioavailable P, %	0.20	0.20
P intake, g/d	11.89	8.66
P retained, g/d	5.31	4.92
P excreted, g/d		
Feces§	5.85	3.64
Urine§	0.73	0.10
Total§	6.58	3.74
Reduction in P excretion, %	--	43

† Five-day balance experiment involving five pigs per treatment at approximately 100 kg body weight. Diets were 14% crude protein and 0.65% lysine. The low phytate corn diet had no supplemental inorganic P, Pierce and Cromwell (1999a).
‡ Both diets contained 14% protein and 0.65% lysine. The normal corn-soybean meal diet contained 0.12% added P from dicalcium phosphate. The low phytate corn-soybean meal diet contained no dicalcium phosphate.
§ Normal vs. low phytate corn ($P < 0.01$).

Table 4–17. Phosphorus excretion by pigs fed corn-soybean meal diets containing normal or low phytate corn and supplemented with microbial phytase.†

	Normal corn		Low phytate corn	
Phytase, 600 units kg^{-1}	–	+	–	+
Total P, %	0.55	0.45	0.45	0.35
Available P, %	0.29	0.19	0.29	0.19
Femur strength, kg	272	267	282	283
P excretion, g d^{-1}‡§	6.80	5.24	4.43	3.34
Reduction in P excretion, %	--	23	35	51

† Each treatment consisted of six pens of two pigs each from 18 to 48 kg body weight, Pierce and Cromwell (1999b).
‡ Normal vs. low phytate corn ($P < 0.05$).
§ Effect of phytase ($P < 0.05$).

ing of Zn and other trace minerals is less likely to occur in the intestinal tract when this type of corn is fed. Thus, dietary levels of some of the trace elements such as Zn also can be reduced, which also will reduce excretion of the trace minerals. Optimum Quality Grains estimates that seed will be commercially available for planting in year 2001.

Soybeans that are low in both phytic acid and oligosaccharides (Table 4–18) have been produced by mutagenesis (Optimum Quality Grains, L.L.C.). Dehulled soybean meal produced from this type of soybean has recently been evaluated at the University of Kentucky. Using slope-ratio procedures similar to those described in the evaluation of normal and low phytate corn, we found that the P in low phytate soybean meal was considerably more bioavailable for both pigs (49 vs. 19%) and chicks (58 vs. 28%) than the P in soybean meal from near-isogenic conventional soybeans (Cromwell et al., 2000a,b). In another study, young pigs fed diets consisting of low phytate corn and low phytate soybean meal with no supplemental inorganic P grew as fast and efficiently, had similar bone traits, and excreted 53% less P than pigs fed diets containing conventional corn and soybean meal supplemented with sufficient inorganic P to meet the pigs P requirement (Table 8–19;

Table 4–18. Composition of normal and low phytate, low oligosaccharide, dehulled soybean meal.†

	Normal soybean meal	Low phytate soybean meal
Total P, %	0.70	0.77
Phytate P, %	0.48	0.22
Nonphytate (inorganic) P, %	0.22	0.55
Bioavailability of P, %‡	19	49
Carbohydrates, %		
Sucrose	7.22	12.32
Galactinol§	0.17	0.00
Oligosaccharides		
Raffinose	0.91	0.55
Stachyose	5.20	0.53

† Source of soybeans was Optimum Quality Grains, L.L.C., a joint venture of Pioneer Hi-Bred International and DuPont Agricultural Products. The two soybean meals were from near-isogenic soybeans.
‡ Determined by slope-ratio procedures. The bioavailabilities of P in soybean meal are relative to that in monosodium phosphate that is considered 100% bioavailable.
§ Galactinol is a precursor to raffinose, stachyose, and *myo*-inositol (a precursor of phytic acid).

Table 4–19. Performance of growing pigs fed diets with normal corn and normal soybean meal or low phytate corn and low phytate soybean meal with various amounts of supplemental P.†

	Normal corn + normal soybean meal			Low phytate corn + low phytate soybean meal		
Added P, %	0.20	0.10	0.00	0.20	0.10	0.00
Total P, %:	0.56	0.46	0.36	0.59	0.49	0.39
Bioavailable P, %:‡	0.27	0.17	0.07	0.44	0.34	0.24
Daily gain, kg§	0.80	0.75	0.63	0.81	0.80	0.79
Feed/gain§	2.22	2.30	2.62	2.24	2.16	2.19
Bone strength, relative§¶	100	74	52	98	102	98
P excreted, g d^{-1}#	7.0	6.2	5.3	5.1	4.0	3.3
Reduction in P excretion, %††				27	43	53

† Each treatment consisted of eight individually penned pigs from 23 to 49 kg body weight. Dietary lysine was 0.95% and Ca was 0.65%. Corn and soybean meal sources are shown in Tables 4–12 and 4–18, Cromwell et al. (2000c).

‡ Based on the following bioavailabilities: normal corn, 20%; low phytate corn, 75%; normal soybean meal, 20%; low phytate soybean meal, 50%; dicalcium phosphate, 100%.

§ Linear effect of added P in normal corn-normal soybean meal diets ($P < 0.01$).

¶ Based on mean of femurs and metatarsal-metacarpals.

Linear effect of added P in both diets ($P < 0.01$).

†† Reduction in P excretion compared with pigs fed the normal corn-soy diet with 0.56 total P [approximate requirement (NRC, 1998)].

Cromwell et al., 2000c). Similar findings also were reported by Spencer et al. (2000a,b).

VI. REDUCTION OR ELIMINATION OF ANTINUTRITIONAL COMPOUNDS

A number of factors have been identified in plant seeds, especially in oilseeds, that are known to have a deleterious effect on growth in animals (Thacker & Kirkwood, 1990). In some instances, certain of these growth inhibitors have been reduced significantly or even nearly completely eliminated by breeding.

A. Glucosinolates and Erucic Acid in Rapeseed

Rapeseed (*Brassica napus* L.) is one of the oldest oilseeds known to mankind. For many years rape was raised as a pasture crop for grazing. Limited amounts of rapeseed were used for animal feeding because of the presence of strong goitrogenic compounds (glucosinolates) and a toxic fatty acid (erucic acid) in the seed (Bell, 1984). A highly successful breeding program was implemented in the 1960s by the Canadian Department of Agriculture to reduce the glucosinolates and erucic acid in rapeseed. After several years of intensive breeding and selection, plant scientists were able to produce a "low-low" or "double 0" (refers to the two toxic factors) rapeseed, and it was officially named "canola" in 1979 (Bell, 1982). Today, canola is grown throughout Canada and in many other areas of the world. Next to soybean meal, canola meal is one of the more commonly used protein supplements in animal feeds. Canola oil is widely used in cooking and its popularity stems from the perception that it is a "healthier" oil due to its relatively high proportions of monounsaturated fatty acids.

B. Trypsin Inhibitors in Soybean

Soybean contains several factors that inhibit the activity of trypsin, one of the major proteolytic enzymes in the intestinal tract of animals (Liener & Kakade, 1980). Trypsin is an endopeptidase secreted by the pancreas into the upper portion of the small intestine (duodenum). Trypsin, along with another pancreatic enzyme, chymotrypsin, digests dietary protein that has already been partially degraded by hydrochloric acid and pepsin in the stomach. The end-products of trypsin-chymotrypsin digestion are small peptides that are further hydrolyzed to dipeptides and free amino acids, which are then absorbed from the gut. Without full activity of trypsin, dietary protein is poorly digested and inefficiently utilized.

Soybean contains several trypsin inhibitors. The two inhibitors most widely studied are the Kunitz and Bowman-Birk trypsin inhibitors (Rackis, 1972). Fortunately, these inhibitors are destroyed by moist heat. The practice in the soybean crushing industry of "toasting" the meal after extracting the oil from the bean is routinely done to destroy these inhibitors. If unprocessed, full-fat soybean is fed to pigs or poultry, growth is severely depressed. Subjecting soybean to heat (roasting) for an appropriate period of time to destroy the inhibitors greatly improves the nutritional properties of the bean. Proper heat treatment is generally evaluated by the urease test, in which the amount of urease in the soybean or soybean meal is measured. Because urease and trypsin inhibitors are both destroyed by heat treatment, trace amounts of urease indicate that soybean or soybean meal have been properly heated.

Plant breeders have successfully produced soybean with reduced amounts of Kunitz trypsin inhibitor (Singh et al., 1969; Hymowitz, 1986). The inhibitor is reduced to about one-half the level found in conventional soybean. While the feeding of unheated low trypsin inhibitor soybean to pigs or chicks does not depress performance to the same extent as when unheated conventional soybeans are fed, growth performance is not optimized unless some heat treatment is applied (Herkelman et al., 1991, 1992). Some research suggests that approximately one-half as much heat is required to maximize the utilization of low trypsin inhibitor soybean as compared with conventional soybean (Herkelman et al., 1991). Further reduction or complete elimination of the Kunitz trypsin inhibitor, as well as the Bowman-Birk and other inhibitors presents a further challenge to soybean breeders.

C. Oligosaccharides in Soybean

Soybean possesses several oligosaccharides; namely raffinose, stachyose, and verbascose, which constitute approximately 7% of the soybean (Rackis, 1975). This family of carbohydrates is characterized by α-1,6-galactosyl linkages that are not digested in mammalian systems due to the lack of α-1,6-galactosidase activity (Cristofaro et al., 1974). These oligosaccharides have been shown to depressed growth performance in young pigs. Genes have now been identified that substantially reduce or eliminate these oligosaccharides in soybean (Table 4–18). Interestingly, the metabolic pathways involved in the synthesis of oligosaccharides in soybeans are the same as those involved in the synthesis of phytic acid. Successful incorporation of these genes into commercial cultivars in order to produce low levels of oligosaccharides should have a major impact on the nutritional properties of soybean, especially when used in diets for young pigs.

D. Gossypol in Cottonseeds

Gossypol, a phenolic compound found in the pigment glands of cottonseed (*Gossypium hirsutum* L.), is toxic to nonruminants, but not to ruminants (Cheeke & Shull, 1985). Glandless cottonseed was developed several decades ago with the intent of producing an acceptable meal for nonruminants, but the concept was not successful due to low lint yield of the cottonseed and to an increased susceptibility of the plant to insects. Plant breeders are now attempting to reduce the gossypol in the seed but increase it in the rest of the plant in order to overcome the agronomic problems. Unfortunately, the lysine and energy content of cottonseed meal is considerably less than in soybean meal (1.72 vs. 2.83%, 2315 vs. 3180 kcal ME kg^{-1}), so it is not likely to be a popular substitute for soybean meal in pig or broiler diets even if the gossypol problem is solved. Cottonseed meal is not used in layer diets because gossypol causes a bluish-green discoloration and egg yolk along with severe mottling (Scott et al., 1976). Also, small amounts of cyclopropene fatty acids, sterculic acid and malvalic acid, in cottonseed meal cause a pink discoloration of the egg white (Scott et al., 1976). Cottonseed meal is still widely used in cattle feed.

E. β-Glucans in Barley

Barley contains β-glucans, polymers of glucose with β-(1-3),(1-4) linkage. The β-glucans are poorly digested by nonruminants and they result in the formation of gels that increase the viscosity of the intestinal tract contents and interfere with the digestion of other nutrients (Cheeke, 1991). An additional problem with poultry is that the hygroscopic nature of glucans makes the excreta wet and sticky. In young chicks, the sticky excreta may dry on the feathers around the vent and inhibit defecation. Low β-glucan barley cultivars have been developed that appear to have improved nutritional value. The enzyme, β-glucanase is commercially available and can be added to barley diets. This enzyme improves the properties of poultry excreta and improves performance. However, in swine, the addition of the enzyme to barley diets has not given consistent positive results (Thacker, 1993).

F. Tannins in Grain Sorghum

Certain types of grain sorghums contain high amounts of polyphenolic compounds, broadly referred to as tannins (Cheeke & Shull, 1985). These compounds form resistant bonds with both dietary proteins and digestive enzymes in the intestinal tract and thus interfere with the digestion of proteins and carbohydrates (Reddy et al., 1982). Due to the unpalatable nature of tannins, grain sorghums with increased tannin content were developed a number of years ago to protect the grain from bird damage in areas where wild birds were prevalent. These grain sorghums were called "bird resistant" sorghum. However the high tannin, bird resistant sorghums are inferior in their nutritional value and have not been widely accepted in animal feeding programs. Breeding programs are presently underway to develop improved cultivars of grain sorghum that have enhanced nutrient digestibility.

VII. GENETIC ENGINEERING TO IMPROVE NUTRITIONAL VALUE

A number of transgenic plants have come about in the past five years as a result of recombinant DNA procedures. The first transgenic plant addressed by the Food and Drug Administration (FDA) was the Flavr Savr tomato (*Lycopersicon esculentum* Mill.), developed by Monsanto Company (St. Louis, MO), that has improved ripening characteristics. Since 1994, the FDA has finalized consultations with developers of 41 transgenic plants (W.D. Price, 1999, personal communications). Among these are insect protected corn, cotton, potato (*Solanum tuberosum* L.), and tomato containing the *cryIA(c)* or *cryIA(b)*gene from *Bacillus thuringiensis* (BT); herbicide resistant corn, soybean, cotton, potato, squash (*Cucurbita* spp.), canola, sugar beet (*Beta vulgaris* L.), and flax (*Linum unsitatissimum* L.); disease-resistant squash and papaya (*Carica papaya* L.); and male sterile corn and canola. In addition, high oleic acid soybean and high lauric acid canola have been developed by foreign gene insertion using recombinant DNA strategies. Genes encoding phytase and α-amylase enzymes have been inserted into tobacco (*Nicotiana tabacum* L.) as a model system (Pen et al., 1993), and similar applications are being investigated with cereal grains and oilseeds. Already, the phytase gene has been successfully introduced into canola such that it can be used as a feed source of the enzyme (Anderson, 1998).

Brazil nut (*Bertholletia excelsa* H.B.K.) 2S albumin is a protein exceptionally high in methionine and cystine (18 and 8% on a mole % basis), and a gene with the coding sequence for this protein has been introduced into tobacco, canola, and narbon bean (*Vicia narbonensis* L.) producing 30% or greater increases in methionine in the transgenic recipients. This gene also has been introduced into soybean with a 50% increase in the methionine content (Anderson, 1998). Unfortunately, the Brazil nut protein has significant allergenic properties that are transferred to the target transgenic plants. Other genes from legumes have been incorporated into cereals with promising effects of increasing the lysine or methionine of storage proteins in cereals. The bean (*Phaseolus vulgaris* L.) β-phaseolin gene and the pea (*Pisum sativum* L.) legumin gene have been introduced into rice (*Oryza sativa* L.) in an effort to elevate lysine (Anderson, 1998). Other novel storage proteins have successfully been transferred into wheat and other cereals. This new technology could result in major improvements in the nutritional value of cereals in the future.

An alternative approach to manipulating the storage protein is to increase the levels of desired free amino acids in the seed. Geneticists have accomplished this by altering the regulating enzymes that control the biosynthetic pathways of amino acid synthesis, thus, allowing free amino acids to accumulate to significant levels. These techniques have been used to produce cell cultures that overproduce free threonine, tryptophan, or lysine (Anderson, 1998).

Twentyfold increases in free lysine have been produced in tobacco leaves with lysine content as high as 2% of the leaf dry weight. These new techniques have been successfully applied to seeds with reports of twofold increases in free lysine in canola, fivefold increases in free lysine in soybean and two- to threefold increases in free lysine in corn seeds (Anderson, 1998). This approach looks promising in that substantial increases in the free amino acids can be obtained, the amino acid should be highly available to the animals since it is not protein bound, the high amino acid trait is conferred by a single dominant gene, and the approach is applicable to many crop species (Anderson, 1998). It will be interesting to see what lies ahead in this fascinating area of biotechnology.

VIII. RELATIVE MERITS OF VALUE-ADDED CULTIVARS VERSUS DIETARY SUPPLEMENTS

An important issue that should be considered by plant breeders and animal nutritionists is the cost/benefit of producing crops with improved nutritional traits, especially if that same improvement could be attained by dietary supplementation. For example, one could ask if there is an advantage in developing cultivars of high oil corn if a commercial source of fat can be added to the feed when it is manufactured. Similarly, one should determine if there are benefits of using high lysine corn or high methionine soybean as compared with simply supplementing normal corn and soybean meal diets with crystalline lysine and methionine that might be commercially available at competitive prices. Often, the costs of various options can be calculated quite easily. But simple cost comparisons may not always give the best answer.

In many cases, having the value-added traits packaged with the product itself has subtle benefits beyond those of economic considerations. For example, high oil corn with a premium in price may still have benefits over normal corn with supplemental fat. One reason is that less dust from the corn itself is lost in the grinding process due to the added fat. As previously discussed, fat additions to pig diets reduces dust in buildings and results in healthier pigs with fewer respiratory problems. Furthermore, the improved air quality benefits people working in confinement buildings. Medium or small producers who mix their own feed could rely on the additional fat in high oil corn to achieve these benefits rather than having to purchase or lease a heated fat tank in order to add supplemental fat to their feeds.

Another example where diets with value-added cultivars may have an advantage is the situation where pig feed produced from corn or soybean meal with improved amino acid profiles (e.g., high lysine corn, high methionine soybean meal) may have an advantage over feed produced from normal corn and soybean meal with added crystalline lysine and methionine. The feed containing corn and soybean meal with improved amino acid profiles may have an advantage of the amino acids being in the intact protein because all of the amino acids are digested, absorbed, and available for use at approximately the same time. This is important because all amino acids must be simultaneously available at the cellular level for protein synthesis to occur. Crystalline lysine and methionine in the diet may be absorbed more quickly from the gut than the amino acids from intact protein and, thus, an ample supply of all amino acids may not be available in the cells for a maximum rate of protein synthesis to occur. In this situation, some of the more rapidly absorbed amino acids, lysine and methionine, may be deaminated and not utilized for protein synthesis.

IX. SUMMARY

Recent advances in plant genetics and breeding have resulted in grains and oilseeds with superior nutritional properties. Among these, high oil corn and low phytic acid corn have received the most attention in the past few years. These and other genes can be combined or "stacked" to produce even further improved cereal grains and oilseeds. The use of recombinant DNA technology has resulted in improved agronomic traits in crops and in improving the nutritional quality of grains and forages fed to animals. Improved amino acid profiles and reduced phytate content of cereal grains have a potential of reducing N and P excretion by 25 to 50% in pigs. Preserving the identity of these new value-added products in the market place will be a major challenge. Plant breeders and animal nutritionists should continue to work together to advance the development of crops with enhanced nutritional value.

REFERENCES

Anderson, P.C. 1998. Crop biotechnology for feed improvement. p. 265–277. *In* K.Wiseman (ed.) Progress in pig science. Proc. Conf. 58th Easter School Agric. Sci. Nottingham Univ. Press, Nottingham, UK.

Araba, M., and N. Dale. 1998. High oil corn: True metabolizable energy, amino acids and fat digestibility. Poult. Sci. 77(Suppl. 1):154.

Bell, J.M. 1982. From rapeseed to canola: A brief history of research for superior meal and edible oil. Poult. Sci. 61:613–622.

Bell, J.M. 1984. Nutrients and toxicants in rapeseed meal: A review. J. Anim. Sci. 58:996–1010.

Carter, S.D., G.L. Cromwell, M.D. Lindemann, L.W. Turner, and T.C. Bridges. 1996. Reducing N and P excretion by dietary manipulation in growing and finishing pigs. J. Anim. Sci. 74(Suppl. 1):59.

Cheeke, P.R. 1991. Applied animal nutrition feeds and feeding. Macmillan, New York.

Cheeke, P.R., and L.R. Shull. 1985. Natural toxicants in feeds and poisonous plants. AVI, Westport, CT.

Cristofaro, E., F. Mottu, and J.J. Wuhrmann. 1974. Involvement of the raffinose family of oligosaccharides in flatulence. p. 313. *In* H. Sipple and K. McNutt (ed.) Sugars in nutrition. Acad. Press, New York.

Cromwell, G.L. 1993. An assessment of the bioavailability of phosphorus in feed ingredients for nonruminants. p. 146–158. *In* Proc. Maryland Nutrition Conf., Baltimore, MD. Univ. Maryland, College Park, MD.

Cromwell, G.L. 1993. A feed composition overview. *In* Designing crops for specific uses. Pap. Symp. 85th Annu. Meet. Crop Sci. Soc. Am., Cincinnati, OH.

Cromwell, G.L., R.A. Pickett, W.M. Beeson, E.T. Mertz, and O.E. Nelson. 1966. Nutritional value of *opaque-2* corn for swine. J. Anim. Sci. 25:890.

Cromwell, G.L., R.A. Pickett, and W.M. Beeson. 1967a. Nutritional value of *opaque-2* corn for swine. J. Anim. Sci. 26:1325–1331.

Cromwell, G.L., J.C. Rogler, W.R. Featherston, and R.A. Pickett. 1967b. Nutritional value of *opaque-2* corn for the chick. Poult. Sci. 46:705–712.

Cromwell, G.L., J.C. Rogler, W.R. Featherston, and T.R. Cline. 1968. A comparison of the nutritive value of *opaque-2*, *floury-2* and normal corn for the chick. Poult. Sci. 47:840–847.

Cromwell, G.L., T.S. Stahly, and H.J. Monegue. 1984. Effects of processing (grinding versus rolling) of normal and mutant corn hybrids on performance of growing pigs. J. Anim. Sci. 59:875–882.

Cromwell, G.L., J.L. Pierce, T.E. Sauber, D.W. Rice, D.S. Ertl, and V. Raboy. 1998a. Bioavailability of phosphorus in low-phytic acid corn for growing pigs. J. Anim. Sci. 76(Suppl. 2):54.

Cromwell, G.L., J.L. Pierce, H.L. Stilborn, D.W. Rice, D.S. Ertl, and V. Raboy. 1998b. Bioavailability of phosphorus in low-phytic acid corn for chick. Poult. Sci. 77(Suppl. 1):117.

Cromwell, G.L., S.L. Traylor, M.D. Lindemann, T.E. Sauber, and D.W. Rice. 2000a. Bioavailability of phosphorus in low-oligosaccharide, low-phytate soybean meal for pigs. *In* Abstract 206. Proc. Midwest Sect. Meet. Am. Soc. Anim. Sci., Des Moines, IA. 13–15 March.

Cromwell, G.L., S.L. Traylor, M.D. Lindemann, H.L. Stilborn, T.E. Sauber, and D.W. Rice. 2000b. Bioavailability of phosphorus in low-oligosaccharide, low-phytate soybean meal for chicks. *In* Abstract 199. Proc. South. Poult. Sci. Soc., Atlanta, GA. 17–18 January.

Cromwell, G.L., S.L. Traylor, L.A. White, E.G. Xavier, M.D. Lindemann, T.E. Sauber, and D.W. Rice. 2000c. Effects of low-phytate corn and low-oligosaccharide, low-phytate soybean meal in diets on performance, bone traits,, and phosphorus excretion by growing pigs. *In* Abstract 208. Proc. Midwest Sect. Meet. Am. Soc. Anim. Sci., Des Moines, IA. 13–15 March.

Crum, R.C., and H.L. Stilborn. 1997. Valuing high oil corn. Feed Manage. 48(12):16–20.

Curtis, S.E., J.G. Drummond, D.J. Grunloh, P.B. Lynch, and A.H. Jensen. 1975. Relative and quantitative aspects of aerial bacteria and dust in swine houses. J. Anim. Sci. 41:1512–1520.

Dudley, J.W., and R.J. Lambert. 1992. Ninety generations of selection for oil and protein in maize. Maydica 37:81–87.

Glenn, S., B. Glenn, C.E. Rieck, D.G. Ely, and L.P. Bush. 1981. Chemical quality, in vitro cellulose digestion, and yield of tall fescue forage affected by mefluidide. J. Agric. Food Chem. 29:1158–1161.

Gordon, W.A.M. 1963. Environmental studies in pig housing. IV. The bacterial content of air in piggeries and its influence on disease incidence. Br. Vet. J. 119:263.

Herkelman, K.L., G.L. Cromwell, and T.S. Stahly. 1991. Effects of heating time and sodium metabisulfite on the nutritional value of full-fat soybeans for chicks. J. Anim. Sci. 69:4477–4486.

Herkelman, K.L., G.L. Cromwell, T.S. Stahly, T.W. Pfeiffer, and D.A. Knabe. 1992. Apparent digestibility of amino acids in raw and heated conventional and low-trypsin inhibitor-soybeans for pigs. J. Anim. Sci. 70:818–826.

Hymowitz. T. 1986. Genetics and breeding of soybeans lacking the Kunitz trypsin inhibitor. Adv. Exp. Med. Biol. 199:291–298.

Kleese, R.A. 1993. Opportunities to genetically improve the nutritional value of feed grains. Invited Pap. Nonruminant Nutr. I Sess. Annu. Meet. Am. Soc. Anim. Sci., Spokane, WA. 7 July.

Liener, E.E., and M.L. Kakade. 1980. Protease inhibitors. p. 7–71. *In* E.E. Liener (ed.) Toxic constituents of plant foodstuffs. Acad. Press, New York.

Lohrmann, T.T., J.D. Hahn, and M. Araba. 1998. Effect of Optimum® high oil corn as a replacement for typical corn and choice white grease in finishing pig diets. J. Anim. Sci. 76(Suppl. 2):56.

McDonald, T.A., and G.A. Mente. 1973. Waxy corn and sources of protein for finishing steers. J. Anim. Sci. 37:350.

Mertz, E.T. 1963. Corn proteins—a chemical and nutritional prospective. p. 1–6. *In* Proc. 18th Ann. Hybrid Corn Res.-Indust. Conf. Publ. No. 18.

Mertz, E.T., L.S. Bates, and O.E. Nelson. 1964. Mutant gene that changes protein composition and increases lysine content of maize endosperm. Science (Washington, DC) 145:279–280.

Mertz, E.T., O.A. Veron, L.S. Bates, and O.E. Nelson. 1965. Growth of rats fed on *opaque-2* maize. Science (Washington, DC) 148:1741–1742.

Nelson, T.S., L.W. Ferrara, and N.L. Storer. 1968. Phytate phosphorus content of feed ingredients derived from plants. Poult. Sci. 47:1372–1376.

Nelson, O.E., E.T. Mertz, and L.S. Bates. 1965. Second mutant gene affecting the amino acid pattern of maize endosperm proteins. Science (Washington, DC) 150:1469–1470.

National Research Council. 1994. Nutrient requirements of poultry. 9th ed. NRC, NAS, Natl. Acad. Press, Washington, DC.

National Research Council. 1998. Nutrient requirements of swine. 10th ed. NRC, NAS, Natl. Acad. Press, Washington, DC.

Pen, J., R.C. Verwoerd, P.A. van Paridon, R.F. Beudeker, P.J.M. van den Elzen, K. Geerse, J.D. van der Klis, H.A.J. Versteegh, A.J.J. van Ooyen, and A. Hoekema. 1993. Phytase-containing transgenic seeds as a novel feed additive for improved phosphorus utilization. Bio/Technology 11:811–814.

Pierce, J.L., and G.L. Cromwell. 1999a. Performance and phosphorus excretion of growing-finishing pigs fed low-phytic acid corn. J. Anim. Sci. 77(Suppl. 1):60.

Pierce, J.L., and G.L. Cromwell. 1999b. Effects of phytase on bioavailability of phosphorus in normal and low-phytic acid corn. J. Anim. Sci. 77(Suppl. 1):60.

Pierce, J.L., G.L. Cromwell, T.E. Sauber, D.W. Rice, D.S. Ertl, and V. Raboy. 1998a. Phosphorus digestibility and nutritional value of low-phytic acid corn for growing pigs. J. Anim. Sci. 76(Suppl. 2):54.

Pierce, J.L., G.L. Cromwell, and V. Raboy. 1998b. Nutritional value of low-phytic acid corn for finishing pigs. J. Anim. Sci. 76(Suppl. 1):177.

Raboy, V., D.B. Dickinson, and M.G. Neuffer. 1990. A survey of maize kernel mutants for variation in phytic acid. Maydica 35:385–390.

Rackis, J.J. 1972. Biologically active components. p. 158–202 *In* A.K. Smith and S.J. Circle (ed.) Soybeans: Chemistry and technology. Vol. I. AVI, Westport, CT.

Rackis, J.J. 1975. Oligosaccharides of food legumes: Alpha-galactosidase activity and flatus problem. p. 207 *In* A. Joanes and J. Hodge (ed.) Physiological effects of food carbohydrates. ACS Symp. Series. ACS, Washington, DC.

Reddy, N.R., S.K. Sathe, and D.K. Salunkhe. 1982. Phytates in legumes and cereals. Adv. Food Res. 28:1–92.

Risley, C.R. and N.L. Bajjalieh. 1996. Apparent digestibility of high oil corn in weanling pigs and growth performance of grower-finisher pigs fed high oil corn. J. Anim. Sci. 74(Suppl. 1):176.

Scott, M.L., M.C. Nesheim, and R.J. Young. 1976. Nutrition of the chicken. 2nd ed. M.L. Scott and Assoc., Ithaca, NY.

Singh, L., C.M. Wilson, and H.H. Hadley. 1969. Genetic differences in soybean trypsin-inhibitors separated by disc electrophoresis. Crop Sci. 9:489.

Spencer, J.D., G.L. Allee, T.E. Sauber, D.S. Ertl, and V. Raboy. 1998. Digestibility and relative bioavailability of normal and genetically modified low phytate corn (LP) for pigs. J. Anim. Sci. 76(Suppl. 2):54.

Spencer, J.D., G.L. Allee, J.W. Frank, and T.E. Sauber. 2000a. Nutrient retention and growth performance of pigs fed diets formulated with low-phytate/low-oligosaccharide soybean meal. *In* Abstract 211. Proc. Midwest. Sect. Meet. Am. Soc. Anim. Sci., Des Moines, IA. 13–15 March.

Spencer, J.D., G.L. Allee, J.W. Frank, and T.E. Sauber. 2000b. Relative phosphorus availability and retention of low-phytate/low-oligosaccharide soybean meals for growing pigs and chicks. *In* Abstract 207. Proc. Midwest. Sect. Meet. Am. Soc. Anim. Sci., Des Moines, IA. 13–15 March.

Thacker, P.A. 1993. Novel approaches to growth promotion in the pigs. p. 295–306. *In* D.J.A. Cole et al. (ed.) Recent developments in pig nutrition. Nottingham Univ. Press, Nottingham, UK.

Thacker, P.A., and R.N. Kirkwood. 1990. Nontraditional feed sources for use in swine production. Butterworth, Stoneham, MA.

Villeagas, E., S.K. Vasal, and M. Bjarnason. 1992. Quality protein maize—what it is and how was it developed. p. 27–48. *In* E.T. Mertz (ed.) Quality protein maize. Am. Assoc. Cer. Chem., St. Paul, MN.

5 Industrial and Other Uses

KENNETH ESKINS, GEORGE F. FANTA,
AND FREDERICK C. FELKER

USDA-ARS
Peoria, Illinois

Designing plants for increased industrial utilization has at least two aspects: (i) modification of the plant to produce specific new products or altered ratios of normal products, and (ii) the adoption and utilization of plant derived products in a wide variety of commercial applications to create new markets for agricultural commodities. Methods of modifying plant products include insertion of new genes to direct production of economically important secondary products, or chemical/environmental regulation of development and alteration of specific pathways of production. On the industrial side, new products from plants or value added modifications of commodities may be produced that can alter the economics of a process, replace more expensive or unavailable products, or even create new technologies around themselves. Although rare secondary products and expensive pharmaceuticals are exciting potential candidates for production by designed plants, the production of new materials by chemically and/or physically modifying the starch, protein, lipid, and fiber components of plants is an equally attractive option. These components are already in the mainstream of industrial utilization, and modifications can be readily adapted and speedily accepted, provided that properties and selling prices are competitive with those of existing products.

The modification of starch and the utilization of value-added starch products has a well-documented history and a potentially bright future. This chapter describes two different modifications of starch carried out at our laboratory and the efforts to transfer these technologies to the private sector. The first of these projects was carried out in the 1970s and led to the development of the water-absorbent starch graft copolymer commonly referred to as "Super Slurper". The second is a current project dealing with the preparation, properties and end-use applications of a new class of starch-oil composites referred to as Fantesk. The 20-yr time period separating these projects saw a drastic change in the U.S. Government's philosophy regarding technology transfer. During that 20 yr, a new philosophy of utilization research and interaction between government and industrial scientist arose. This evolution of thought culminated in the Technology Transfer Act of 1986 that allowed government scientists to enter into cooperative research and development agreements (CRADA's) with industrial partners to co-develop specific products for commercialization. Under this new legislation, exclusive licenses to government patents could be granted, and for the first time, the public, private industry, and the inventors would share in the profits of new technology.

It is hoped that this chapter will not only provide an understanding of the technical aspects of these two very different research projects but also will provide some insight into how the transfer of technology from government laboratories to private industry is carried out today, as compared with 20 yr ago.

I. STARCH GRAFT COPOLYMER ABSORBENTS

A. Technical Aspects of Starch

Starch is a polysaccharide composed of repeating 1,4-α-D-glucopyranosyl units (often called anhydroglucose units, or AGU) and is isolated from the seeds, roots, and tubers of living plants as discrete granules ranging from about 5 to 40 μm in diameter. Starch granules are generally composed of a mixture of linear and branched polymers (amylose and amylopectin, respectively). The molecular weight of the amylose component is on the order of several hundred thousand, whereas amylopectin has a molecular weight of several million. Although the amylose/amylopectin in normal cornstarch is about 25:75, starch varieties are commercially available that have amylose contents ranging from nearly 0 to about 70% of total starch.

Starch is insoluble in water at room temperature due to H bonding and crystallinity within the granule; however, granules will swell and lose their birefringence (i.e., gelatinize) when water dispersions are heated. At roughly 80 to 90°C, the highly swollen and partially fragmented granules form a smooth dispersion or paste, and a significant fraction of the starch (largely amylose) goes into solution. Complete solubility is not easily achieved, however, unless an autoclave or steam jet cooker is used to obtain temperatures greater than 100°C. Research on the preparation and properties of starch graft copolymers began at our Center in the early 1960s, and during the past 30 to 35 yr, numerous monomers and initiator systems have been studied (Fanta & Doane, 1986). Although a number of these graft copolymers have potentially useful properties, the water absorbent polymers prepared by graft polymerization of acrylonitrile followed by alkaline hydrolysis have generated the most commercial interest.

B. Graft Copolymers

Early work on the synthesis and characterization of starch-*g*-polyacrylonitrile (S-*g*-PAN) formed the basis for the absorbent polymer technology. Mino and Kaizerman (1958) showed that graft polymerization could be easily initiated by adding ceric salts (e.g., ceric ammonium nitrate) to a suspension of starch and monomer in water at room temperature. Ceric ion is a powerful oxidizing agent and complexes rapidly with starch. The initially formed starch-Ce(IV) complex further reacts to yield Ce(III) and a starch macroradical, which can then initiate polymerization of grafted PAN. Evidence has been published to support the reaction path shown in Fig. 5–1 (Hintz & Johnson, 1967; Pottenger & Johnson, 1970). Ceric ammonium nitrate initiated grafting of acrylonitrile onto starch takes place with nearly complete conversion of monomer to polymer and with little formation of ungrafted PAN. Copolymers containing 50 to 60% grafted PAN are easily obtained. Scanning electron microscopy showed that graft polymerization takes place not only on the surfaces of starch granules but within the granule interior as well (Fanta et al., 1977). The appearance of grafted starch granules resembles that of ungrafted starch (Brockway & Seaberg, 1967; Fanta et al., 1977). Also, the outward appearance of starch granules is retained, even after removal of the starch moiety by acid or enzyme hydrolysis.

The structure of S-*g*-PAN prepared from gelatinized starch is quite different than that of analogous copolymers prepared from starch in its native granule state (Burr et al., 1967; Fanta et al., 1970). Gelatinized starch (obtained by heating an aqueous starch suspension to about 85°C) produced PAN grafts with an average molecular weight of about 800 000, and the average separation between grafts was about 4000 AGU. Use of granular starch as the substrate for grafting reduced both the graft molecular weight and the number of AGU per graft by a factor of six to seven.

Treatment of starch-*g*-PAN with hot alkali (Gugliemelli et al., 1969; Weaver et al., 1971) results in hydrolysis of the nitrile substituents of PAN to a mixture of

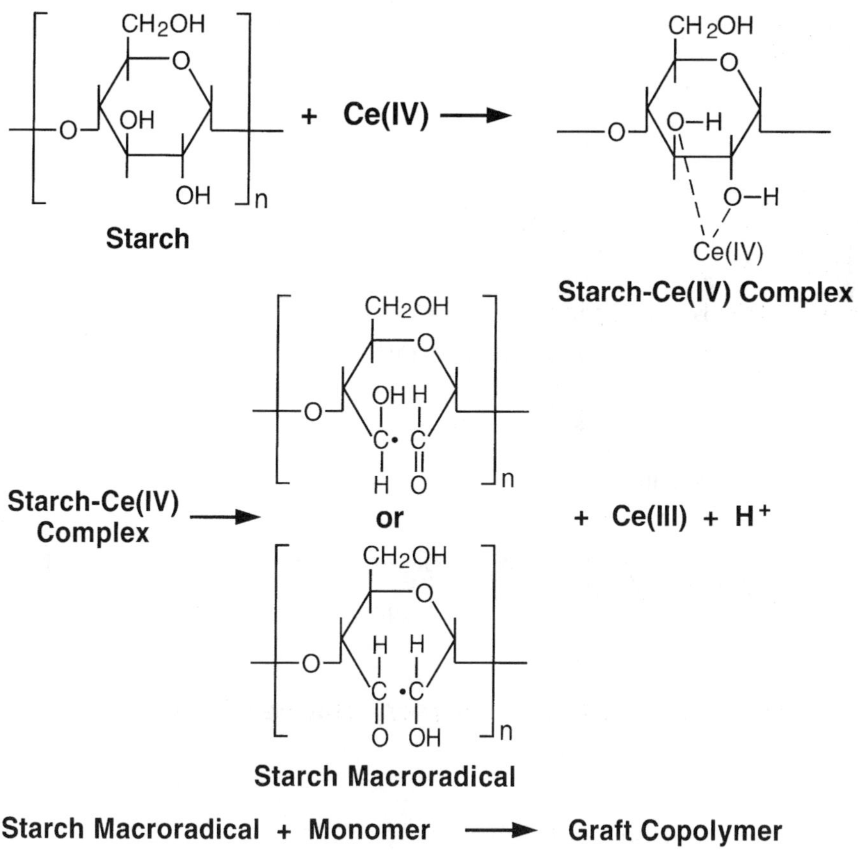

Fig. 5–1. Graft polymerization unto starch initiated by Ce (IV).

carboxamide and alkali metal carboxylate (see Fig. 5–2). Aqueous dispersions produced by the hydrolysis reaction are extremely viscous and were initially evaluated as commercial thickeners. Ratios of carboxamide to carboxylate vary with hydrolysis conditions but are typically on the order of 1:2. Complete hydrolysis of PAN to the sodium salt of poly(acrylic acid) does not occur, even under the most stringent reaction conditions.

Research at our Center showed that starch-*g*-PAN prepared from gelatinized starch dissolves to only a minor extent during alkaline hydrolysis and remains largely in the form of highly swollen but insoluble gel particles (Taylor & Bagley, 1974). It was correctly concluded that the thickening action of the graft copolymer in water is due to nearly complete absorption of water by the gel to give a system consisting of highly swollen, deformable gel particles that are closely packed and in intimate contact. When the ionic strength of the medium is increased or the swollen gel is diluted to the point where solvent is in excess, gel particles no longer remain tightly packed, and the viscosity of the system drops sharply.

Since starch and hydrolyzed PAN are both soluble in aqueous alkali, the insolubility of hydrolyzed S-*g*-PAN (HSPAN) suggests the occurrence of cross-linking during the synthesis of these polymers. Cross-linking also is indicated by the similarity in outward appearance between HSPAN gel particles and particles of un-

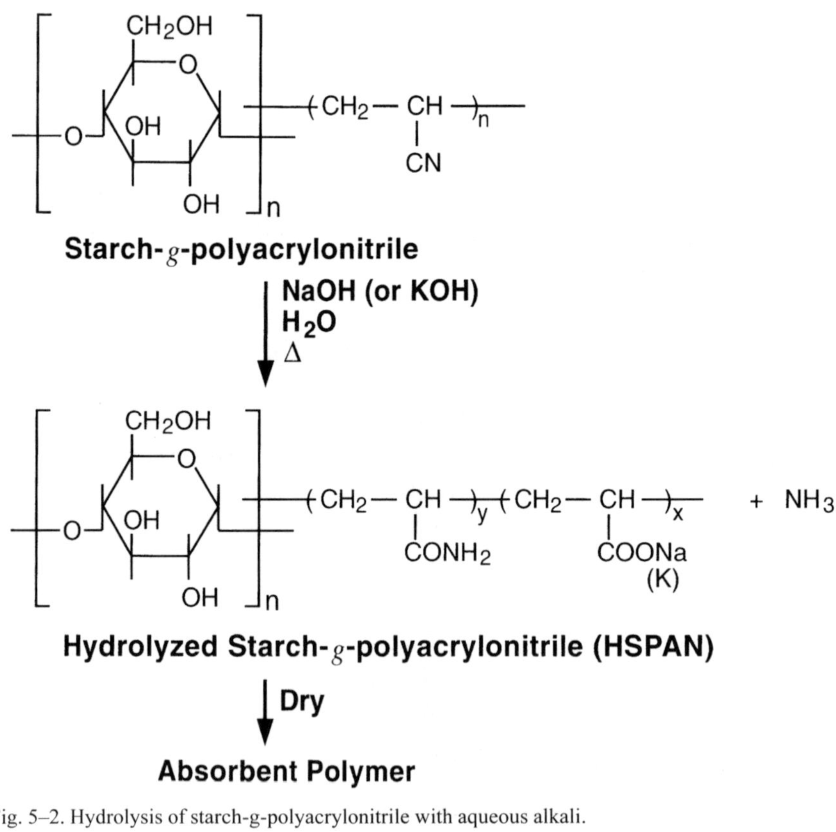

Fig. 5–2. Hydrolysis of starch-g-polyacrylonitrile with aqueous alkali.

hydrolyzed starch-*g*-PAN when examined by scanning electron microscopy (Fanta et al., 1977). Evidence has indeed been found that cross-linking reactions occur during both graft polymerization and alkaline hydrolysis (Fanta et al., 1982).

Although HSPAN can be easily isolated from water dispersions by air drying or precipitation, we were surprised to observe that the resulting films and particles of dry HSPAN did not revert back to the original smooth gel dispersion when placed back into water. Instead, the polymer swelled without disintegrating and absorbed an amount of water equal to many times its original weight. This unexpected property formed the basis for the practical use of these polymers as water absorbents. Our initial observations were made with continuous HSPAN films prepared by air drying (Weaver et al., 1974). When placed in water, these films swelled as a single entity and produced continuous sheets of highly swollen gel with sufficient integrity to be carefully manipulated without breaking The exact shape of the dry film was retained by the swollen gel sheet. The HSPAN polymers prepared from gelatinized starch showed a greater tendency to form large particles and films of continuous gel than the analogous graft copolymers prepared from granular starch. Moreover, gelatinized starch-based products absorbed over twice as much water as those prepared from granular starch.

Precipitation of HSPAN as a grainy solid by addition of a water-miscible nonsolvent, such as methanol, has been extensively used as a method of isolation and yields the polymer in pure form, since excess alkali and inorganic salts remain soluble in methanol-water. The deionized water absorbency of methanol-precipitated HSPAN is on the order of 1000 g g^{-1} of dry polymer. Since HSPAN polymers are polyelectrolytes, ion-containing fluids are absorbed to a much lesser extent. For example, the absorbency of 1% sodium chloride solution is lower than that of deionized water by roughly a factor of 10.

When maximum purity and absorbency are not needed (e.g., for agricultural applications), the viscous reaction mass produced by alkaline hydrolysis may be drum dried to yield HSPAN in the form of coarse flakes. This process is both inexpensive and convenient, and generates no by-products other than steam (Weaver et al., 1977). It is necessary, however, to avoid excess alkali during hydrolysis so that the final product will not be contaminated by large amounts of inorganic salts. We observed that the use of 0.6 to 0.8 mole of sodium or potassium hydroxide per mole of acrylonitrile repeating unit allows the hydrolysis to proceed at a reasonable rate without the disadvantages of excess alkali. Although the deionized water absorbency of drum-dried HSPAN is only about 300 g g^{-1}, water is rapidly absorbed, and gel blocking does not occur.

The absorbent properties of HSPAN have suggested a number of practical applications. In the area of personal care, these include: (i) addition to disposable soft goods (e.g., diapers, adult incontinent pads, hospital underpads and feminine napkins) to increase the absorbance of body fluids and to decrease bulk; (ii) addition to ostomy and urinal bags to prevent leaks and odors; (iii) application to wounds to promote healing (Spence, 1980; Valdez, 1980; Geronemus & Robins, 1982); and (iv) addition to body powders as a perspiration absorbent (Anon., 1977; Spence, 1981). In the agricultural area, HSPAN has been successfully tested as: (i) a seed coating to absorb and hold water at the seed surface to promote germination (Deterling, 1981; Tanzy, 1981; Kirkland et al., 1985); (ii) a transplanting aid to hold water near the root zones of plants (Hamilton & Lowe, 1982); (iii) a soil additive to increase the water holding capacity of poor or marginal soils (Shrader & Mustejeran, 1977; Miller, 1979); and (iv) a tackifier component in hydroseeding formulations (Kay, 1980). Some other applications tested include removal of water from organic solvents and fuels (Fanta et al., 1980), addition to fire fighting fluids to increase viscosity, and preparation of gels to be used as agar substitutes (Leppla, 1976; Cooke, 1977).

C. Technology Transfer of Starch Graft Copolymers

The factors influencing technology transfer 20 to 25 yr ago were quite different than they are today under the Technology Transfer Act (1986). Although USDA-ARS has always encouraged the filing of patents by its scientists and the licensing of these patents by the private sector, patent licenses in the 1970s could only be granted on a nonexclusive basis. This meant that no single company could be given the exclusive right to market and to derive profit from a technology resulting from government-funded research. As a result, companies were reluctant to invest in the development and manufacture of a product, when their competitors could easily obtain government licenses to make, use, or sell the same material. Despite the fact that licensing of government patents required no payment of royalties, companies felt that they would rather invest in their own research and development programs, which, if successful, would yield technology that they would exclusively own. Although a law passed in 1980 permitted the government to issue exclusive licenses under certain conditions, most licenses granted were still on a nonexclusive basis. Moreover, this law was passed too late to affect licensing of the HSPAN technology.

Since there was no strong emphasis on technology transfer and little incentive to aggressively seek licenses for government patents, scientists generally relied upon technical journals, trade publications, and presentations at scientific meetings to publicize their results and to generate interest in their work within the private sector. Following discovery of the water absorbent properties of HSPAN in late 1972, our initial observations were disclosed in two trade publications (Anonymous, 1973a,b) in mid-1973. Experimental results also were presented in the same year at a meeting of the American Chemical Society (Weaver et al., 1973). These disclosures resulted in dozens of suggestions for potential applications and hundreds of requests for additional information and samples. Apparatus in our pilot plant was used to prepare multi-pound quantities of HSPAN for sampling, and it soon became necessary to assign one of our employees the nearly full time task of responding to inquiries.

Our patent application, covering the composition of HSPAN, its method of preparation, and a number of practical uses, was filed in April 1974 (within 1 yr of our first disclosure). Because of its length and diversity, the application was divided into four separate patents, all of which issued in 1976. The first patent in this series (Weaver et al., 1976a) covered the addition of HSPAN to aqueous emulsions, suspensions, and dispersions as a method for selectively absorbing water and thus reducing the water contents of these systems. The second patent (Weaver et al., 1976b) covered: (i) the use of HSPAN as a soil conditioner, (ii) the addition of

HSPAN to soil to increase its water holding capacity, and (iii) the use of HSPAN to entrap seeds within a water absorbent matrix to facilitate their germination. Our third patent (Weaver et al., 1976c) dealt with the use of HSPAN for immobilization of enzymes, while the final patent in this series (Weaver et al., 1976d) covered composition of matter and methods used for HSPAN preparation.

During the time period between patent filing and issuance, the many possible uses of HSPAN as a water absorbent stimulated the imagination of the private sector and resulted in a large amount of publicity. The name "Super Slurper," (coined by Dean Mayberry of our USDA-ARS Information Office) also stimulated company interest. Thousands of inquiries were received, and a multimillion pound market for the polymer was projected by some analysts. The HSPAN received *Industrial Research Magazine*'s IR-100 award in 1975. The following year, the HSPAN research team received USDA's Distinguished Service Award and also the Inventor of the Year Award, sponsored by the Association for the Advancement of Invention and Innovation. About 40 nonexclusive licenses to make, use, or sell the polymer were granted.

As soon as a patent license was granted, USDA scientists began working actively with the licensee and provided information on polymer properties and on all aspects of the polymerization process. Licensees also were encouraged to visit our Center, observe our pilot plant facility and, if necessary, make trial experiments. In addition, companies producing HSPAN were officially listed by our Laboratory as suppliers of the polymer, and their names were provided to all parties requesting information. In return for this listing, suppliers were requested to respond to all inquiries received and to supply small samples of HSPAN free of charge upon request. A major chemical company in Minnesota was the first company to license our patent application. Their scientists were already familiar with the technology, since they were at the time carrying out a USDA-funded contract study to evaluate practical uses for a number of starch graft copolymer products. This company opened a pilot plant in Illinois, and began producing HSPAN in 1977 to 1978. Although the company was sold to a large, multinational corporation shortly after the plant was constructed, production continued and research and development efforts were not interrupted. Super Absorbent Co., in North Carolina, and a medical supply company in Texas also opened pilot plant facilities at about this same period of time. The pilot plant in Illinois produced a methanol precipitated polymer, whereas the other two companies used drum drying as their method of isolation.

In spite of the many patent licenses granted, sales and new markets for HSPAN in the late 1970s and early 1980s were slow to develop. Companies with marketing plans that included large purchases of HSPAN were discouraged by the relatively small quantities of polymer being produced in pilot plant facilities. On the other hand, manufacturers were reluctant to increase production capacity unless they were assured of a continued strong demand for their product. Neither purchaser nor manufacturer wished to make the first real commitment. Also, polymers having the capacity to absorb hundreds of times their weight in water were at that time a novelty, and markets for such materials were not firmly established. Sales of HSPAN thus did not only depend upon having a superior product for a particular market, but also required the creation of new and uncertain markets that did not yet exist.

To this picture must be added the major disadvantage associated with all nonexclusive government licensing prior to the Technology Transfer Act (1986), i.e., the possibility that a competitor also might obtain a government license and begin to compete for profits as soon as markets became established. Because of this concern, some U.S. and overseas companies began research programs to develop their own absorbent polymers outside of our USDA patents. Many of these efforts proved successful, and some excellent products were introduced into the marketplace. Some of these new products were based on starch and cellulose; while others were totally synthetic (e.g., polymers prepared by copolymerizing acrylamide and acrylic acid in the presence of cross-linking agents). Although the successful marketing of these absorbents was initially perceived as a threat to HSPAN commercialization, we soon recognized that the number of applications for water absorbent polymers was sufficiently large to provide markets for a variety of different products.

The single occurrence that irreparably damaged our efforts toward HSPAN commercialization was the decision made by the major supplier of HSPAN to cease production and to abandon all research and development efforts related to the polymer. This action, which took place in 1982, was apparently the result of a management decision pertaining to all new commercial ventures and was not due to any technical difficulties specifically related to HSPAN. Nevertheless, the action of this company could not avoid the perception by the growing superabsorbent industry as a lack of confidence in the ability of HSPAN to compete in the marketplace. In 1983, the company's product inventory and production equipment were sold to a major starch company in the Midwest. Although this starch company continued to supply HSPAN in small quantities, their marketing interests were centered on their own granular starch based superabsorbent, which they themselves had developed and patented.

The closing of the production facility in Illinois left only two companies as HSPAN suppliers. Although the medical supply company in Texas produced and marketed HSPAN for a few years, mainly for addition to ostomy and urinal bags and as a component in body powders, they too ceased production. Fortunately, Super Absorbent Co. maintained their production efforts and continues to market HSPAN today, largely for agricultural applications. Sales of HSPAN to the disposable soft goods industry (e.g., for use in diapers and tampons) is limited, since many of the newly developed superabsorbents hold body fluids more tenaciously and yield firmer gels after fluid absorption.

Although the amount of HSPAN sold at the present time by the remaining supplier is only a fraction of the annual sales predicted for the polymer 15 to 20 yr ago, these sales are still significant and are continuing to increase. The commercial development of HSPAN is thus a very real success story for USDA-ARS scientists, especially in view of the rather poor climate for technology transfer that existed in the 1970s and the many difficulties and setbacks that occurred. Perhaps the most rewarding aspect of this research, however, is the realization that the discovery of HSPAN and our early efforts to bring this material to the marketplace was a major factor in the development of the multimillion dollar, worldwide superabsorbent industry that we see today.

II. FANTESK, A MULTIPLE USE STARCH /OIL COMPOSITE

A. Invention

Circumstances leading to the discovery of starch-oil composites demonstrate the synergy between basic and applied research. The initial discovery was the result of our attempts to synthesize new lipid membrane systems and also new starch derivatives with improved compatibility with synthetic plastics. After initial attempts to chemically attach lipids to the solid starch polymers proved unsatisfactory, the idea arose to physically incorporate lipids into the starch helix prior to chemical bonding. We knew from past experience that jet-cooking was capable of solubilizing starch, and we assumed that co-cooking starch and lipid together would provide the intimate mixture needed for helical complex formation. This proved to be true, and co-cooking starch, water, and soybean fatty acids produced a creamy, lubricous, gel-like composite that did not separate upon standing or exhibit the usual retrogradation characteristic of cooked starch solutions. When the material was heated in a microwave oven, it melted and flowed freely, but still did not phase separate upon cooling. The composite also was freeze/thaw stable and could be dried into a film or powder that could be reconstituted with water. At this stage of the investigation, we were convinced that we were dealing with a simple fatty acid inclusion complex. We decided to prove that point by jet cooking starch with triglycerides, which reportedly do not form inclusion complexes because of the sterically hindered triglyceride structure. To our considerable surprise, soybean [*Glycine max* (L.) Merr.] oil, olive (*Olea europeae* L.) oil and lard also formed similar composites, and subsequent experiments with many fats, oils, and starches demonstrated the wide range of composites possible. We now know that the fats and oils are microencapsulated (1–10 μ in size) in a starch/water matrix and are surrounded by an adhering starch film that keeps the oil droplets from coalescing. The unusual nature of these compositions led us to seek patent protection and to explore the many possible practical and commercial applications of the technology. Two patents were issued for Fantesk (a trademark issued to the USDA for the technology and the composites, and derived from combining the names of the inventors). The first (Eskins & Fanta, 1997) covered "Non-Separable Starch-Oil Compositions" prepared by co-jet cooking. The second, "Non-Separable Compositions of Starch and Water-Immiscible Organic Materials" (Eskins & Fanta, 1999) extended the claims to include lipids added to starch after jet-cooking.

B. Physical Properties

1. Process Description and General Physical Properties

Stable and nonseparable compositions comprised entirely of starch, water and oil are prepared without dispersants or emulsifiers by passing a continuously stirred suspension of granular, unmodified starch and oil in water through a steam jet cooker operated under excess steam conditions (Eskins & Fanta, 1994). We also have determined that the oil and starch/water may be separately coinjected into the hydroheater portion of the cooker to yield a similar product. The jet cooking process solubilizes and reduces the molecular weight of starch and intimately mixes the oil, starch and water under conditions of high temperature and turbulence. The resulting products are stable and do not phase separate into water and lipid components even after prolonged standing. A portion of the jet-cooked product also may be recycled into the original mixing chamber to pre-emulsify the starch, water, and oil components. This "addback" procedure results in a product with smaller and more uniformly dispersed oil droplets. A variety of starches and lipids may be used to produce compositions having starch to lipid ratios ranging from 100:5 to 100:50. Starch sources include those from corn (*Zea mays* L.), wheat (*Triticum aestivum* L.), rice (*Oryza sativa* L.) and potato (*Solanum tuberosum* L.), and amylose/amylopectin may be varied over a broad range. Lipids (or lipidlike materials) include vegetable oils and fats, animal oils and fats, resins, waxes, and various petroleum-based products.

The viscosity of jet cooked dispersions depends on solids content and, at room temperature, ranges from free flowing liquids at about 5% solids to soft gels at 20% solids. Increases or decreases in temperature cause corresponding changes in fluidity and gelling properties. The dispersions may be drum dried to yield solid compositions that are not oily to the touch. These dry products can be readily rehydrated to provide compositions similar to the original (Fanta & Eskins, 1995).

The smoothness and lubricity of these compositions immediately suggested their use as thickeners, emulsifiers and fat substitutes in food products. The presence of oil in these composites also facilitated the incorporation of additional amounts of water-insoluble materials such as oils, volatile essential oils and food flavoring materials, antioxidants, pharmaceuticals, and agricultural chemicals. These properties also lend themselves to the formulation of seed coatings, glues and adhesives, cosmetics, and drug delivery systems.

2. Microproperties

Ultrastructural properties of starch-oil composites were investigated (Eskins et al., 1996) by scanning electron microscopy (SEM) of dried thin films and by light microscopy (LM). Direct observation of the composite with a light microscope reveals a random distribution of oil droplets in the starch dispersion (Fig. 5–3). When thin films of the cooked or reconstituted dispersions are allowed to dry at room temperature, fractured, extracted with hexane, and examined with SEM, the resulting micrographs show a smooth starch surface interspersed with many cavities previously filled with oil, ranging in size from 1 to 10 μm (Fig. 5–4). The size distribution of oil droplets is determined by a number of factors, including amount of oil, number of cook cycles, and addition of protein or back additions of jet-cooked starch-oil composites. The effects of changing the starch/oil are shown in Fig. 5–4a,b,c. When large amounts of oil (starch/oil of 100:50 or greater) are used, the oil droplets are large and not as well dispersed as with smaller amounts of oil. The addition of protein, or recycling a portion of the jet-cooked starch-oil composite to the physical mixture of starch, water and lipid prior to jet-cooking improves the dispersion and reduces the size of microencapsulated oil droplets. Recycling of cooked material through the steam jet cooker changes the size distribution of oil droplets in the "twice-cooked" composite (Fig. 5–5).

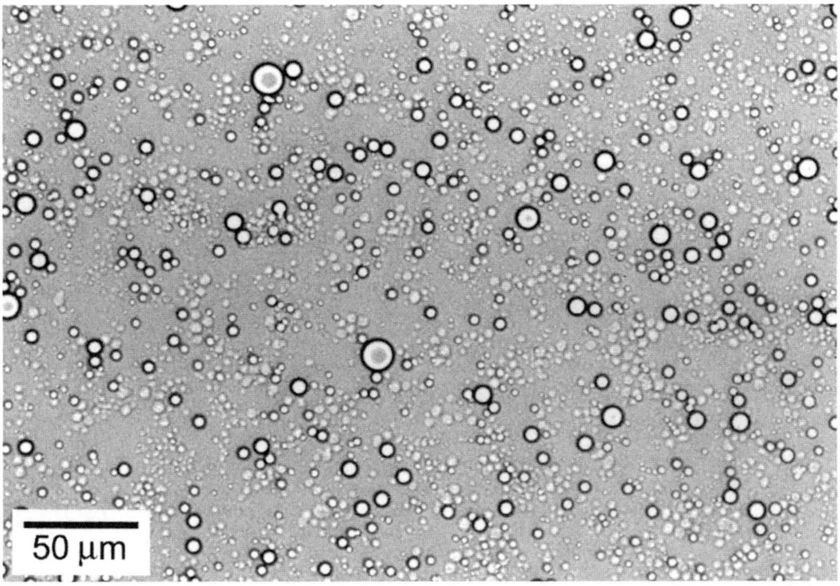

Fig. 5–3. Light micrograph of jet-cooked starch-oil composite made from normal food grade starch (100 parts) and soybean oil (20 parts).

While conducting experiments aimed at determining the lowest starch concentration at which stable composites could be made, it was noticed that very low solids composites formed a white surface layer upon standing. Centrifugation of any diluted Fantesk preparation allowed the isolation of a bouyant opaque surface layer, which had a very high oil content but remained water dispersible, even after numerous washings. Further experiments showed that even though all soluble and suspended microgel particles of starch had been removed, the oil droplets were still covered by an adhering starch film, which could be stained with iodine if it contained amylose. Transferring a small amount of washed surface layer into excess ethanol fixed and dehydrated the surface film of starch while extracting the soy oil. It was found that critical point drying preserved the structure of the hollow starch

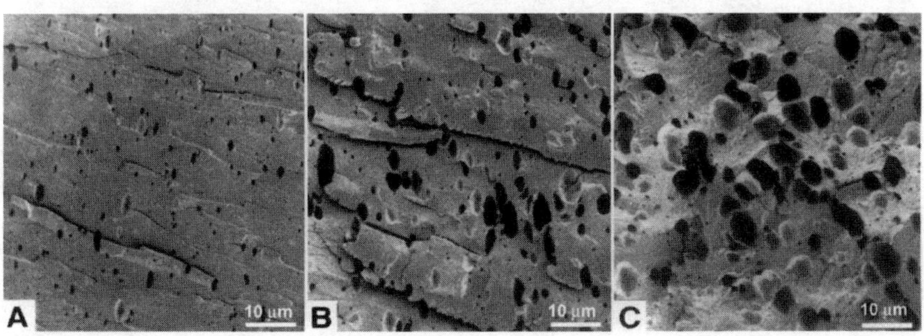

Fig. 5–4. The SEM of fracture face of starch-soy oil composites dried to film and extracted with ethanol followed by hexane, containing: (*A*) 100 parts starch, 5 parts soy oil; (*B*) 100 parts starch, 20 parts soy oil; (*C*) 100 parts starch, 40 parts soy oil.

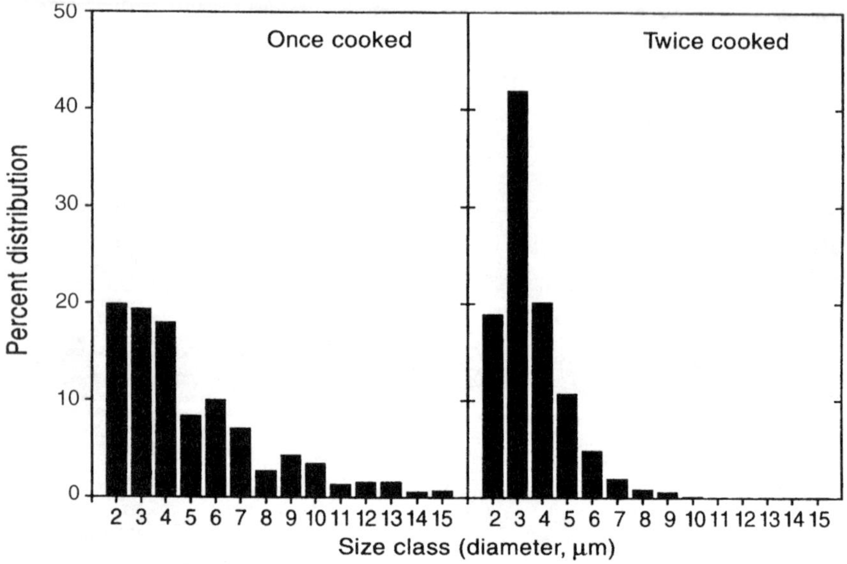

Fig. 5–5. Change in percentage distribution of oil droplet size in single and twice cooked composites.

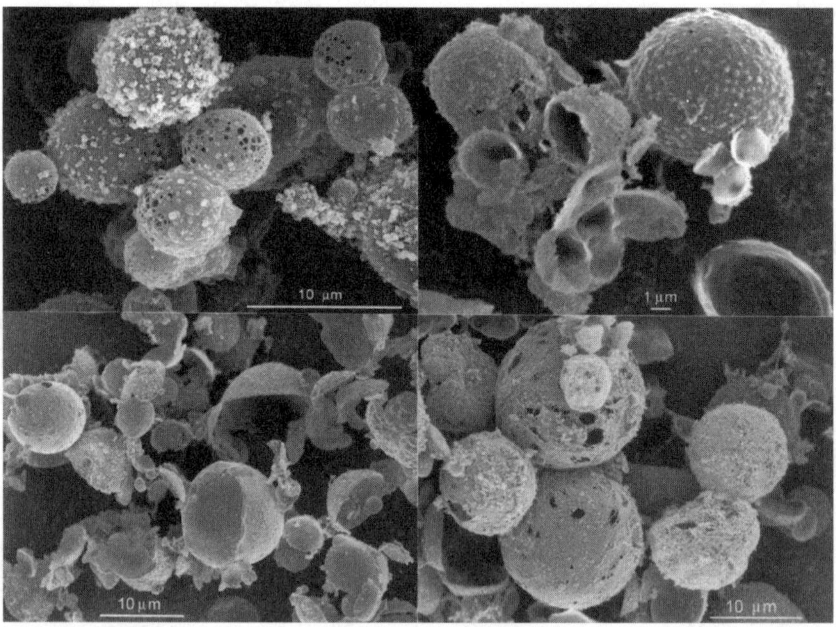

Fig. 5–6. The SEM of hollow starch spheres precipitated with ethanol and critical point dried. Samples prepared with different starches and dilution procedures show varied textures.

Table 5–1. Recovery of oil from jet-cooked starch-oil composites.

	Oil added		Recovered oil, percentage of product wt.		
Starch	g (100 g)$^{-1}$ starch	Percentage of product wt.	Loosely bound	Tightly bound	Total
Normal	20	16.7	7.5	4.5	12.0
Normal	40	28.6	16.8	5.9	22.7
Waxy	40	28.6	19.1	4.3	23.4
Amylomaize	40	28.6	14.3	8.4	22.7

spheres, and that the texture of the prepared spheres varied with starch type and dilution treatment (Fig. 5–6). Using this technique, the formation of an adhering starch shell was shown to occur with a wide variety of lipid phases (Fanta et al., 1999). These shells were not observed in earlier SEM studies because the starch shells are indistinguishable from the interstitial starch network at the high solids concentrations necessary for that technique. Only when the composite is diluted does the starch at the oil droplet surface take on the appearance of a shell.

Microscopic demonstration of the starch layer adhering to oil droplets provides the best explanation so far of why the composite is so stable, and why the oil in the surface layer of diluted composites does not coalesce. The adhesion of starch films to the oil droplets tends to counteract their bouyancy by increasing the net specific gravity of the droplets. The starch shells also prevent direct contact between oil droplets. Evidence for this interpretation is that amylase treatment of washed surface layer suspensions quickly causes the oil to coalesce and phase separate.

3. Extraction Studies

The composites were studied for stability and resistance to separation of the hydrophilic and hydrophobic phases by solvent extraction in the solid state and over a wide range of aqueous dilutions (Knutson et al., 1996). Dry samples were extracted with hexane and the hexane evaporated to determine amount of extractable oil. The extracted samples were then dispersed, hydrolyzed with amylase, and re-extracted with hexane to determine the amount of tightly bound oil. The amounts of "loosely bound" and "tightly bound" oil in various composites determined by hexane extraction of the drum dried powder are shown in Table 5–1. Content of tightly bound oil from composites having 20 or 40 parts of oil per 100 parts of normal corn starch were similar (4–6% of total sample weight). However, in various starches or mixtures of starches, the proportion of tightly bound oil increased with increased amylose content.

Aqueous dispersions were then reconstituted from drum-dried material and extracted eight times with hexane. The results from these extractions were compared with extractions of simple mixtures of oil and gelatinized starch containing equivalent proportions of oil. More than 90% of the oil was extracted from gelatinized starch-oil mixtures with five extractions, whereas the jet cooked starch-oil composite containing 40 parts per 100 parts of starch of oil lost less than 30% of its oil in eight extractions. This retention of oil in aqueous composites increases with the amylose content of the starch and with the solids content of the aqueous composite (Figs. 5–7, 5–8).

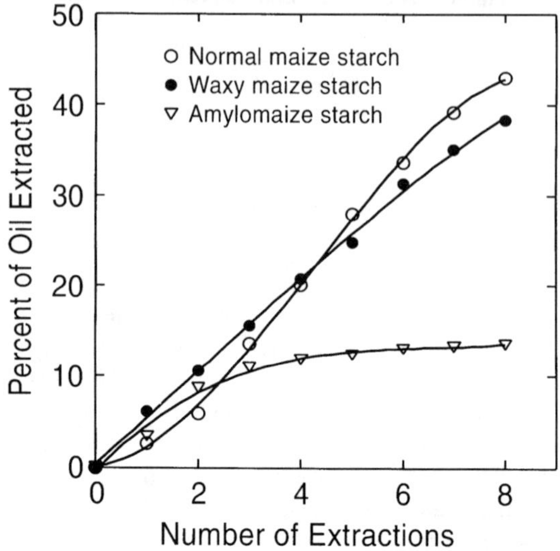

Fig. 5–7. Percentage of oil extracted from reconstituted (1.25% solids) composites of starch-oil (100:40) as a function of starch type.

C. Technology Transfer

An important aspect of utilization research is bringing new discoveries to the market place (Villet, 1994; Knauth, 1991). From the time of the initial invention of the Fantesk process in 1993, experimental composites with an extremely broad range of ingredients have been made. Many of these formulations have resulted in

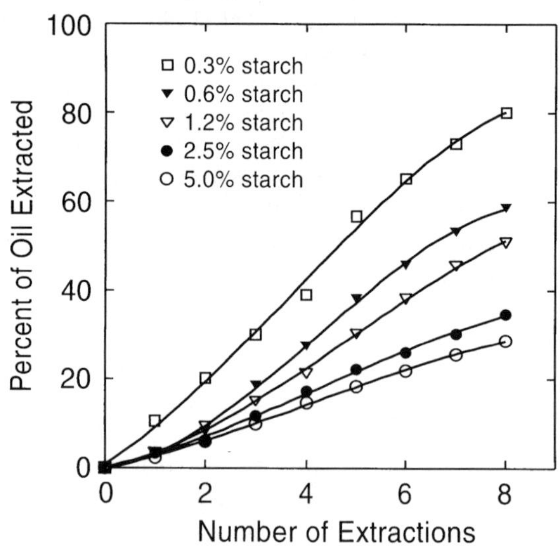

Fig. 5–8. Percentage of oil extracted from reconstituted starch-oil composites (100:40) as a function of percentage solids of the composite.

products showing very promising physical characteristics, including some with surprisingly appropriate and useful properties. Four of our most successful technology transfer experiences are described below: food applications, seed coatings, skin care products, and water-based lubricants. In addition to these, other applications such as drug delivery systems, coatings, and fragrance delivery have been successfully demonstrated but have not yet been developed as far with industrial partners.

1. Food Applications

Our first contact with a large food company was at once instructive and frustrating. We contacted the vice president of sales of a large food company and explained in general terms the concept of our products and possible applications. We were invited to make a presentation at their technology center and prepared suitable viscosity data and micrographs of structure. We learned, however, that they were not particularly interested in the basic science of our discovery but only in fully formulated end-use products. Our efforts at cooperation were stymied because we failed to anticipate the manner in which the product would be evaluated. Some reasonably sophisticated application work was necessary to capture the initial interest of these companies.

At about the same time, we began to present the results of some of our work to another large food company and also stepped up our own evaluation of the product. Through a subsidiary company, we arrived at a potential partner who was interested in finding applications for our composites in low fat meats, sauces and gravies. We also induced them to look at low fat microwave popcorn as another application. Popcorn seeds, coated with our starch-butter composite and dried, popped very well and retained a mild butter aroma and flavor delivered on crisp and non-oily popped kernels. A CRADA was eventually signed with this company to pursue these applications.

Another contact was made with a company primarily interested in dairy applications, and this new partner also progressed from nondisclosure agreements, through preliminary applications to a CRADA with well-defined areas of cooperation and use. Preliminary products included an extremely well received low-fat ice cream, yogurts, sour creams, water soluble vitamins and salad dressings.

A local meat processing company cooperated with us in making very low-fat sausages and hamburger patties containing starch/lard and starch/beef tallow composites. The water-holding property of the starch component combined with the flavor dispersing effect of the microscopic fat droplets resulted in surprisingly juicy and flavorful cooked meats. In-house taste panel experiments demonstrated that very low-fat preparations made with Fantesk composites were similar in texture and taste to much higher-fat control preparations.

Despite early promising results and initial enthusiasm, none of the CRADA partners or cooperating companies showed a sustained interest. One company that had obtained a license for food applications never developed products for commercialization, and let the license expire. Thus, after several years of promising food application demonstrations, technology transfer in this area seemed stalled.

To stimulate renewed interest in food applications, we formulated a soft-serve ice cream product with 0.5% butterfat, with the butter supplied in the form of starch/butter Fantesk. Microscopy revealed that our product contained butterfat droplets in the 1- to 10-μ range, presumably coated with starch. In contrast, commercial low-fat soft-serve mix contained 3.5% butterfat in the form of submicron-sized droplets, presumably coated with milk proteins. Despite the much lower butterfat content, our product had a buttery taste and a smooth, creamy feel to the mouth. This product was demonstrated at a major national food technology conference and was accepted as a quality product by a sophisticated and critical audience. We also have incorporated oil-based flavorings into Fantesk, and these efforts have resulted in new contacts with better possibilities for CRADA research and licensing opportunities.

2. Seed Coatings

It almost was immediately evident that any system that could put oil and water together into a stable composite would have multiple uses, even outside the food industry. One of the first nonfood applications that we explored was use of the composite as a coating for agricultural and turf seeds, and we found that its performance was excellent. We searched trade magazines for potential users of such seed-coating technology and contacted several. A company interested in turfgrasses responded to our initial inquiry and visited our laboratory to examine the use of starch-oil composite to deliver oil or water based chemicals to the germinating seed and the soil surrounding it, as well as protect the seeds from excessive moisture during early planting. This initial investigation was conducted under a nondisclosure agreement and rapidly produced some positive results. Over a period of 10 mo, increasingly close cooperation led to the establishment of a CRADA to study these coating systems in more detail.

A period of intense development and optimization research followed during which we learned from our cooperator how to coat seeds properly and evaluate the physical performance of the coating. A very broad range of ingredients and combinations was tried, along with new strategies for introducing melted lipids into the processing stream. This cooperation resulted in a successful formulation that enhanced germination and stand establishment in a university field test arranged by our CRADA partner. A license was issued, and we then worked with our in-house machinist to construct a much larger-scale jet cooker than the ones we had used in our laboratory. After delivering this equipment to the company, setting it up, demonstrating its use, and working with them further, an effective and economical seed coating formula was optimized. This product represented the first commercial sale of a Fantesk composite product, which occurred less than 7 yr after the initial discovery of the process.

3. Skin Care Products

The first jet-cooked starch-oil composite had the outward appearance of a hand lotion: it was a thick, free-flowing, opaque, white, slippery liquid that could be easily spread over the skin. However, its performance as a hand lotion left much to be desired, such as less stickiness, less gritty starch particles left on the skin, and a less malodorous lipid phase than soybean oil. But the idea of delivering moisturizing ingredients and lipids to the skin without using chemical emulsifiers was unusual enough to interest one company in the skin care products area. Under a CRADA, a series of cooperative experiments were performed, resulting in the development of a specific formulation that both parties agreed was an excellent moisturizing lotion. A similar product with a higher lipid content was developed as a barrier cream. Independent testing arranged by our CRADA partner demonstrated the efficacy of the barrier product as a skin protectant. This company is very close to commercialization and has expressed an interest in using our recently renovated pilot plant facility for the initial scale-up production. They also will be developing Fantesk-based skin products with active ingredients for specific medical purposes.

4. Water-Based Lubricants

The lubricity of Fantesk liquid is immediately obvious, but what distinguishes the starch-oil composite from a simple jet-cooked starch dispersion is the aqueous distribution of oil droplets. These droplets are small enough to be uniformly distributed in the aqueous phase, and, as we have seen, do not phase separate. But they are large enough to do some good when a droplet gets caught between two metal surfaces, for example. Presumably the oil droplet is compressed, breaks out of the starch shell (this can be demonstrated under a light microscope by pressing down forcefully on the coverslip), and flows between the surfaces. Even after the free oil is released to the aqueous medium, new starch films can form because of the soluble starch present. Two companies have expressed interest in water-based lubricant systems and have established CRADA's, one involving metal-working fluids, and another interested in oil drilling mud additives. The latter company submitted our composite formulations to an independent commercial testing laboratory, and the results showed superior lubricating performance. This company is likely to pursue commercialization, and has begun negotiating with a third party who will mass-produce the large quantities required for further field testing.

III. CONCLUSION

The ability of the government to issue exclusive licenses to practice the Fantesk technology provided a good incentive for companies to work with us to develop end-use products. Not all such efforts were successful, for various reasons. In general, those instances where it was left to the companies to develop products were less successful than those in which we played an ongoing, active role in the research efforts. Several end-use patents have been applied for during this period, and one, covering starch-based seed coatings, has been issued. Both the starch graft copolymer absorbent and the starch-oil composite technologies have provided new commercial opportunities for utilizing agricultural commodities. These technologies are flexible enough to take advantage of any new starch or oil commodities made possible by genetic engineering of crop plants. We have shown that postharvest processing technology is an important and essential step in getting plant products from the field to the marketplace.

REFERENCES

Anonymous. 1973a. Chemical Engineering News. 21 May, p. 16.

Anonymous. 1973b. Chemical Marketing Reporter. 21 May, p. 5.

Anonymous. 1977. Specialties give better grip on the game. Chemistry Week 6 April, p. 34–35.

Brockway, C.E., and P.A. Seaberg. 1967. Grafting polyacrylonitrile to granular corn starch. J. Polym. Sci. 5 (A-1):1313–1326.

Burr, R.C., G.F. Fanta, C.R. Russell, and C.E. Rist. 1967. Influence of swelling and disruption of the starch granule on the composition of the starch-polyacrylonitrile copolymer. J. Macromol. Sci. Chem. A1:1381–1385.

Cooke, R.C. 1977. The use of an agar substitute in the initial growth of Boston ferns in vitro. Hort. Sci. 12:339.

Deterling, D. 1981. "Super Slurper" gets your crop moving earlier. Progr. Farm. 96(2):56K–56P.

Eskins, K., and G.F. Fanta. 1997. Non-separable starch-oil compositions. U.S. Patent 5 676 994.

Eskins, K., and G.F. Fanta. 1999. Non-separable compositions of starch and water immiscible organic materials. U.S. Patent 5 882 713.

Eskins, K., G.F. Fanta, F.C. Felker, and F.L. Baker. 1996. Ultrastructural studies on microencapsulated oil droplets in aqueous gels and dried films of a new starch-oil composite. Carbohydr. Polym. 29:233–239.

Fanta, G.F., F.L. Baker, R.C. Burr, W.M. Doane, and C.R. Russell. 1973. Polyacrylonitrile distribution in grafted starch granules determined by scanning electron microscopy. Stärke 25:157–161.

Fanta, G.F., F.L. Baker, R.C. Burr, W.M. Doane, and C.R. Russell. 1977. Scanning electron microscopy of saponified starch-g-polyacrylonitrile. Stärke 29:386–391.

Fanta, G.F., R.C. Burr, and W.M. Doane. 1982. Cross-linking in saponified starch-g-polyacrylonitrile. p. 195–215. *In* D.N.S. Hon (ed.) Graft copolymerization of lignocellulosic fibers. ACS Symp. Ser. no. 187.

Fanta, G.F., R.C. Burr, W.L. Orton, and W.M. Doane. 1980. Liquid phase dehydration of aqueous ethanol-gasoline mixtures. Science (Washington, DC) 210:646–647.

Fanta, G.F., R.C. Burr, C.R. Russell, and C.E. Rist. 1970. Copolymers of starch and polyacrylonitrile. Influence of granule swelling on copolymer composition under various reaction conditions. J. Macromol. Sci. Chem. A4:331–339.

Fanta, G.F., and W.M. Doane, 1986. Grafted starches. p. 149–178. *In* O.B. Wurzburg (ed.) Modified starches: Properties and uses. CRC Press, Boca Raton, FL.

Fanta, G.F., and K. Eskins. 1995. Stable starch-lipid compositions prepared by steam jet cooking. Carbohydr. Polym. 28:171–175.

Fanta, G.F., F.C. Felker, K. Eskins, and F.L. Baker. 1999. Aqueous starch-oil dispersions prepared by steam jet cooking. Starch films at the oil-water interface. Carbohydr. Polym. 39:25–35.

Geronemus, R.E., and P. Robins. 1982. The effect of two new dressings on epidermal wound healings. J. Dermatol. Surg. Oncol. 8:850–852.

Gugliemelli, L.A., M.O. Weaver, C.R. Russell, and C.E. Rist. 1969. Base-hydrolyzed starch-polyacrylonitrile (S-PAN) graft copolymer. S-PAN-1:1, PAN M.W. 794,000, J. Appl. Polym Sci. 13:2007–2017.

Hamilton, J.L., and R.L. Lowe. 1982. Use of a water absorbent in tobacco seedling production and transplanting. Tob. Sci. 26:17–20.

Hintz, H.L., and D.Johnson. 1967. The mechanism of oxidation of cyclic alcohols by cerium (IV). J. Org. Chem. 32:556–564.

Kay, B.L. 1980. Mulch choices for erosion control and plant establishment. Weeds Trees Turf 19(8):16–24.

Kirkland, E.R., E. Palanuik, and H.V. Ingmar. 1985. U.S. Patent 4 495 724.

Knauth, K. 1991. Sharing the technology wealth. Governing Executive. November, p. 34–39.

Knutson, C.A., K. Eskins, and G.F. Fanta. 1996. Composition and oil-retaining capacity of jet-cooked starch-oil composites. Cer. Chem. 73:185–188.

Leppla, N.C. 1976. Starch compounds as adjunct gelling agents in larval diet of the cabbage looper. J. Georgia Entomol. Soc. 11:251–254.

Miller, D.E. 1979. Effect of H-SPAN on water retained by soils after irrigation. Soil Sci. Soc. Am. J. 43:628–929.

Mino, G., and S. Kaizerman, 1958. A New Method for the preparation of graft copolymers. Polymerization initiated by ceric ion redox systems. J. Polym. Sci. 31:242–243.

Pottenger, C.R., and D. Johnson. 1970. Mechanism of cerium (IV) oxidation of glucose and cellulose. J. Polym. Sci. 8 (A-1):301–318.

Shrader, W.D., and A. Mustejeran. 1977. Potentials of a polyacrylonitrile-starch polymer, Super Slurper, to modify water holding properties of soils. ACS Div. Org. Coatings & Plast. Chem. 37(1):683–687.

Spence, W.R. 1980. U.S. Patent 4 226 232.

Spence, W.R. 1981. U.S. Patent 4 272 514.

Tanzy, K. 1981. Is there a Super Slurper in your future? Soybean Dig. 41(3):306.

Taylor, M.W., and E.B. Bagley. 1974. Dispersions or solutions? A mechanism for certain thickening agents. J. Appl. Polym. Sci. 18:2747–2761.

Technology Transfer Act. 1986.

Valdez, H. 1980. A hydrogel preparation for cleansing and protecting equine wounds. Equine Pract. 2:33–36.

Villet, R.H. 1994. Increasing the value of agricultural feedstocks. Chemtech 24(9):44–48.

Weaver, M.O., L.A. Gugliemelli, W.M. Doane, and C.R. Russell. 1971. Hydrolyzed starch-polyacrylonitrile graft copolymers: effect of structure on properties. J. Appl. Polym. Sci. 15:3015–3024.

Weaver, M.O., E.B. Bagley, G.F. Fanta, and W.M. Doane. 1973. Gel sheets produced by hydration of films from the potassium salt of hydrolyzed starch-polyacrylonitrile graft copolymers. *In* SSYM 009. Abstr. Am. Chem. Soc. 166th Natl. Meet. 26–31 August.

Weaver, M.O., E.B. Bagley, G.F. Fanta, and W.M. Doane. 1974. Gel sheets produced by hydration of films from the potassium salt of hydrolyzed starch-polyacrylonitrile graft copolymer. Appl. Polym. Symp. 25:97–102.

Weaver, M.O., E.B. Bagley, G.F. Fanta, and W.M. Doane. 1976a. U.S. Patent 3 935 099.

Weaver, M.O., E.B. Bagley, G.F. Fanta, and W.M. Doane. 1976b. U.S. Patent 3 981 100.

Weaver, M.O., E.B. Bagley, G.F. Fanta, and W.M. Doane. 1976c. U.S. Patent 3 985 616.

Weaver, M.O., E.B. Bagley, G.F. Fanta, and W.M. Doane. 1976d. U.S. Patent 3 997 484.

Weaver, M.O., R.R. Montgomery, L.D. Miller, V.E. Sohns, G.F. Fanta, and W.M. Doane. 1977. A practical process for the preparation of Super Slurper, a starch-based polymer with a large capacity to absorb water. Stärke 29:413–422.

6 Breeding Cereal Small Grains for Value-Added Uses

WILLIAM A. BERZONSKY AND HERBERT W. OHM

Purdue University
West Lafayette, Indiana

Broadly defined, a value-added trait encompasses a myriad of genetically determined plant components. Furthermore, genetic improvement that increases the economic and utilization value of a crop is at the core of plant breeding. Therefore, to effectively illustrate how plant breeders have added value to crops, logic dictates that this chapter cover a finite group of crops and value-added traits. It will describe breeding cereal small grain crops for end-use quality improvement.

Whether the end-use quality of a crop is enhanced directly or indirectly, the source of such improvement is genetic variability. In wheat (*Triticum* sp.), for example, quality might be improved directly by elevating grain protein, or it might be improved indirectly by increasing resistance to head scab, caused by *Fusarium graminearum* Schwab. Once a breeder has identified an end-use trait for improvement, germplasm must be screened for promising genotypes.

Cultivars represent the primary gene pool, and the wild, related species represent the secondary and tertiary gene pools as described by Harlan and de Wet (1971). Sufficient variability may be present in cultivars, or it may only exist in wild, related species. A breeder's use of the secondary and tertiary gene pools for crop improvement is dependent on many considerations, not the least of which are possible crossability barriers and transfers of undesirable genes (Harlan 1976). Nevertheless, as a result of advances in technologies and methodologies, plant breeders presently have the genetic tools to tap virtually any of the three gene pools. Indeed, there are many recent examples demonstrating success in doing so. Fundamental to these successes has been the application of classical genetics, cytogenetics, cell and tissue culture and other biotechnologies, including use of DNA markers for plant selection. To illustrate methodologies and technologies, this chapter will review some of these examples and speculate on future possibilities for adding end-use qualities.

I. WHEAT GENETIC VARIATION FOR QUALITY

Although some diploid wheats ($2n = 2x = 14$) are cultivated, most wheat products are derived from durum wheat (*Triticum turgidum* var. *durum*) ($2n = 4x = 28$,

---

 Designing Crops for Added Value, Agronomy Monograph no. 40.

AABB) and common wheat (*Triticum aestivum* L.) ($2n = 6x = 42$, AABBDD) (Orth & Shellenberger, 1988). Although it also may be used for breadmaking, durum wheat is used in the production of pasta products, and because of the additional viscoelastic properties of its flour, common wheat is primarily used to produce pastry and bread products. Common wheat is classified according to grain hardness (soft or hard), bran color (white or red), and growth habit (spring or winter). Differences in many grain characteristics determine whether the flour is best suited for the production of cakes, biscuits, crackers, cereals, pastries, soup thickeners, noodles, flat breads, pan breads, etc. Durum wheat typically is the hardest of all wheats, and it has a high protein content (Dick & Matsuo, 1988). Although there are exceptions, hard red spring or winter wheats, characterized by high protein, make the most suitable bread flours. Soft white and red winter wheats by comparison, have low protein and are most suitable for the production of cakes, biscuits, crackers, and pastries (Yamazaki et al., 1981). Other common wheats, such as club wheat and hard white wheat, have qualities that make them desirable for important specialized markets. Club wheat flour has low α-amylase activity and a superior gluten for the production of Japanese sponge cake (Nagao, 1995). In the USA under an "identity preserved grain" marketing system, hard white wheat is being used to produce bread products nutritionally equal to red wheat products, but lower in sugar content (Freiberg, 1991).

The industrial uses of wheat are based mainly on the properties of wheat starch and gluten (Eliasson, 1989). Wheat starch is used for sizing of paper, for producing textiles and adhesives, and for producing syrups and sugars (Bashuk, 1986). Gluten is used to fortify flour and add rigidity and volume to specialty baked goods. It also is used as a meat and poultry binder; a protein additive in pet foods, cereals, and snack foods; and as an additive to improve the texture of synthetic cheeses. Nonfood applications for gluten include its use in chewing gum, mascara, and pharmaceuticals (Magnuson, 1985). In addition, there is an impressive list of potential nonfood uses for gluten (Bietz & Lookhart, 1996). Wheat gluten could be an essential ingredient in biodegradable plastics or as an additive in inks, laundry detergents, cosmetics, adhesives, and rubber products.

A decade ago, it was estimated that only 2 to 4% of the total world wheat production was used for industrial applications, and no breeding programs were dedicated to developing cultivars strictly for industrial uses (Bushuk, 1986). The focus on production of wheat starch and gluten industrial products varies between countries. It depends on the existing economic policies and the availablility of low-cost starch alternatives. In the European Community, where industry subsidies have existed, starch and gluten production has been rapidly expanding. In North America, without subsidies and with lower-cost corn starch available, there has been a less rapid expansion (Bietz & Lookhart 1996, Leuck, 1989). Nevertheless, given favorable agricultural and economic conditions, breeders might someday target improvement of wheat traits specifically for nonfood usage. Two examples would be adding fiber to wheat straw for paper production, or increasing grain starch content for ethanol production (Wrigley & McMaster, 1989).

Despite the many subtle quality differences between the various wheat types, each has important quality standards for flour color, grain hardness, protein content, and flour yield. Fortunately, whether or not the end-use product is for food, genetic variation exists for each of these traits, so they may be manipulated to improve quality.

A. Bran and Flour Color

Both wheat endosperm color and bran color determine flour color. A white flour is preferred for the production of most bread and pastry products, but an amber yellow semolina color is an especially strict quality requirement for durum wheat (Cantrell, 1987). This color is determined by xanthophylls, and the end-use pasta product color is a result of interaction of these xanthophylls with lipoxygenase during processing. Genotypes with low lipoxygenase have been selected (Joppa & Williams, 1988), and since semolina color is highly heritable (Braaten et al., 1962), early generation selection for color is frequently a breeding practice (Cantrell, 1987). Jain et al. (1975) determined that much variation exists for kernel color between entries in the USDA World Durum Collection. Variation also exists in related wheat species. For example, endosperm pigmentation was threefold greater in domesticated *Triticum monococcum* L. ($2n = 14$, AA) accessions compared to durum (D'egidio et al., 1993). Some spelt wheat (*Triticum aestivum* subsp. *spelta*) cultivars had a twofold higher β-carotene content compared to hard red spring wheat (Abdel-Aal et al., 1995).

The fact that common wheat is categorized on the basis of bran color indicates that genetic diversity exists for this trait. Genetic differences account for a large proportion of the variation in flour color between cultivars (Miskelly, 1984; Shuey, 1976). Three partially dominant genes determine red bran pigmentation, while white bran results from the presence of recessive alleles (Heyne & Johnson, 1954). Interestingly, both wheat bran pigmentation and endosperm color can be altered by introducing different chromosome segments of *Agropyron elongatum* (Host) P.B. ($10x$, E genome). A dark-blue grain is produced when a chromosome segment of *Agropyron* replaces a portion of wheat chromosome 4B (Jan et al., 1981); whereas, a yellow endosperm results from the replacement of a portion of chromosome 7D (Knott, 1980).

B. Preharvest Sprouting and α-Amylase Activity

Preharvest rains can activate α-amylase in wheat kernels leading to a degradation of grain color, a reduction in grain hardness, and a lowering of flour yield, especially in durum and white wheats (Soper et al., 1989; Sorrells et al., 1989). In fact, there is a relatively tight association between white grain color and preharvest sprout susceptibility (Freed et al., 1976). Within each color class however, there are genotypes exhibiting a range of reactions to preharvest rain. Everson and Hart (1961) were able to classify red wheat cultivars into dormant, moderately dormant, and nondormant types; whereas, they found most white wheat cultivars to be nondormant. Kratochvil and Sammons (1991) distinguished between susceptible and tolerant red wheat cutivars, and Czarnecki (1987) identified a red variety with seventeen-fold lower α-amylase activity compared to a sprout-susceptible cultivar. 'Clark's Cream' and its derivatives are white wheats with a higher level of resistance to preharvest sprouting (Morris & Paulsen, 1987). Since transferring sprout resistance from hexaploid wheat to durum wheat may adversely affect kernel color (Soper et al., 1989), it is fortuitous that a durum relative, *Triticum dicoccoides* L. ($2n = 28$, AABB), also has preharvest sprout resistance (Mares, 1987).

C. Grain Hardness

Because grain hardness corresponds closely to milling and flour quality, market classes are in part defined by this trait. A fine flour consisting of small particles is desirable for the production of pastries. Flour composed of larger particles is associated with kernel hardness. Therefore, soft wheat is best suited for pastries, and hard wheat, which has large flour particles, is best suited for breads. In fact, hard wheat flour exhibits starch degredation when hard wheats are milled to produce small particles (Hoseney, 1987). Durum wheat, having the hardest endosperm, is milled into a coarse product called semolina. In addition to the extreme hardness of the endosperm, the durum grain must be highly vitreous to produce a high-quality, transluscent pasta product (Cantrell, 1987). Therefore, a reduction in grain vitreousness, commonly referred to as "yellow berry", is undesirable. The expression of yellow berry is genetically determined (Dhaliwal et al., 1986), and *T. monococcum* exhibits considerable genetic variability for expression of yellow berry (D'egidio et al., 1993).

Grain hardness is determined by a major gene and several minor genes (Symes, 1965; Symes, 1969). The major gene for soft grain is carried on the D-genome of wheat (Williams, 1988). Several studies have demonstrated genetic variability for kernel hardness within soft and hard wheat classes and the high heritability of this trait (Yamazaki & Donelson, 1983; Pomeranz et al., 1985). There are various theories that may explain biochemical endosperm differences between soft and hard wheats (Anjum & Walker, 1991), but one in particular seems most feasible. Friabilin, a starch granule protein, is present in higher amounts in soft wheats compared to hard wheats (Greenwell & Schofield, 1986). A subunit of this protein was bound more to soft wheat starch. In addition, its accumulation corresponded well with grain texture differences in near-isogenic wheat lines (Jolly et al., 1993). Apparently, friabilin reduces the tight association between endosperm starch granules and proteins, resulting in a soft endosperm.

D. Seed Endosperm Proteins

Wheat grain is an important source of energy, nutrition, and fiber for livestock and humans. A wide variety of wheat food products are rich in carbohydrates, proteins, vitamins, and minerals (Betschart, 1989). Wheat proteins provide many essential amino acids in the human diet, a notable exception being lysine, and they are intrinsically tied to baking quality. Because of this, wheat breeders have concentrated considerable effort into identifying genetic materials with quantitative and qualitative protein differences.

Most wheat grain proteins reside in the endosperm and are traditionally classified according to whether they are salt-soluble (albumins and globulins), ethanol-soluble (gliadins), or acid/alkali-soluble (glutenins). Approximately 83% of the wheat grain proteins are gliadins and glutenins (Wall, 1990). The gliadins consist of monomeric polypeptide chains; whereas, the glutenins are made of high molecular weight complex polypeptide chains. When water is added to wheat flour, these proteins form a substance called gluten, which can expand under yeast fermentation and hold its form when baked. The ability to expand is likely attributable to glutenins, while the ability to hold form appears due to gliadins (Pritchard & Brock, 1994), more generally referred to as the "strength" and "volume" characteristics of the dough, respectively. Expression of wheat protein genes, perhaps more than any other quality component, is influenced by environment (Graybosch et al., 1996; Mariani et al., 1995). Even so, evolution and domestication have given rise to a plethora of wheat genotypes with different glutenin and gliadin allelic profiles.

Searching for a means to identify protein variation between genotypes that also would associate with end-use quality, geneticists found they could separate the various endosperm proteins by molecular weight on electrophoretic gels. In doing so, certain combinations of glutenin and gliadin subunits have been identified as markers for desirable or undesirable wheat quality attributes (Bietz, 1987). Some 40% of the soft red wheat and 90% of the soft white wheat surveyed have a glutenin allele with 2 + 12 subunits, which is associated with poor breadmaking quality; whereas, 62 to 91% of the U.S. hard wheat surveyed has an allele with 5 + 10 subunits, which is associated with good breadmaking quality (Lookhart et al., 1993). Payne et al. (1987) used the association of these subunits with breadmaking quality to develop glutenin quality scores for 84 wheats developed in the UK. They estimated that 47 to 60% of the breadmaking quality variation between the cultivars was due to glutenin subunit variation. Subunits of both combinations are determined by alleles of the *Glu-D1* gene located on chromosome 1D (Payne & Lawrence, 1983). Hou et al. (1996b) recently identified another group of glutenin subunits that are correlated to Japanese sponge cake and sugar-snap cookie baking qualities. Cross and Guo (1993) studied a wheat collection for the presence of glutenin alleles. Among the 1889 accessions they examined, 147 had previously undocumented glutenin alleles or allelic combinations. Although, a high percentage of U.S. wheats have glutenin alleles associated with high quality (Lookhart et al., 1993), substantial glutenin differences between cultivars of different classes exist (Graybosch et al., 1994; Rogers et al., 1989).

Durum wheat lacks chromosome 1D, considered to carry the major gluten quality genes for breadmaking (Liu et al., 1994; Kerber & Tipples, 1969), but durum cultivars still exhibit glutenin allelic variability as a result of genes on chromosome 1B (Pena et al., 1994). In fact, D-genome genes often suppress the expression of A and B-genome endosperm genes (Bittel et al., 1991; Howes et al., 1990; Galili & Feldman, 1984). Compared to common wheat, it has been difficult to pinpoint specific qlutenin gene subunits in durum wheat that are closely associated with improved end-use quality; however, the 5 + 10 and 2 + 12 subunits appear to also have a similar effect on durum end-use quality (Liu et al., 1996).

Gliadin variability between wheat cultivars also exists. Huebner et al. (1995) detected large cultivar differences between the gliadin profiles of hard red winter wheat. Likewise, Wieser et al. (1994) discovered that there was a great deal of gliadin subunit variation between 16 diverse wheat cultivars originating from several different countries. But, 66% of the durum cultivars from a large population had one of only three gliadin profiles (Nieto-Taladriz, 1994). In some of these cases, gliadin variability might be due to structural gene mutations (D'Ovidio et al., 1991).

As was the case with glutenin, specific gliadin allele subunits also appear to be associated with wheat quality for soft white wheat, club wheat, and soft red winter wheat (Hou et al., 1996a). In durum wheat, resistance to pasta degredation during cooking is due to the presence of a specific gliadin designated as γ-gliadin 45. Poor pasta-making is due to the presence of a gliadin referred to as γ-gliadin 42 (Damidaux et al., 1980; Taha & Sagi, 1987). Breeders have selected for the presence of γ-gliadin 45. The result has been the development of durums with increased gluten strength having improved pasta-making characteristics (Liu et al., 1996). The gene for γ-gliadin 45 is located on chromosome 1B (Joppa et al., 1983).

Protein variability among the wild and related wheat species augments that lacking among the cultivated tetraploid and hexaploid types. A number of wheat species, especially *T. monococcum*, *T. dicoccoides*, and *Triticum tauschii* ($2n = 14$, DD), are highly variable for the quantity and quality of their seed storage proteins compared to cultivars (William et al., 1993; Ciaffi et al., 1992). More distantly related genera, such as *Agropyron* (wheatgrass) and *Secale* (rye), also can be important sources for protein variability. A single wheatgrass chromosome added to a wheat complement increased endosperm seed protein 15% (Soliman et al., 1980), while chromosome arm 1RS of rye translocated to wheat chromosome arm 1BL altered the glutenin/gliadin seed protein balance resulting in "sticky" dough (Dhaliwal & MacRitchie, 1990). The expression of the sticky dough quality itself however, is dependent on the wheat genotype into which the rye segment is inserted (Graybosch et al., 1990).

E. Seed Starch

Wheat starch granules are composed of two high-molecular-weight polymers, amylose and amylopectin (Lineback & Rasper, 1988). Amylose content is a determinant of end-product quality, because cultivars with lower amylose content are superior for Japanese noodle production (Oda et al., 1980). Amylose synthesis is controlled by a granule-bound synthase, which is encoded by three *waxy* genes (Miura et al., 1994). The genes have been located to chromosomes 7AS, 4AL, and 7DS (Nakamura et al., 1993). Yamamori et al. (1992) have shown that Japanese wheat cultivars with lower amylose content also have less *waxy* protein. They indicated that 'Kanto 107', a cultivar with low amylose content, was being used in crosses to breed superior genotypes for Japanese noodle production. Soulaka and Morrison (1985) isolated amylose from 23 common and 26 durum wheats and found mean starch amylose content varied from 26.3 to 30.6%. They concluded that a major portion of this variation was attributable to genotype, considering an estimated environmental variation of ±0.6% amylose. Nemeth et al. (1994) reported a range for starch amylose content of 19.2 to 22.5% when they analyzed 11 soft wheat cultivars from Canada, Australia, and the USA.

F. Flour Yield

Flour yield is of primary importance to the wheat miller. According to Briggle et al. (1968), flour yield is highly heritable. Although flour yield is not a visually selectable trait for breeders in the field, Yamazaki and Andrews (1982) found significant genotpe-dependent flour yield differences between 37 U.S. soft wheat cultivars. A 10% wheat flour yield difference, attributable to genotype, was observed by Baenziger et al. (1985) after milling 22 soft and hard winter wheats. Nemeth et al. (1994) concluded that U.S. soft wheat cultivars gave higher flour yields compared to Canadian and Australian wheat. Accessions of *T. monococcum* had flour yields similar to common wheat cultivars, but by comparison, their yields greatly exceeded those for durum wheat cultivars milled and tested in the same manner (D'egidio et al., 1993).

A significant genotype × environment interaction for flour yield can be a factor complicating selection for this quality trait. Bassett et al. (1989) calculated a significant genotype × environment interaction for the flour yields of four U.S. soft white wheat cultivars grown in 63 site-years, but they also calculated a 0.72 to 0.81 correlation between flour yield and grain hardness. One way breeders may be able to identify high-yielding flour genotypes is by screening for kernel morphological characteristics or flour particle size, which is related to hardness (Wu et al., 1990). A softer wheat kernel will generally produce more break flour and have a smaller flour particle size (Hoseney, 1987). Using image analysis of kernels, Berman et al. (1996) were able to account for 66% of the flour yield variation they observed between eight wheat lines. However, Carver (1994) was unable to successfully develop high-yielding flour types by basing selection solely on near-infrared reflectance spectroscopy measurements (measurements of kernel hardness).

II. BREEDING WHEAT FOR ADDED QUALITY

Over years, breeders have substantially improved several end-use wheat quality traits, while also managing to improve other agronomic traits (Souza et al., 1993; Cox et al., 1989; Patterson & Allan 1981). Much of this improvement is the result of recombining quality traits from improved wheat parents, and then selecting for the desired genotype based on various quality test data (Everson, 1987). This is still the basis of wheat end-use quality improvement, and it is likely to remain so. But, because a number of plant breeding discussions cover classical breeding methodologies for end-use cereal quality (Patterson & Allan, 1981; Brown & Patterson, 1992; Joppa & Williams, 1988; Anderson & Reinbergs, 1985), this chapter will describe breeding biotechnologies that augment these classical techniques, and this section will discuss their application to wheat breeding.

A. Marker-Assisted Selection and Breeding

The most rapidly advancing area of plant genetics and breeding is that of molecular genetics and its use to identify DNA markers for important plant traits. Many of the techniques associated with breeding wheat on the molecular level have lagged behind other organisms and crops, mainly because of the complex genetic make-up of wheat (Talbert, 1993). Presently, this obstacle is being overcome with the institution of new research approaches and the adaptation of successful techniques once applied only to other crops.

Breeding efficiency is maximized if the genetic component of an end-use trait is large compared to the environmental component. If possible, a breeder selects for end-use quality improvement on the basis of phenotype, which includes both components. Herein is the dilemma for breeders. Many of the important quality traits are determined by more than one gene each with a small incremental effect (quantitative trait loci, QTL), or by a single gene that can be difficult to select for, especially if seasonal weather is not favorable. In both scenerios, the large environmental effect can be a hindrance to progress. Breeders continue to use morphological markers, isozyme markers, and various statistical treatments to minimize the environmental component of phenotype (Stuber, 1992). However, a DNA marker closely linked to the gene and independent of the environment is invaluable to a breeder for the assessment of the genetic component of end-use quality. In addition, accumulation of such markers will aid in the construction of linkage maps so that favorable end-use gene combinations can be produced and breeding progress can be predicted.

B. Deoxyribonucleic Acid Markers

Restriction fragment length polymorphism (RFLP) and polymerase chain reaction (PCR)-based DNA analyses have been the two most widely employed techniques to identify molecular markers. Each technique has its advantages and disadvantages (Ragot & Hoisington, 1993), but both have been utilized to identify markers for important wheat quality traits.

The RFLP analysis (Helentjaris et al., 1985) involves isolating genomic DNA from segregating individuals. The segregating individuals are developed by hybridizing two lines that differ for a trait of interest. The DNA is digested with a restriction endonuclease, which cleaves the DNA at enzyme-specific sites throughout the genome. The fragment byproducts of the digestion are separated by electrical charge according to length on a gel. They are then denatured and bound as single-stranded DNA to a membrane. The bound fragments are allowed to hybridize with plant DNA fragments (probes) that have been radioactively or chemically labeled. The probe fragments are cloned in bacteria and may represent single or repetitive DNA sequences from the plant genome. They also may be localized within the genome (Appels et al., 1989) or dispersed throughout (McIntyre et al., 1988). Any band that results from probe hybridization and cosegregates with the trait of interest among progeny from the original cross is a possible DNA marker. Inherent problems with this technique are that it is labor-intensive and often requires the use of radioactivity; however, it has significant advantages over morphological markers and its use has already generated much data (Helentjaris, 1992; Tanksley et al., 1989).

The PCR-based analyses include: random amplified polymorphic DNA (RAPD) (Williams et al., 1990), amplified fragment length polymorphism (AFLP) (Vos et al., 1995), sequence tagged site (STS) (Olson et al., 1989), and sequence tagged microsatellite site (STMS) (Beckmann & Soller, 1990) techniques. Each method is based on the polymerase chain reaction (Saiki et al., 1988). In this reaction, template DNA is denatured to single-stranded DNA, and primers defining a DNA segment are hybridized to specific sites on complementary DNA strands. The

DNA polymerase then catalyzes an extension of each of the primers for the defined segment, duplicating the complementary DNA strands. Cycling of these events leads to a significant amplification of the DNA segment delimited by the primers. These amplified products are then separated on an electrophoretic gel to visualize possible genomic polymorphisms. The RAPD, AFLP, STS, and STMS techniques differ because the amplification might be random or nonrandom, or because the primer sequences are of different lengths or anneal to different genomic sites.

Compared to RFLP analysis, RAPD analysis is relatively quick, does not require the processing of probes or the handling of radioactivity, and requires much less template DNA (Tingey et al., 1992). It also should be noted that the RFLP and RAPD technologies are not necessarily mutually exclusive. Some RAPD fragments can be converted into RFLPs, and RFLPs, after sequencing, can be made into STSs.

Since preharvest sprout tolerance in wheat is a quantitative trait and has low heritability, Anderson et al. (1993) sought to identify QTL associated with tolerance in two populations of white wheat. Using RFLP analysis, they identified QTL that accounted for up to 51% of the genetic variance for sprout tolerance in a segregating population. Smith et al. (1994) demonstrated the utility of using PCR-based DNA markers when they distinguished between closely related glutenin alleles for high quality among diverse wheat cultivars. They predicted that these markers might be used to develop soft wheat cultivars with strong dough characteristics for cracker production. Similarly, D'Ovidio (1993) found PCR markers linked to glutenin genes associated with low and high quality in durum wheat cultivars. Using nondestructive techniques, he was able to detect differences between the genomic DNA originating from individual plants. Thus, the high-quality plants were still available to be used in crosses.

Several improved technologies, hybridization schemes, and sampling methodologies have been devised to circumvent limitations associated with the marker assisted breeding of wheat. He et al. (1992) ran their wheat RAPDs on denaturing gradient gels and detected higher levels of polymorphisms compared to using conventional agarose gels. Denaturing gradient gels enable fragment separation based on size and sequence differences (Fisher & Lerman, 1983), Dweikat et al. (1993) were able to detect differences between soft and hard wheats using this same system.

Development of recombinant inbred and near-isogenic lines are attempts to minimize differences between all loci except for the trait of interest. These methodologies have long been practiced in breeding programs, but their use is increasing in importance because of the search for selectable DNA markers (Sorrells, 1991). Rousset et al. (1992) developed a wheat recombinant inbred population from which they were able to identify glutenin genes for improved baking quality. Michelmore et al. (1991) introduced a DNA-pooling strategy, bulked segregant analysis, to more efficiently detect genomic polymorphisms in segregating populations. An exciting, rapidly developing approach to finding valuable selection markers in wheat takes advantage of the concept of conserved genomic homoeology ("synteny") across grass genera. There is an ever-increasing body of evidence to indicate that grasses such as rice (*Oryza sativa* L.), wheat, and maize (*Zea mays* L.) share a similar gene constitution and order, despite significant disparities in their genome sizes (Bennetzen & Freeling, 1993). Utilizing this concept of synteny, Sorrells et al. (1995) are mapping gene loci in rice, which has the simplest genome of all the grasses, to determine gene relationships and identify valuable markers in wheat.

C. Cytogenetic Manipulations

Another approach to breeding for improved end-use quality is to alter the chromosome constitution of wheat. The chromosomal locations of many genes influencing wheat quality have been located to the group-1 and group-6 chromosomes (Law & Payne, 1983). "Chromosome-engineering" has resulted in the fusion of alien and wheat chromosome segments. The goal of these fusions is to transfer desirable genes from the alien species to wheat, while minimizing the transfer of undesirable genes. Aside from recombination between homologous chromosomes, this is one way to introduce valuable genetic variation for quality. For example, segments of rye chromosomes have been fused with wheat chromosomes to increase disease and insect resistances (Zeller & Hsam, 1983), thereby indirectly improving quality. Unfortunately, replacement of chromosome 1BS of wheat with 1RS of rye can negatively impact wheat quality (Lee et al., 1995). Other wheat-rye translocations however, can have a positive effect on quality (Knackstedt et al., 1994), and wheat-rye group-1 translocations may nevertheless be genetically altered to mitigate poor quality attributes (Lukaszewski & Curtis 1992; Koebner & Shepherd, 1988). Similar chromosome manipulations involving wheat and wheatgrass (*Agropyron* sp.) have been successful. Not only do certain *Agropyron* genes confer disease and insect resistances to wheat (Friebe et al., 1991), some also elevate endosperm protein levels (Soliman et al., 1980). Wheat-*Agropyron* translocations have been induced that maintain the *Agropyron* resistance genes but eliminate possible undesirable genes (Friebe et al., 1994). Yellow flour, the result of a wheat-*Agropyron* translocation, actually improves the quality of some wheats used for pasta production. Wheat-*Agropyron* translocations also have been altered to reduce their negative impact on quality (Knott, 1980).

A very powerful technique, which promises to improve the efficiency of manipulating alien chromosomes for end-use wheat quality, is in situ chromosome hybridization (Miller et al., 1995). With this technique, species-specific DNA probes can be hybridized to critical chromosomes in situ. The result is that geneticists can physically quantify the amount of alien chromosomal DNA introduced into wheat, helping a breeder to determine if it can be tolerated. Another technique that holds considerable promise in quantifying and limiting the amount of alien DNA introduced into wheat is chromosome flow cytometry (Dolezel et al., 1994). In flow cytometry, the amount of DNA present in chromosomes is determined by the amount of fluorescence and the scatter pattern as individual stained chromosomes pass through a laser excitation beam. Furthermore, flow cytometry offers a chance to sort chromosomes based on DNA content. This may enable the production of flow-sorted chromosome libraries (Wang et al., 1992), which in turn could facilitate the cloning of wheat quality genes.

Knowing that many of the wheat endosperm glutenin and gliadin genes are carried on the group-1 chromosomes, cytogeneticists have successfully altered endosperm protein profiles by manipulating the dosage of chromosomes 1A, 1B, and

1D in wheat (Rogers et al., 1990; Galili et al., 1986). The consensus from these studies is that genes on homoeologous group-1 chromosomes in higher doses compensate for the loss of genes on other chromosomes. In addition, the compensating effect sometimes results in an improvement in end-use quality. These facts, combined with the fact that hexaploid wheat tolerates chromosome substitutions and deletions, indicate that group-1 chromosome composition might be tailored to maximize quality for specific end-use applications. Substitution of chromosome 1D from common wheat for 1A or 1B in durum wheat has improved the gluten strength of semolina, and additional improvements can be expected if specific glutenin subunits (e.g., 5 + 10) are identified and transferred (Liu et al., 1996). Existing RFLP probes specific for each homoeologous group-1 chromosome arm of wheat would be very valuable in these attempts (Sharp et al., 1989).

Dosage manipulation of wheat chromosomes carrying *waxy* genes might be a way to improve soft wheats for Japanese noodle production or a way to improve the pasting properties of wheat starch for industrial purposes. Nakamura et al. (1993) discovered that the three *waxy* genes carried on chromosome arms 7AS, 4AL, and 7DL had different effects on amylose production. Loss of chromosome 4AL, and the *Wx-B1* gene it carried, resulted in the greatest reduction in amylose content. *Waxy* wheats lacking all three gene loci and deficient for amylose production were produced by crossing with cultivars having the null alleles (Nakamura et al., 1995). Cytogeneticists could eliminate the *waxy* genes by producing chromosome deletions. Using chromosome translocations, they also might try to effect gene expression by separating the functional gene from possible regulator genes. Miura et al. (1994) speculated that chromosome 7B suppresses expression of the *Wx-b1* gene, because a primative translocation between 7B and 4A placed the functional gene on 4A and its regulator on 7B.

Synthetic hexaploid wheats, produced by hybridizing durum with *T. tauschii*, exhibit considerable variation in glutenin composition and end-use quality (Pena et al., 1995; Lagudah et al., 1987). These synthetic hexaploids can be readily hybridized to cultivated hexaploids in efforts to introduce high-quality genes from *T. tauschii* . In addition, *T. tauschii* can be directly hybridized to hexaploid wheat. The recovery of hybrids by embryo rescue and the subsequent backcrossing to hexaploid wheat has advantages compared to traditional backcrossing (Gill & Raupp, 1987). The technique could be employed by breeders to rapidly transfer end-use quality genes from *T. tauschii*.

Another wild species candidate for improving the gluten strength of wheat is *Triticum dicoccoides*. Steiger et al. (1996) reported that several chromosomes from *T. dicoccoides* significantly increased protein quality and content in durum wheat. Sharma et al. (1981) identified *T. dicoccoides* accessions that had a flour protein content of about 31% compared to 14 to 18% for cultivars, and Waines et al. (1987) found a significant level of variation between accessions for seed lysine content. Similarly, Srivastava et al. (1988) found *T. dicoccoides* accessions among 1800 collected in Ethiopia to have high seed protein content. After examining the genetic variability present in 12 *T. dicoccoides* populations in Israel, Nevo et al. (1982) concluded this species was underutilized as a germplasm source for improving either durum or bread wheat. Perhaps future breeding efforts will remedy this by focusing on transferring end-use quality genes from *T. diccoccoides* to wheat. Breeders can expect normal recombination between the A- and B-genomes of *T. diccocoides* and the A- and B-genomes of the cultivated wheats (Gupta, 1991).

D. Genetic Transformation, Tissue Culture, and Doubled-Haploid Breeding

One of the limitations to exploiting all of the wheat gene pools is obtaining sexual hybrids between species and genera. It is not feasible to obtain all types of wide hybrids in a breeding program, although hybridization success depends on many genetic factors associated with both putative parents (Sharma, 1995). Genetic transformation has considerable promise as a means of augmenting the sexual transfer of valuable genes from distantly related species and genera to wheat. Transformation of wheat has been accomplished using three different methods; *Agrobacterium tumefaciens*-mediated transfer (Woolston et al., 1988), protoplast electroporation (Lee et al., 1988), and microprojectile bombardment (Vasil et al., 1991). To date, microprojectile bombardment has been the most successful and repeatable tranformation technique for wheat. In microprojectile bombardment, metal particles are coated with the DNA to be transferred, and are then propelled into wheat tissue, usually embryonic calli. Transfer of a gene by microprojectile bombardment is typically proven by the activity of a reporter gene, β-glucuronidase (GUS), fused to the putative transfer gene.

Because wheat transformation is in its infancy, few as yet have directly applied it to the improvement of wheat end-use quality. However, Blechl (1993) and McKinnon and Henry (1995) defined how it might be applied. They point out that endosperm proteins are the most likely targets for molecular manipulation. For example, combining quality-determining glutenin subunits in novel combinations, and then inserting them by transformation might increase the dough strength in bread wheats. Incorporating high-lysine and threonine-coding genes from distantly related species also could vastly improve wheat protein quality for human consumption. Even the detrimental quality genes from rye, for example, could be reduced if antisense secalin genes, those that reduce the protein product of rye, are introduced via transformation. Blechl and Anderson (1996) demonstrated the processes involved when they produced a hybrid wheat glutenin gene by combining the coding regions of two glutenin subunits. The hybrid gene was expressed in three generations of wheat plants originally transformed by microprojectile bombardment. In addition, expression was controlled by the native glutenin gene regulatory sequences, and a single gene copy produced protein levels equivalent to those of the native genes. Preliminary indications were that an increase in the hybrid gene copy number also would result in an increase in glutenin accumulation relative to other seed storage proteins, or transgene expression would result in a decrease in the expression of endogenous glutenin genes.

A *waxy* gene product of wheat has been cloned and sequenced (Clark et al., 1991). Now that genetic materials exist that will help to understand the relationship between amylose-amylopectin ratios and starch quality (Yasui et al., 1996), perhaps the genetic engineering of *waxy* genes for diverse end-use wheat products is possible. Newly engineered *waxy* genes would likely be inserted by transformation, just as was done for the aforementioned hybrid glutenin gene.

Once a gene has been transferred to wheat by transformation, regeneration of plants from tissue cultures and embryos is necessary. The inability to regenerate wheat tissues in vitro to complete plants has been a hindrance to studying the expression of alien genes, but recent advances have made this less of a bottleneck to wheat improvement (Maheshwari et al., 1995). Wheat tissue culture itself promotes somaclonal variation that could produce mutants for improved wheat quality traits. Sharma et al. (1992) and Larkin et al. (1984) observed variants from tissue culture for yellow berry formation, bran color, α-amylase activity, and gliadin protein composition. Selection for somaclonal variants among elite genotypes might add another dimension to the improvement of these quality traits. However, the production of many unfavorable variants and the high cost of screening for such desirable ones will certainly limit use of somaclonal variation for improving quality (Larkin & Scowcroft, 1981).

Doubled-haploid (DH) breeding entails producing haploid plants, plants with the gametophytic chromosome number, and doubling the chromosome number of the haploids, usually by chemical treatment. Wheat haploids may be produced by culturing anthers in vitro, by combining certain nuclear genes with specific wild wheat cytoplasms, or by crossing with a barley species, *Hordeum bulbosum* L. (Baenziger et al., 1984) . Wheat breeders are exhibiting renewed interest in DH breeding with the discovery that wheat plants pollinated with maize also frequently produce haploids (Laurie & Bennett, 1988). Baenziger et al. (1984) reviewed the benefits derived from DH breeding. The principle benefit is that the technique produces homozygous lines quicker than for most other breeding methodologies. This is an especially important benefit in winter wheat breeding, due to the relatively long growing season. Many wheat breeding methodologies, including DH breeding, begin with an initial hybridization. Theoretically, DH individuals best represent the gametophytic variation from the hybrid. Bedo et al. (1992) hybridized two Hungarian bread wheat lines and developed DH lines from the hybrid. They identified transgressive DH lines that exceeded the parents for both protein and gluten content, suggesting DH lines with improved quality could be selected. The benefits of DH line production also can be extended to genetic studies designed to map end-use quality genes. For example, because DH production quickly fixes recombinant lines in a homozygous state, the technique could be used to quickly develop a recombinant inbred population for molecular analyses of quality traits.

III. BARLEY GENETIC VARIATION FOR QUALITY

Unlike cultivated wheat, cultivated barley (*Hordeum vulgare* L.) ($2n = 14$) is a diploid species and it is primarily produced for malt and animal feed, though it also is sometimes consumed directly as a human food in flat breads, gruels, and soups (Poehlman, 1985). Barley is called "2-rowed", "4-rowed", or "6-rowed", depending on whether one, two, or three of the florets in a given spikelet are fertile. The various end-use quality traits depend on whether the grain is to be used for malt or feed.

Burger and LaBerge (1985) aptly described the malting process as "a modified germination and growth process". When barley seeds are germinated for brewing, they produce α-amylase and β-amylase enzymes that hydrolyze the endosperm starch to sugars. The β-glucanases also are produced to breakdown β-glucan, a main constituent of barley endosperm cell walls (Fincher & Stone, 1993). The activities of these three enzymes are very important to malt quality, and they are regulated in the initial grain steeping and germination process. The end-use product is produced when sugars in the malt are fermented to produce beer or whisky, and the barley genotype greatly influences the final color and taste of the beverage (Burger & LaBerge, 1985).

"Malting-grade" barley differs from "feed-grade" barley for many quality characteristics. Most notably, malting-grade barleys have low grain protein and low dormancy. They also typically have hulls, since the hulls help filter the wort in the brewing process (Bamforth & Barclay, 1993). Malt barley grain with high protein content and high dormancy is unsatisfactory, because beer produced from such grain has degraded color and an undesirable haze. Another important quality, the amount of fermentable sugars extracted, also is reduced when high protein barley is used in malting (Burger & LaBerge, 1985). Ideally, the best malting barley cultivars have white, plump kernels and high test weights (grain weight per unit volume). Plump kernels and high test weights are often used as indicators of high malt extract potential (Peterson & Foster, 1973).

Feed barley is an important source of energy, vitamins, fiber, and protein to animals, especially poultry and swine (Newman & McGuire, 1985). The energy value of barley to animals is usually expressed as digestible energy (DE) or metabolizable energy (ME). This energy value generally defines feed quality, and several different interrelated physical and biochemical factors determine the value of a feed barley (Bhatty, 1993). Hulless cultivars, for example, had a higher DE compared to ones with hulls when fed to mice and swine (Bhatty et al., 1975; Bhatty et al., 1979). Grain dormancy, starch, protein, and lipid content all impact the ME of feed barley. High dormancy, high starch content, high protein content (≈15%), and high lipid content are desirable in feed barley (Bhatty, 1993). Protein quality also is of greater significance in feed barley. A balanced protein complement with all of the essential amino acids for animals is the preference (Ostrowski, 1977). Unfortunately, like many cereal grains, barley is naturally deficient in threonine and lysine (Eggum, 1978).

Genetic variation exists for both malt and feed barley end-uses. In fact, for end-use quality, barley has been extensively studied and genetically manipulated. As a result, barley germplasm collections have many naturally occurring, end-quality variants as well as many induced mutants (Van Hintum, 1994; Moseman & Smith, 1985). This means feed and malt barley breeders have access to diverse germplasm collections for the improvement of end-use quality. This is important when one considers that there is some crossover usage of barley. For example, malt barley grains are also frequently used as animal feed after their use in brewing (Peterson & Foster, 1973).

A. Seed Enzyme Activity

The activity of α-amylase and β-amylase barley grain enzymes is measured by what is termed "diastatic power" (Bamforth & Barclay, 1993). A diallel analysis revealed that specific combining ability and general combining ability were both important factors in the expression of diastatic power (Hayter & Riggs, 1978). Useful variants for α-amylase activity and diastatic power among North American barley cultivars were identified by Horsley et al. (1995). Natural variation for α-amylase activity also was determined to be high for the progenitor to cultivated barley, *Hordeum vulgare* L. *ssp. spontaneum* ($2n = 14$) (Brown & Jacobsen, 1982). The PCR-based analyses of wild barley accessions confirmed that substantial α-amylase variation exists (Weining et al., 1994). Furthermore, Allison and Ellis (1973) were able to group barley cultivars into three distinct classes based on their β-amylase polymorphisms. Goering and Eslick (1976) described a barley with an α-amylase that could hydrolyze its own starch as well as an additional four times as much corn starch by weight.

Other variants for amylase activity are those with mutated regulatory systems. Ho and Shih (1980) described barley with altered sensitivity to gibberellic acid, a hormone that has significant effects on α-amylase synthesis. The β-glucanase activities also vary significantly for some wild barley (Ahokas & Naskali, 1990) as well as for cultivars (Berbigier et al., 1981; Morgan et al., 1983a).

B. Seed Endosperm Proteins

Because barley endosperm proteins, hordeins, influence malt and feed quality (Ullrich et al., 1981), extensive germplasm searches have been made to find natural protein variants (Muntz et al., 1979; Johansson, 1978). The endosperm protein composition of barley also has been successfully altered by mutagenesis (Walther & Seibold, 1979; Balaravi et al., 1976). 'Hiproly', a natural high-lysine barley variant, and 'Riso 1508', a high-lysine mutant, have been useful as breeding sources in this respect (Eggum, 1978; Chaterjee et al., 1975; Helm et al., 1974; Newman et al., 1990; Ullrich & Eslick, 1978). Alternatively, 'Karl' barley has been useful as a low-protein source (Goblirsch et al., 1996; Burger et al., 1979)

Unfortunately, increasing barley grain protein levels often results in a 20 to 30% reduction in yield (Hagberg, 1978), which is likely the result of less grain carbohydrate accumulation (Doll & Koie, 1978). It has been very difficult to break this association by conventional breeding methodologies, but Stolen (1979) reported on the early selection of high-protein barley lines with apparently normal yields. *Hordeum spontaneum* K. Koch has a significantly higher protein content compared to cultivars (Friedman & Atsmon, 1988). Jaradat (1991) concluded that this wild species had great potential for breeding high protein cultivars, particularly since he had identified high-protein variants of the wild species that also had heavy kernels.

C. Seed Starch and Lipids

Similar to the barley protein situation, there are natural variants as well as mutants for seed starch and lipid composition. MacGregor and Fincher (1993) concluded that the natural range for grain starch content was 58 to 64% in barley. Torp (1980) reported a similar range for cultivars, but exotics exhibited from 43 to 62% starch and mutant barley types from 17 to 48% starch. Walker and Merritt (1969) described the identification of a spontaneous mutant for high amylose content. It contained about two times the normal amount of amylose naturally present in the barley seed. Two induced starch mutants, 'waxy Hector' and 'chalky Glenn', significantly influenced malt quality (Swanston, 1991). Chalky Glenn, exhibiting better malt quality, had a softer endosperm and a lower β-glucan level; whereas, waxy Hector had greater water uptake and loss of milling energy.

Compared to starch, less effort has been spent in examining barley collections for grain lipid variation (Morrison, 1993). Nevertheless, grain lipids are important components of malt and feed quality. They influence beer head retention and flavor (Morrison, 1978), and they contain more energy by weight compared to starch. Parsons and Price (1974) and Fedak and De La Roche (1977) disagreed as to the significance of the lipid variation they discovered in different barley collections. Wild-type barley has an average seed lipid content of 2% (Bhatty & Rossnagel, 1989). In a USDA collection, Parsons and Price (1974) found one plant introduction with a lipid content 35% higher than a control cultivar, and they suggested that this was sufficient to initiate breeding for high lipid content. However, Fedak and De La Roche (1977) found insignificant lipid variation between 21 Canadian barley lines, and they concluded that breeding should not be initiated. Riso 1508 barley, a high-lysine mutant, also has high lipid content (Bhatty & Rossnagel, 1979). Crude lipid content of Riso averaged 5.3% compared to that of a wild-type cultivar that averaged 3.4% (Bhatty & Rossnagel, 1980).

D. Malt Extraction and Feed Energy

There are many interacting quality parameters that influence the amount of malt extracted and the amount of energy available to animals, but relatively few directly impact both (Ullrich et al., 1981). Barley "gum" content, the viscous material associated with β-glucans (MacGregor & Fincher, 1993), in part determines malt extract amounts and the amount of usable energy available to animals. A high β-glucan and gum composition in barley lowers malt extraction (Morgan et al., 1983b; Barrett et al., 1973) and decreases digestibility for poultry (Campbell et al., 1989). However, gum content is highly heritable (Greenberg, 1977), and it varies between cultivars (Horsley et al., 1995; Molina-Cano & Conde, 1982) and wild barley (Henry & Brown, 1987).

Hulled barleys are less digestible to monogastric animals compared to hulless cultivars (Bhatty et al., 1979). The hull decreases digestibility because it increases grain fiber content (Foster & Prentice, 1987). Hulless or "naked grain" barley cultivars have been produced (Bhatty, 1986); however, they are not ideal for malting because they have comparatively lower yields (Bhatty, 1987). Some hulless cultivars have value for human food preparation (Berglund et al., 1992), because hulless cultivars with *waxy* starch have a higher β-glucan content compared to malt or feed barleys (Fastnaught et al., 1996). The β-glucan lowers cholesterol levels in humans (Davidson et al., 1991).

IV. BREEDING BARLEY FOR ADDED QUALITY

Improvement of barley quality is complicated by its use for both malt and feed products; however, compared to wheat, barley is less complex genetically. Thus, much is already understood about the genetic basis of barley quality. The barley breeder's task is to apply this knowledge to the continual improvement of malt and feed quality. Another equally important task is to continue to elucidate the genetic mechanisms of quality. Considering the simplicity of the barley genome and the array of biotechnological tools available, research directed towards addressing these tasks is producing exciting results.

As one of the earliest crops to be genetically manipulated by man, barley has been bred with improved grain yield, disease resistance, and lodging resistance (Cattivelli et al., 1994; Anderson & Reinbergs, 1985; Wych & Rasmusson, 1983). But, there are quality improvements yet to be made. Already, genes or portions of genes important for quality, such as; α-amylase, β-glucanase, hordien, and starch synthetase genes, have been cloned (Leah & Mundy, 1989; Loi et al., 1988; Kreis et al., 1986; Sorenson, 1989; Rohde et al., 1988). Kasha et al. (1993) suggest that many of these genes will someday be combined with specific promoters to enable greater expression and regulation in transformed barley plants.

A. Marker-Assisted Selection and Breeding

Recognizing the breeding significance of a molecular map for barley, The North American Barley Genome Mapping Project (Blake et al., 1991) was inititiated to develop an extensive RFLP map. Using wheat-barley addition lines, another group has been working to generate a similar map (Jahoor et al., 1991; Graner et al., 1990). Becker et al. (1995) augmented the map with AFLP markers, and Qi et al. (1996) combined the individual RFLP maps into one. The consensus map has useful breeding markers on every barley chromosome. In addition, the diploid nature of barley makes it an excellent genetic model for comparative mapping of quality traits to the more complex polyploid and related cereal species (Sherman et al., 1995).

The usefulness of the barley molecular map has been demonstrated by Becker and Heun (1995) and Tragoonrung et al. (1992), who employed known primer sequences based on the maps to generate potential PCR markers for various barley gene coding sequences, including those for α-amylase, endosperm protein, and starch. Although the level of polymorphism detected for cultivated barley has been relatively low, probably as a consequence of domestication (Saghai Maroof et al., 1995), molecular end-use quality markers have been identified. Molnar and McKay (1991), for example, identified RFLPs associated with hordein seed storage proteins. Using PCR-based technology, Weining et al. (1994) synthesized DNA primers for the *α-amy1* gene of barley, and Chalmers et al. (1993) found RAPD markers linked to QTL determining the energy required to mill barley grain. Genotypes with high energy requirements typically do not have high malt yield extracts, because their endosperm starches resist enzymatic hydrolysis. Using a doubled-haploid population and RFLP mapping, QTL also were identified for grain protein content, diastatic power, α-amylase activity, and malt extraction (Hayes et al., 1993).

Another use for DNA polymorphisms is to provide breeders with a means to assess the relatedness of potential parents. This would be valuable for selecting divergent parents for quality in either a beginning barley breeding program or an established one (Melchinger et al., 1994).

Unfortunately, two major problems are associated with altering barley quality, either by classical or molecular breeding techniques. First, several quality traits are coded by interrelated multigene families (Doll, 1983). Consequently, more than one gene must be manipulated to effect a major change in quality. Second, changing one quality trait often alters another trait, frequently in a way detrimental to improving overall quality. This is evident by the fact that high-lysine mutants have significantly reduced starch content (Kreis & Doll, 1980). A solution to these problems appears to require a better molecular understanding of the gene regulatory processes involved. Both Kasha et al. (1993) and Shewry and Kreis (1987) suggest that important quality genes be identified, sequenced, and cloned. After that, the expression of these genes and the processes underlying regulating expression can be studied. In the future, gene regulation, and not necessarily structure, might be altered to improve barley quality.

Gibberellic acid and its Ca-dependent secondary messenger are central to barley germination (Fincher & Stone, 1993). Therefore, they are excellent candidates for genetic manipulation. Several aspects of barley α-amylase gene expression and regulation also could be altered to improve malt or feed quality (Svensson, 1994). The pathway to the production of proanthocyanidins might be amenable to genetic manipulation. It is desirable to limit the production of proanthocyanidins, because they are associated with the hazy quality of beer. Erdal (1986) reported on the discovery of mutants that block the synthesis of proanthocyanidins. Perhaps "fine-tuning" these mutants by enhancing expression of the blocking genes would further limit quality problems (Horsley et al., 1991).

As Kasha et al. (1993) point out, a key element to studying and modifying multigene families is manipulating relatively large pieces of DNA. As a result, many future genetic improvements for barley quality will likely develop from cloning gene families into a system that accepts kilobase-size DNA. The work of Kleine et al. (1993) illustrates the first step in this approach. Using barley c-DNA, they produced a yeast artificial chromosome (YAC) library and identified a 120-kb insert from barley that corresponded to a segment of the barley *Hor1* gene. The *Hor1* gene codes for an endosperm storage protein. Study of the expression of this isolated gene, and its regulation with respect to other hordein genes, may lead to the production of barley with improved malting or feed characteristics.

B. Mutation Breeding and Cytogenetic Manipulations

Inducing genetic mutations and chromosome changes in barley remains a viable means to improve barley quality, despite the fact that barley does not tolerate genetic deletions as well as do polyploid crop species. Duplications of whole chromosomes (trisomics) or segments of chromosomes are most useful in barley cytogenetics and breeding (Ramage, 1991; Hagberg, 1986). On a basic level, duplications allow quality genes to be mapped to specific chromosomes (Tsuchiya, 1991), and they may enable better characterization of proximally located genes (Hagberg & Hagberg, 1991).

Segmental chromosome duplications can be produced by mutation using ionizing irradiation or chemicals (Hagberg & Hagberg, 1991). If duplicated chromosome segments contain quality genes, gene expression could be altered, something breeders might take advantage of to improve quality. Jensen (1984) described duplications for chromosome 5, which carries genes for grain protein. Unfortunately, like approximately 90% of such duplications, these proved inferior in vigor to their genetically normal parents (Hagberg & Hagberg, 1991). But, some barley protein mutants have been selected that not only have an improved protein composition, but also exhibit stable, high-yield (Munck et al., 1986). This indicates that it might be possible to generate selected chromosome duplications carrying quality genes without some of the associated adverse effects. Certainly, the utility of this approach would depend on duplicating only those chromosome DNA stretches that code for the quality gene of interest. Targeted duplication might be accomplished cytologically by the combined use of chromosome banding (Linde-Laursen, 1988) with flourescence in situ hybridization employing *Hordeum*-specific DNA sequences (Anamthawat-Jonsson & Heslop-Harrison, 1993). With barley, it may also be feasible, using microdissection, to produce chromosome-specific DNA libraries (Schondelmaier et al., 1993). One can, for example, envision portions of chromosomes from desirable quality mutants being dissected out, cloned, and inserted back into low-quality genotypes with high yield.

As has been described, there is ample genetic variation for quality in wild barley species. However, with the exception of *Hordeum spontaneum* K. Koch, the use of wild species has been limited because of hybridization barriers with *Hordeum vulgare* L. (Von Bothmer & Jacobsen, 1991). The primary attraction to the use of wild barley species has been to improve disease resistance (Fedak, 1985). In fact, at least for some malt quality characteristics, it appears that breeders have sought to maintain a narrow genetic base and select for a stable, consistent cultivar performance (Wych & Rasmusson, 1983). Reluctance to use wild species may be well-founded considering the fact that some barley species not only improve disease resistance but also reduce malt quality (Swanston, 1987).

Breeders still realize the value of genetic variability, and the use of wild species to improve barley quality continues to be explored. Although the hybrids they produced were aneuploid, Linde-Laursen and Von Bothmer (1986) demonstrated that successful crosses between *Hordeum vulgare* L. and three distantly related species were possible. Intergeneric hybrids with barley also are possible if embryo rescue is practiced (Fujigaki, 1995). Unfortunately, the practical production of such wide hybrids is difficult and very rare (Fedak, 1985). Thus, the direct transfer of quality genes from wild barley species or other genera is not currently feasible. In addition, Von Bothmer and Jacobsen (1991) concluded that a better understanding of crossability barriers and genomic interrelationships are prerequisites for future success by direct transfer. Until this happens, developing transformation methodologies might facilitate the transfer of end-use quality genes from wild to cultivated barley.

C. Tissue Culture, Genetic Transformation, and Doubled-Haploid Breeding

As with wheat, nontraditional methods for introducing improved quality genes into barley are becoming reality. Presently, barley plants can be regenerated from anthers, embryos, or protoplasts derived from embryonic tissue (Mannonen et al., 1994). Used in conjunction with particle bombardment, the regenerating tissues can serve as targets for genetic transfer of quality genes. Aside from that, they sometimes exhibit somaclonal variation (Ullrich et al., 1991). This variation may be detrimental to the overall breeding goal or it may be useful for quality improvement. For example, Bregitzer et al. (1995) compared tissue-culture derived genotypes to normal barley for malt quality. They found that the tissue-culture derived genotypes generally had lower quality with elevated protein, reduced kernel weight, and reduced malt extract. Conversely, Bright et al. (1982) took advantage of in vitro variation by culturing barley embryos in the presence of threonine and lysine. Selections from surviving embryos had altered amino acid biosynthesis and produced grain with approximately 6% higher threonine and methionine levels. Such mutants would be desirable as feed-type barleys, but they generally have poor vigor and low yield.

Genetic transformation of barley can be achieved by *Agrobacterium* infection, electroporation/chemical-mediated DNA uptake, or particle bombardment, but the most successful means appears to be particle bombardment (Mannonen et al., 1994). A quality gene candidate for transformation is a thermostable β-glucanase coding sequence from a fungus, *Trichoderma reesei*. This gene has been successfully inserted into barley (Mannonen et al., 1994), and Lee et al. (1989) demonstrated that certain promoter/gene constructs can be expressed in aleurone cells. Transgenic plants with β-glucanase and α-amylase genes have already been patented (Mannonen et al., 1994).

Hybridization with *Hordeum bulbosum* L. ($2n = 14$ and $2n = 28$) has facilitated the production of haploids in barley genetics and breeding for years (Kasha & Reinbergs, 1981). After hybridizing a barley species to *H. bulbosum* L., the *bulbosum* chromosomes are selectively eliminated during embryo formation and haploids are produced (Subrahmanyam & Kasha, 1973). As with other haploid techniques, the *H. bulbosum* technique has its breeding advantages. It can be used to fix polymorphic genotypes in a homozygous state for genetic mapping (Graner et al., 1991). It also can facilitate a more rapid establishment of homozygosity in breeding programs (Cattivelli et al., 1994). After the chromosomes are doubled, the population theoretically represents a sampling of the hybrid gametes. Therefore, it is one way for a breeder to determine the potential value of a barley hybrid combination. Pickering and Devaux (1992) described the breeding uses of barley doubled-haploids, including the release of commercial cultivars from DH breeding.

Improvement of barley quality by DH breeding will likely result from identifying mutants for increased enzyme activity or protein levels. Unlike euploid cell culture, recessive and dominant mutants are detectable in haploids. On the other hand, compared to anther culture, somaclonal variation is reduced in *H. bulbosum*-produced haploids (Devaux et al., 1993; Snape et al., 1988). It would be interesting to compare tissue culture, anther culture, and the *H. bulbosum* technique for progress in improving a barley quality trait.

V. OAT GENETIC DIVERSITY FOR QUALITY

Oat exists as diploid ($2n = 14$), tetraploid ($2n = 4x = 28$), or hexaploid ($2n = 6x = 42$) species (*Avena* sp. L.), and they may have either black, grey, red, yellow or white hulls. However, the white-hulled hexaploid species, *Avena sativa* L. (AACCDD), is most widely cultivated (Youngs & Forsberg, 1987). Cultivated oat is used in a wide-variety of ways. Although approximately 75% of world oat production is used as animal feed (Murphy & Hoffman, 1992), grain also is processed into "steel-cut" or, more commonly, "rolled" oat for human consumption (Burnette et al., 1992). The versatility of the crop is evident by use of both vegetative and grain portions of the oat plant for animal feed, use of both whole oat grain and flour for human food, and even use of milling byproducts for industrial purposes (Webster, 1986).

The importance of oat as an animal feed means that many growers do not directly sell their oat products. Rather, they utilize oat on the farm to nourish livestock and improve livestock quality. The oat grain without the hull, the "groat", provides animals with protein, energy, minerals, and vitamins (Youngs & Forsberg, 1987). Compared to the grains of other crops, groats are an excellent substitute or supplement in the diets of many animals, and they are particularly desirable in the starting diets of nonruminants (Schrickel et al., 1992; Youngs & Forsberg, 1987). The fibrous oat hull is not of consequence to ruminants, but it limits the nutritive value of oat as a feed for nonruminant animals. Consequently, it is more nutritious to feed nonruminants with oat that has a high groat percentage, or oat that has their kernels easily threshed from the hull, "naked oat". Doyle and Valentine (1988) found naked oat to have the highest digestible energy compared to other cereal grains; however, naked oat also was considerably lower yielding compared to oat with hulls. Doyle and Valentine (1988) suggest a high-yielding, naked oat would be an economical alternative to expensive maize and soybean (*Glycine max* L.) feeds in Europe. Similar to barley, the chemical properties determining high-quality feed groats are: high protein content balanced for amino acids, and a high starch and lipid content (Youngs & Forsberg, 1987).

Oat is the raw material for a wide variety of human food products (McKechnie, 1983), and as the additional health benefits of oat to humans were discovered (Shinnick et al, 1991; Anderson & Lin Chen, 1986), use of groats in human foods expanded (Burnette et al., 1992). Use of oat as a fat substitute in some foods (Inglett & Grisamore, 1991) is a good example of its expanded use. After the seed is cleaned and the hull is removed from the kernel, groats are steamed and rolled into flakes or ground to produce flour (Deane & Commers, 1986). High-quality oat for for this process has a high groat-to-hull ratio, and they also have a high test weight. On a nutritional level, high-quality oat has a high protein content with a well-balanced amino acid composition. They also have a high starch and lipid content that does not lead to a bitter taste or spoilage when the grain is stored (Forsberg & Reeves, 1992).

Compared to kernels of the other cereal small grains, the oat kernel has some unique physical and biochemical features. For example, the oat bran is higher in soluble fiber than other cereal small grains (Peterson, 1992). This is due to the high β-glucan content of the bran, which has been linked to reducing blood cholesterol in humans (Davidson et al., 1991). Groats also contain a higher concentration of proteins than other cereal small grains, and more importantly, oat protein is the most nutritious. This is because there is a high lysine-containing globulin fraction present in oat protein (Lockhart & Hurt, 1986). Although many cereal proteins are involved in the allergic reaction of celiac disease, oat proteins only induce symptoms in the most severe celiac cases (Baldo & Wrigley, 1984). Finally, oat oil has valuable antioxidant properties, a desirable feature in limiting the breakdown of food products during cooking (Dimberg et al., 1993; Kahlon, 1989).

Unique properties attributable to oat have led to use of processed oat products and byproducts in nonfood applications. In cosmetics, oatmeal or oat flour has been used in facial masks, soaps, and shampoos (Webster, 1986). Because oat hulls have a high pentosan content, oat hulls, a byproduct of processed groats, are a valuable raw material in the production of furfural. Furfural has its uses in the petroleum and rubber industry (Caldwell & Pomeranz, 1973). Other uses for oat products and byproducts have been envisioned, although they have not yet been commercially developed. Oat starch grains have properties that make oat attractive as a raw material in paper manufacture (Wilhelm et al., 1989). Wu (1990) suggested that oat would be a good source for the production of ethanol and protein-rich food products; whereas, Rosenberg et al. (1978) suggested that oat hulls be used as a substrate to produce nutritious fungal protein. An anticaries feature of oat hulls has been well-documented (Madsen, 1981). Thus, there might someday be manufacture of an oat food additive to prevent tooth decay in diets that warrant it (Lorenz, 1984).

Opportunities exist to breed for animal feed use, human food use, and industrial use of oat. However, many of the same groat characteristics determine the quality of different end-use products. Consequently, whether a breeder is to develop a cultivar for rolled oat production, swine production, or furfural production, genetic diversity is required for groat protein content, starch and lipid content, and β-glucan content. Oat germplasm collections, consisting of both wild species and cultivars, contain genotypes that exhibit variation for these traits (Wesenberg et al., 1992).

A. Groat Endosperm Proteins

The endosperm storage proteins of oat are significantly different than those of other cereal small grains. First, overall seed protein content is consistently higher than in wheat or barley (Peterson & Brinegar, 1986). Second, although threonine and lysine are still limiting, the amino acid composition is more balanced by comparison; therefore, groat proteins are more nutritious. Finally, unlike wheat and barley, when groat protein content increases, lysine proportions do not significantly change, apparently because most of the protein is contained in the lysine-rich globulin fraction (Peterson & Brinegar, 1986). This is important because it means that nearly the full nutritive value of oat can be retained when breeding for high protein.

A survey of oat cultivars in the world collection showed that protein content varied from approximately 12 to 24% (Robbins et al., 1971). Frey (1977) also reported that groat protein varied considerably among oat cultivars. The result of breeding for high protein has been the release of cultivars with approximately a 16% higher protein content compared to normal cultivars (Forsberg & Reeves, 1992). Wild oat species also are important sources of high protein. Pomeranz et al. (1973) examined 11 oat species plus *A. sterilis* L. ($2n = 42$, AACCDD) for protein content, and found that on average all species had higher protein content than *A.*

sativa. One wild tetraploid oat species, *A. murphyi* Ladiz. ($2n = 28$, AACC), has a seed protein content as high as 27% (Ladizinsky & Fainstein, 1977). *Avena fatua* L. ($2n = 42$, AACCDD)—a wild oat considered a weed in the western USA—also has potential as a source of high groat protein (Rines et al., 1980). The value of wild oat germplasm to high-protein breeding has been repeatedly demonstrated (Reich & Brinkman, 1984; Takeda & Frey, 1979; Pomeranz, 1975).

B. Groat Starch and Lipids

Oat starch and lipids are important sources of energy to animals fed groats. Oat genotypes may contain from 42 to 64% starch in their groats (MacArthur & D'Appolonia, 1979; Paton, 1977; Youngs, 1974) and from 2 to 12% lipids (Frey & Hammond, 1975; Brown & Craddock, 1972). Genetic variation also exists for the saturated and polyunsaturated fatty acid composition of oat lipids (Saastamoinen et al., 1989; Youngs et al., 1977). Polyunsaturated fatty acids are essential in the diets of monogastric animals, but they also may be responsible for the bitter taste that develops in oil-containing foods that have spoiled (Youngs & Forsberg, 1987). To improve the end-use value of groats for animal feed and human food, breeders are sometimes faced with simultaneously altering seed protein and lipid content. Youngs and Forsberg (1979) showed that this was possible. They were able to identify six oat genotypes with high seed protein (≈21%) and high oil (≈10%) and four with high protein (≈19%) and low oil (≈7%). Frey and Hammond (1975) determined that oat needed to have minimum concentrations of 21% protein and 17% oil to become commercially competitive for end-use with soybean in Iowa. Therefore, at least for protein content, there is potential to breed an oat that would be competitive.

C. Groat β-Glucans

When included in human diets, plant fiber, and specifically water-soluble fiber, lowers blood cholesterol (Anderson & Lin Chen, 1986). Oat fiber has this quality because of groat β-glucans (Wood et al., 1989; Klopfenstein, 1988). Oat β-glucans have rheological properties that also make them promising as thickening agents in foods (Wood, 1986). On the other hand, β-glucans also reduce oat feed quality for poultry (Anderson et al., 1978).

The genetic component of oat β-glucan content is significant (Peterson, 1991), but its low heritability indicates it is highly influenced by environment (Humphreys & Mather, 1996). Cho and White (1993) analyzed 212 oat genotypes, including *A. sterilis* L. accessions, for β-glucan content and found significant genetic differences between them. Prentice et al. (1980) described large differences in β-glucan content between two oat cultivars. In fact, oat β-glucan content can vary from a low of 2.5% (Anderson et al., 1978) to a high of 6.6% (Stuart et al., 1987).

Furthermore, the genetic component for β-glucan content is very predictable at these extremes (Lim et al., 1992). In a study of three oat nurseries over 4 yr, Peterson et al. (1995) found genetic variation for β-glucan content sufficient to make breeding for higher β-glucan content feasible. In addition, they concluded that selecting for higher β-glucan content would not negatively impact breeding for other traits. Holthaus et al. (1996) determined that β-glucan content was polygenic in inheritance and that genes with additive effects were involved. They also concluded that breeders could effectively select for lower or higher levels of β-glucan content in oat.

D. Groat Yield and Feed Value

A high proportion of groat-to-hull is an indication of high milling yield (Forsberg & Reeves, 1992). Though the hull is not very nutritious in human foods and it limits digestibility in nonruminant animals, it protects the oat kernel before and after harvest (Youngs & Forsberg, 1987), and it is a good source of dietary fiber (Rasper, 1979). Because of this, a breeder's goal is often to minimize the hull, but not to eliminate it. Marshall and Kolb (1986) found significant cultivar variation for groat fraction over 2 yr at two locations. In addition, they determined that groat fraction, and not grain test weight, was the best indicator of digestible energy for animals. Groat percentage differed markedly between Minnesota cultivars released since 1923 (Wych & Stuthman, 1983), although there was no correlation with date-of-release. Pomeranz et al. (1979) also reported genetic differences for groat-to-hull ratio. Few hulless oat cultivars have been released in the USA compared to Europe (Brown & Patterson, 1992). But, there is genetic variability for the trait (Marshall & Shaner, 1992), and naked oat is potentially quite valuable for animal feed (Doyle & Valentine, 1988). If considered for feed, some hulless oats are similar to maize in value (Burrows, 1986a).

VI. BREEDING OAT FOR ADDED QUALITY

According to Brown and Patterson (1992), adding quality to oat is becoming a higher breeding priority. If research continues to link oat and oat products to improved human health, this breeding emphasis is not likely to change (Forsberg & Shands, 1989). As has been illustrated, sufficient genetic variability exists to enable breeders to meet the demands of this changing market.

When comparing modern oat cultivars (cultivars released after 1965) to old (cultivars released from 1950–1965), worldwide grain yields are improved by some 30 to 40% (Peltonen-Sainio, 1994); however, much quality improvement remains if the potential of oat, both as a feed and a food crop, is to be maximized (Burrows, 1986b). Improving quality without limiting yields will be a major challenge for oat breeders. Generally, modern oat cultivars appear to maintain their photosynthetic tissue longer and have the capability to withstand stress environments better than old cultivars (Lynch & Frey, 1993). But, Schipper and Frey (1992) concluded that increasing the protein and oil content in oat resulted in a higher demand for photosynthate. Thus, one wonders at what cost to yield will added quality have? Whatever the cost, augmenting classical breeding techniques with newer biotechnologies should improve the difficult task of selecting for high-yielding genotypes that also have added quality. In turn, this would enable oat breeders to maintain or improve upon existing cultivar release time frames.

A. Marker-Assisted Selection and Breeding

Because cultivated oat is a hexaploid, attempts to map molecular markers in oat have encountered the same difficulties faced when trying to map any large, complex genome. In a complex genome, the primary problem is identifying sufficient polymorphisms to map useful markers for selection. Nevertheless, one map has been published for oat (O'Donoughue et al., 1995). Furthermore, the oat breeding benefits derived from such efforts may justify additional resources applied to mapping. One approach for cultivated oat has been to first map relatives with simpler genomic compositions. Since the hexaploid oat has three different but homoeologous (similar) genomes (Thomas, 1992), diploid progenitors and putative progenitors have chromosomes and genes that also are homoeologous to cultivated oat. Additionally, having arisen from a common ancestor, related grass genera have homoeology to cultivated oat (Moore et al., 1993). Consequently, genes of related species and grass genera likely map to the same hexaploid oat chromosomes, and the genes generally have an order similar to those in cultivated oat (Van Deynze et al., 1995). In this way, the cultivated oat genome can be indirectly mapped by extrapolating mapping results of simpler genomes to the polyploid, as was demonstrated by O'Donoughue et al. (1992).

Evaluating genetic relatedness among elite oat parents would benefit breeders, especially if the evaluation was independent of the environment and could be conducted without a requirement for pedigree information. Moser and Lee (1994) applied RFLP analysis to crosses of nine elite oat lines. They concluded that sufficient polymorphisms were detectable for breeders to make good use of the technique. Importantly, given the value of *A. sterilis* L. as germplasm (Wesenberg et al., 1992), RFLPs adequately characterized genetic relationships between *A. sterilis* accessions (Goffreda et al., 1992; Beer et al., 1993). RAPDs also have been used to effectively describe relationships between *A. sterilis* accessions (Heun et al., 1994).

On the molecular level, oat seed protein genes and their expression mechanisms are becoming clearer. The production by oat of higher amounts of seed globulins relative to prolamines, rare among cereals (Peterson & Brinegar, 1986), is due to post-transcriptional regulation (Boyer et al., 1992; Fabijanski & Altosaar, 1985). Globulin genes are redundant in the oat genome, and they are homologous to similar genes in rice (Shotwell et al., 1988). Because of several differences between oat globulin and prolamine mRNA sequences, translation of globulin genes may be more efficient (Shotwell et al., 1990).

Future breeding schemes to add quality to oat may capitalize on this understanding of protein gene expression. Globulin genes for oat have already been cloned (Tanchak et al., 1995). Molecular geneticists and breeders may be able to enhance globulin production by modifying the existing gene promoters or by coupling the transcriptional gene seqeunces to more efficient promoters. Because many copies of the globulin genes are present in oat, gene dosages might be manipulated to increase globulin expression. Finally, since globulin genes have been cloned, they could possibly be inserted into different oat genotypes to assess the best "genetic environment" to maximize globulin expression. Because coding regions of oat globulin genes can vary depending on the cultivar (Tanchak et al., 1995), gene expression may depend on the genetic background into which it is inserted.

B. Cytogenetic Manipulations

The nature of genomic relationships between diploid, tetraploid, and hexaploid oat is similar to that between diploid and polyploid wheat. There is homoeology between species having the A, C, and D genomes in common, but in polyploids, there also is a diploidization mechanism that inhibits pairing between homoeologous chromosomes (Thomas, 1992). Depending on the related species used, this has implications for transferring desirable quality genes to cultivated oat. For example, adding genes from wild species with A, C, or D-genome chromosomes can be accomplished by recombination with homologous cultivated oat chromosomes. Thomas et al. (1980) illustrated this when they incorporated higher seed protein into *A. sativa* from *A. maroccana* Gand. ($2n = 28$, AACC), a related wild tetraploid. Similarly, when cultivated oat was hybridized to *A. sterilis*, which primarily exhibits homologous pairing with *A. sativa* (Thomas, 1992), Takeda and Frey (1979) were able to transfer higher seed protein.

With more distantly related species, because of a lack of homologous chromosome recombination, transferring genes to cultivated oat sometimes requires overcoming hybridization barriers or promoting homoeologous pairing. Although homoeologous pairing has not been used to directly enhance oat quality, it has been used, for example, to transfer mildew resistance to cultivated oat (Thomas et al., 1980), and there appears to be no technical reason why it also could not be used to transfer genes for quality.

To be compatible with a breeder's goal to add quality while maintaining other desirable traits, minimal chromatin should be introgressed from related oat species. This is partly why no alien oat chromosome addition or substitution line has been released as a cultivar; whereas, chromosome translocations are present in cultivars (Thomas, 1992). As quality genes are transferred from related species to cultivated oat, monitoring the amount of chromatin involved continues to be important. C-banding oat chromosomes (Hutchinson & Postoyko, 1986) and labelling them in situ (Chen & Armstrong, 1994) aid the breeder in this respect, since both techniques enable the detection of introgressed chromosome segments. In fact, genome-specific oat probes not only enable a breeder to follow the introgression of chromatin into cultivated oat, but they also better define the relationship of related species to oat. Fominaya et al. (1995) showed that a C-genome DNA probe could be used to identify chromosome rearrangements in several related oat species.

Identification of genes responsible for the production of β-glucans in oat is the focus of one effort to improve cultivated oat quality (Hardin, 1990). With the complexity of the cultivated oat genome, this may prove a bit like looking for "a needle in a haystack". However, once these genes are found, cloning them will undoubtedly be the first step to understanding their mode of action and to inserting them into other genotypes. A promising technique on both these fronts is chromosome microdissection. Chen and Armstrong (1995) isolated a single oat chromosome by microdissection, enriched the DNA for low-copy sequences, and cloned a portion of the enriched DNA. Future molecular breeding to add quality to oat also will likely require handling relatively large portions of DNA and some way to limit the search for the critical genes.

C. Tissue Culture, Genetic Transformation, and Doubled-Haploid Breeding

Breeders have made great strides in adding quality to oat by selecting among segregants from *Avena* species hybrids. To date, the genetic variation generated from sexual hybridizations between genotypes within *A. sativa* and between *A. sativa* and related species has proven sufficient. However, in light of food industry demands for a wide variety of oat products (Nelson & Eilertson, 1996), one might justifiably argue that it is essential for breeders to broaden their germplasm base by utilizing secondary and tertiary gene pools. For breeders who want to use these gene pools, new biotechnologies offer considerable promise.

Oat tissue culture produces genetic variation and, with the proper conditions, calli into which genes from distantly related genera and species might be inserted. Therefore, it not only can serve as a source of useful variability, it also can serve as a means to overcome possible crossability barriers. Oat calli were first produced by culturing immature or mature oat embryos (Cummings et al., 1976; Lorz et al., 1976); however, it is now possible to produce calli from immature oat leaf tissue (Chen et al., 1995). Often, the genetic variation associated with oat tissue culture is the result of chromosome aberrations (Rines et al., 1992; Johnson et al., 1987). However, Dahleen et al. (1991) identified tissue-culture derived oat families without chromosome aberrations. One family exhibited genetic variation for up to six agronomic traits compared to controls, and seed protein was one trait for which there was significant variation. Rines et al. (1992) concluded that it was not possible to predict the future value of tissue-culture-derived variation, but it does look more promising than the same type variation observed in other crops (Dahleen et al., 1991). Oat breeding programs may someday focus more on selecting for higher groat protein from somaclonal variants.

As was the case with wheat and barley, inserting genes from distantly related genera and species is becoming more practical for oat. Somers et al. (1992) reported transforming oat tissue with the *β-glucuronidase* gene from *Escherichia coli* and the *Streptomyces hygroscopicus bar* gene with a cauliflour mosaic virus promoter. They discovered that both genes were expressed in seeds of regenerates. Although Somers et al. (1992) produced transgenic oats by microprojectile bombardment, one can expect other DNA delivery mechanisms (e.g., *Agrobacterium*-mediated transfer and protoplast electroporation) to someday be applicable to oat.

Anther culture, a feasible option for producing haploid plants and exploiting possible somaclonal variation for quality in most cereals, has been used very little in oat (Rines et al., 1992). Plant regeneration from oat calli is difficult and highly dependent on genotype (Rines, 1983). However, pollination of oat with maize and other distantly related genera followed by embryo rescue has emerged as another approach to producing haploids. As Rines et al. (1992) noted, this technique has the advantage of being rather genotype-independent. Therefore, elite oat germplasm, and not only easily cultured germplasm, can be used to produce haploids. Rines and Dahleen (1990) obtained 14 haploid oat plants from 3300 maize-pollinated oat florets. In addition, progeny from the selfed haploid plants were both aneuploid and euploid. Among the aneuploids produced in this way, monosomics may prove to be very valuable cytogenetic stocks (Phillips et al., 1995). One can envision using the monosomics produced from crosses with maize to locate any number of oat quality loci to a specific chromosome. Another interesting outcome to this type of haploid production procedure is that not all chromosomes of the pollinator are eliminated after fertilization. Matzk (1996) found that a few plants produced from oat pollinated with pearl millet (*Pennisetum glaucum* L.) and maize retained one to four chromosomes of the pollinator. Phillips et al. (1995) described maize chromosome additions to the oat genome and presented evidence that oat/maize chromosome translocations exist.

Although a complete understanding of the expression of maize genes in oat appears years removed from the present, this technique may someday offer breeders a way to directly incorporate maize quality genes into oat. If this transfer technique is perfected and gene expression is understood, breeders could tap a virtual storehouse of quality traits in maize. For example, maize geneticists and breeders have significantly modified seed protein content and quality, carbohydrate metabolism, oil content and quality, and carotenoid content (Glover & Mertz, 1987).

VII. SUMMARY

Adding value to crops has been the goal of plant breeders since the first unique plant was selected. In order to illustrate new breeding methods and biotechnologies, this chapter has highlighted adding quality to cereal small grains. Unquestionably, this represents only a miniscule sampling of the type of added-value breeding that is taking place, and cereal breeders are adding value in other ways, most notably by improving disease resistance. In addition, breeders of other crops are improving processing traits, shelf life, and the way crops taste, smell, and feel (Jones, 1992).

One could categorize the methods and biotechnologies described in this chapter as "genetic engineering". To do so would be appropriate, but one should realize that this "genetic engineering" is no substitute for the historic genetic engineering, plant breeding, that has already added considerable value to crops. Rather, the newer biotechnologies all augment sexual hybridization of crops coupled with efforts to select whole-plants with added-value. This favorable breeding/biotechnology relationship is not likely to change anytime soon (Duvick, 1996; Jones, 1992).

Compared to just a few years ago, many more tools are available to add value to crops. In embarking on any effort to add value to a crop, a breeder is faced with a series of decisions. First, it must be determined if the trait has economic value in the marketplace. Second, available genetic diversity for the trait must be assessed, and finally, a decision on what type of hybridization and selection approach to take must be made. The economic value of small grains has traditionally been tied to food and feed production (Bashuk, 1988). Because of health interests, there is a general desire to incorporate more grains into the human diet. As a result, the cereal grain specialty market is growing (Nelson & Eilertson, 1996). Moreover, a move towards using renewable, biodegradable resources in industry and changing economic climates will encourage the breeding of cereal small grains for nonfood purposes. Therefore, with the need for an even wider spectrum of end-use products, cereal breeders are likely to choose more of the newer biotechnologies to assess genetic variability and make selections so that their programs are as efficient and productive as possible.

Ultimately, public perception of crops genetically engineered to have added-value may determine the success of breeding efforts which incorporate newer biotechnologies (Duvick, 1996; Barefoot et al., 1994). However, it behooves plant breeders to continue using marker-assisted selection, molecular cytogenetics, genetic transformation, tissue culture, and other newer biotechnologies, because their use in the applied science of breeding also adds to a basic understanding of added-value genetics. The result for end-use consumers will be a continual improvement of product quality.

REFERENCES

Abdel-Aal, E.-S.M., P. Hucl, and F.W. Sosulski. 1995. Compositional and nutrient characteristics of spring einkorn and spelt wheats. Cer. Chem. 72:621–624.

Ahokas, H., and L. Naskali. 1990. Variation of α-amylase, β-amylase, β-glucanase, pullulanase, proteinase and chitinase activities in germinated samples of the wild progenitor of barley. J. Inst. Brew. 96:27–31.

Allison, M.J., and R.P. Ellis. 1973. The inheritance of β-amylases in developing barley grain. Biochem. Gen. 10:165–173.

Anamthawat-Jonsson, K., and J.S. Heslop-Harrison. 1993. Isolation and characterization of genome-specific DNA sequences in Triticeae species. Mol. Gen. Genet. 240:151–158.

Anderson, J.A., M.E. Sorrells, and S.D. Tanksley. 1993. RFLP analysis of genomic regions associated with resistance to preharvest sprouting in wheat. Crop Sci. 33:453–459.

Anderson, J.W., and W.-J. Lin Chen. 1986. Cholesterol-lowering properties of oat products. p. 309–333. *In* F.H. Webster (ed.) Oats: Chemistry and technology. AACC Publ., St. Paul, MN.

Anderson, M.A., J.A. Cook, and J.A. Stone. 1978. Enzymatic determination of 1,3:1,4-β-glucans in barley grain and other cereals. J. Inst. Brew. 84:233–239.

Anderson, M.K., and E. Reinbergs. 1985. Barley breeding. p. 232–268. *In* D.C. Rasmusson (ed.) Barley. Agron. Monogr. 26. ASA, Madison, WI.

Anjum, F.M., and C.E. Walker. 1991. Review on the significance of starch and protein to wheat kernel hardness. J. Sci. Food Agric. 56:1–13.

Appels, R., P. Reddy, C.L. McIntyre, L.B. Moran, O.H. Frankel, and B.C. Clarke. 1989. The molecular-cytogenetic analysis of grasses and its application to studying relationships among species of the Triticeae. Genome 31:122–133.

Baenziger, P.S., R.L. Clements, M.S. McIntosh, W.T. Yamazaki, T.M. Starling, D.J. Sammons, and J.W. Johnson. 1985. Effect of cultivar, environment, and their interaction and stability analyses on milling and baking quality of soft red winter wheat. Crop Sci. 25:5–8.

Baenziger, P.S., D.T. Kudirka, G.W. Schaeffer, and M.D. Lazar. 1984. The significance of doubled haploid variation. p. 385–414. *In* J.P. Gustafson (ed.) Gene manipulation in plant improvement. Proc. 16th Stadler Genet. Symp., Columbia, MO. Plenum Press, New York.

Balaravi, S.P., H.C. Bansal, B.O. Eggum, and S. Bhaskaran. 1976. Characterization of induced high protein and high lysine mutants in barley. J. Sci. Food Agric. 27:545–552.

Baldo, B.A., and C.W. Wrigley. 1984. Allergies to cereals. p. 289–356. *In* Y. Pomeranz (ed.) Advances in cereal science and technology. Vol. 6. AACC, St. Paul, MN.

Bamforth, C.W. and A.H.P. Barclay. 1993. Malting technology and the uses of malt. p. 297– 254. *In* A.W. MacGregor and R.S. Bhatty (ed.) Barley: Chemistry and technology. AACC, St. Paul, MN.

Barefoot, S.F., R.N. Beachy, and M.S. Lilburn. 1994. Labeling of food-plant biotechnology products. Cer. Foods World 39:760–765.

Barrett, J., J.F. Clapperton, D.M. Divers, and H. Rennic. 1973. Factors affecting wort filtration. J. Inst. Brew. 79-407–413.

Bassett, L.M., R.E. Allan, and G.L. Rubenthaler. 1989. Genotype x environmental interactions on soft white winter wheat quality. Agron. J. 81:955–960.

Becker, J., and M. Heun. 1995. Barley microsatellites: Allele variation and mapping. Plant Mol. Biol. 27:835–845.

Becker, J., P. Vos, M. Kuiper, F. Salamini, and M. Heun. 1995. Combined mapping of AFLP and RFLP markers in barley. Mol. Gen. Genet. 249:69–73.

Beckmann, J.S., and M. Soller. 1990. Toward a unified approach to genetic mapping of eukaryotes based on sequence tagged microsatellite sites. Biotechnology 8:930–932.

Bedo, Z., I. Karsai, V. Gy, and L. Lang. 1992. Breadmaking quality of doubled haploid lines derived from wheat anther culture. J. Genet. Breed. 46:263–268.

Beer, S.C., J. Goffreda, T.D. Phillips, J.P. Murphy, and M.E. Sorrells. 1993. Assessment of genetic variation in *Avena sterilis* using morphological traits, isozymes, and RFLPs. Crop Sci. 33:1386–1393.

Bennetzen, J.L., and M. Freeling. 1993. Grasses as a single genetic system: Genome composition, collinearity and compatability. Trends Genet. 9:259–261.

Berbigier, A., J.B. Denis, L. Jestin, and R. Scriban. 1981. Genetic and environmental variation of β-glucanase activity in barley breeding material. p. 204–210. *In* Barley breeding. Vol. 5. Proc. 4th Int. Barley Genet. Symp., Edinburgh, UK. 22–29 July. Redwood Burn Ltd., UK.

Berglund, P.T., C.E. Fastnaught, and E.T. Holm. 1992. Food uses of waxy hull-less barley. Cer. Foods World 37:709–714.

Berman, M., M.L. Bason, F. Ellison, G. Peden, and C.W. Wrigley. 1996. Image analysis of whole grains to screen for flour-milling yield in wheat breeding. Cer. Chem. 73:323–327.

Betschart, A.A. 1989. Nutritional potential and limitations of wheat. p. 85–101. *In* Y. Pomeranz (ed.) Wheat is unique: Structure, composition, processing, end-use properties, and products. AACC, St. Paul, MN.

Bhatty, R.S. 1986.The potential of hulless barley. A review. Cereal Chem. 63:97–103.

Bhatty, R.S. 1987. Miller yield and flour quality of hulless barley. Cer. Foods World 32:268–272.

Bhatty, R.S. 1993. Nonmalting uses of barley. p. 355–417. *In* A.W. MacGregor and R.S. Bhatty (ed.) Barley: Chemistry and technology. AACC, St. Paul, MN.

Bhatty, R.S., J.D. Berdahl, and G.I. Christison. 1975. Chemical composition and digestible energy of barley. Can. J. Anim. Sci. 55:759–764.

Bhatty, R.S., G.I. Christison, and B.G. Rossnagel. 1979. Energy and protein digestibilities of hulled and hulless barley determined by swine-feeding. Can. J. Anim. Sci. 59:585–588.

Bhatty, R.S., and B.G. Rossnagel. 1979. Oil content of Riso 1508 barley. Cer. Chem. 56:586.

Bhatty, R.S., and B.G. Rossnagel. 1980. Lipids and fatty acid composition of Riso 1508 and normal barley. Cer. Chem. 57:382–386.

Bietz, J.A. 1987. Genetic and biochemical studies of nonenzymatic endosperm proteins. p. 215–241. *In* E.G. Heyne (ed.) Wheat and wheat improvement. 2nd ed. ASA, Madison, WI.

Bietz, J.A., and G.L. Lookhart. 1996. Properties and non-food potential of gluten. Cer. Foods World 41:376–382.

Bittel, D.C., G. Hueros, N. Jouve, and J.P. Gustafson. 1991. Changes in expression of seed storage protein genes effected by chromosome 1D of wheat. Genome 34:845–848.

Blake, T., L. Dahleen, J. Dvorak, P. Gustafson, P. Hayes, D. Hoffman, K. Kasha, S. Procunier, W. Kim, A. Kleinhofs, N. Lapitan, A. LaRoche, M. A. Saghai Maroof, S. Molnar, G. Fedak, G. Scoles, R. Skadsen, M. Sorrells, and S. Tanksley. 1991. An RFLP map of barley. p. 245–248. *In* Proc. 6th Int. Barley Genet. Symp., Vol. 1, Helsingborg, Sweden.

Blechl, A.E. 1993. Genetic transformation: The new tool for wheat improvement. Cer. Foods World 38:846–847.

Blechl, A.E., and O.D. Anderson. 1996. Expression of a novel high-molecular-weight glutenin subunit gene in transgenic wheat. Nature Biotech. 14:875–879.

Boyer, S.K., M.A. Shotwell, and B.A. Larkins. 1992. Evidence for the translational control of storage protein gene expression in oat seeds. J. Biol. Chem. 267:17 449–17 457.

Braaten, M.O., K.L. Lebsock, and L.K. Sibbitt. 1962. Intergeneration relations of physical properties of dough and carotenoid pigment content in durum wheat. Crop Sci. 2:277–281.

Bregitzer, P., M. Poulson, and B.L. Jones. 1995. Malting quality of barley lines derived from tissue culture. Cer. Chem. 72:433–435.

Briggle, L.W., W.T. Yamazaki, and W.D. Hanson. 1968. Heritability of three soft wheat quality characters in the F2 and F3 generations. Crop Sci. 8:283–285.

Bright, S.W.J., B.J. Miflin, and S.E. Rognes. 1982. Threonine accumulation in the seeds of a barley mutant with an altered aspartate kinase. Biochem. Genet. 20:229–243.

Brown A.H.D., and J.V. Jacobsen. 1982. Genetic basis and natural variation of α-amylase isozymes in barley. Genet. Res. Camb. 40:315–324.

Brown, C.M., and J.D. Craddock. 1972. Oil content and groat weight of entries in the world oat collection. Crop Sci. 12:514–515.

Brown, C.M., and F.L. Patterson. 1992. Conventional oat breeding. p. 614–656. *In* H.G. Marshall and M.E. Sorrells (ed.) Oat science and technology. ASA and CSSA, Madison, WI.

Burger, W.C., and D.E. LaBerge. 1985. Malting and brewing quality. p. 367–401. *In* D.C. Rasmusson (ed.) Barley. Agron. Monogr. 26. ASA, Madison, WI.

Burger, W.C., D.M. Wesenberg, J.E. Carden, and P.E. Pawlisch. 1979. Protein content and composition of Karl and related barleys. Crop Sci. 19:235–238.

Burnette, D., M. Lenz, P.F. Sisson, S. Sutherland, and S.H. Weaver. 1992. Marketing, processing, and uses of oat for food. p. 248–263. *In* H.G. Marshall and M.E. Sorrells (ed.) Oat science and technology. ASA and CSSA, Madison, WI.

Burrows, V.D. 1986a. Tibor oat. Can. J. Plant Sci. 66:403–405.

Burrows, V.D. 1986b. Breeding oats for food and feed: Conventional and new techniques and materials. p. 13–46. *In* F.H. Webster (ed.) Oat: Chemistry and technology. AACC, St. Paul, MN.

Bushuk, W. 1986. Wheat: Chemistry and uses. Cer. Foods World 31:218–226.

Caldwell, E.F., and Y. Pomeranz. 1973. Oats. p. 393–411. *In* Y. Pomeranz (ed.) Industrial uses of cereals. AACC, St. Paul, MN.

Campbell, G.L., B.G. Rossnagel, H.L. Classen, and P.A. Thacker. 1989. Genotypic and environmental differences in extract viscosity of barley and their relationship to its nutritive value for broiler chickens. Anim. Feed Sci. Technol. 26:221–230.

Cantrell, R.G. 1987. Breeding and genetics of durum wheat. p. 11–40. *In* J. Janick (ed.) Plant breeding reviews. Vol. 5. Van Nostrand Reinhold Co., New York.

Carver, B.F. 1994. Genetic implications of kernel NIR hardness on milling and flour quality in bread wheat. J. Sci. Food Agric. 65:125–132.

Cattivelli, L., G. Delogu, V. Terzi, and A.M. Stanca. 1994. Progress in barley breeding. p. 95–181. *In* G.A. Slafer (ed.) Genetic improvement of field crops. Marcel Dekker, Inc., New York.

Chalmers, K.J., U.M. Barua, C.A. Hackett, W.T.B. Thomas, R. Waugh, and W. Powell. 1993. Identification of RAPD markers linked to genetic factors controlling the milling energy requirement of barley. Theor. Appl. Genet. 87:324–320.

Chatterjee, S.R., N.S. Verma, S.C. Gulati, J.S. Bakshi, and Y.P. Abrol. 1975. Identification of barley strains with improved amino acid balance. Euphytica 24:725–730.

Chen, Q., and K. Armstrong. 1994. Genomic in situ hybridization in Avena sativa. Genome 37: 607–612.

Chen, Q., and K. Armstrong. 1995. Characterization of a library from a single microdissected oat (*Avena sativa* L.). Genome 38:706–714.

Chen, H., G. Xu, D.C. Loschke, L. Tomaska, and B.G. Rolfe. 1995. Efficient callus formation and plant regeneration from leaves of oats (*Avena sativa* L.). Plant Cell Rep. 14:393–397.

Cho, K.C., and P.J. White. 1993. Enzymatic analysis of β-glucan content in different oat genotypes. Cer. Chem. 70:539–542.

Ciaffi, M., L. Dominici, D. Lafiandra, and E. Porceddu. 1992. Seed storage proteins of wild wheat progenitors and their relationships with technological properties. Hereditas 116:315–322.

Clark, J.R., M. Robertson, and C.C. Ainsworth. 1991. Nucleotide sequence of a wheat (*Triticum aestivum* L.) cDNA clone encoding the waxy protein. Plant Mol. Biol. 16:1099–1101.

Cox, T.S., M.D. Shogren, R.G. Sears, T.J. Martin, and L.C. Bolte. 1989. Genetic improvement in milling and baking quality of hard red winter wheat cultivars, 1919 to 1988. Crop Sci. 29:626–631.

Cross, R.J., and B. Guo. 1993. Glutenin variation in a diverse pre-1935 world wheat germplasm collection. p. 287–298. *In* A.B. Damania (ed.) Biodiversity and wheat improvement. ICARDA, J. Wiley & Sons, New York.

Cummings, D.P., C.E. Green, and D.D. Stuthman. 1976. Callus induction and plant regeneration in oats. Crop Sci. 16:465–470.

Czarnecki, E. 1987. Breeding and selection for pre-harvest sprouting resistance in red wheats. p. 45–53. *In* D.J. Mares (ed.) Proc. 4th Int. Symp. Pre-harvest Sprout. Cereals. Westview Press, Boulder, CO.

Dahleen, L.S., D.D. Stuthman, and H.W. Rines. 1991. Agronomic trait variation in oat lines derived from tissue culture. Crop Sci. 31:90–94.

Damidaux, R., J.-C. Autran, and P. Feillet. 1980. Gliadin electrophoregrams and measurements of gluten viscoelasticity in durum wheats. Cer. Foods World 25:754–756.

Davidson, M.H., L. Dugan, J.H. Barns, J. Bova, K. Story, and K.B. Drennan. 1991. The hypocholesterolemic effects of β-glucan in oatmeal and oat bran: A dose controlled study. J. Am. Med. Assoc. 265:1833–1837.

Deane, D., and E. Commers. 1986. Oat cleaning and processing. p. 371–412. *In* F.H. Webster (ed.) Oats: Chemistry and technology. AACC, St. Paul, MN.

D' egidio, M.G., S. Nardi, and V. Vallega. 1993. Grain, flour, and dough characteristics of selected strains of diploid wheat, *Triticum monococcum* L. Cer. Chem. 70:298–303.

Devaux, P., A. Kilian, and A. Kleinhofs. 1993. Anther culture and *Hordeum bulbosum*-derived doubled haploids: Mutations and methylation. Mol. Gen. Genet. 241:674–679.

Dhaliwal, H.S., and F. MacRitchie. 1990. Contributions of protein fractions to dough handling properties of wheat-rye translocation cultivars. J. Cer. Sci. 12:113–122.

Dhaliwal, H.S., S.K. Sharma, D.S. Multani, and A.S. Randhawa. 1986. Genetics of yellow berry in wheat (*Triticum aestivum*). Theor. Appl. Genet. 72:548–550.

Dick, J.W., and R.R. Matsuo. 1988. Durum wheat and pasta products. p. 506–547. *In* Y. Pomeranz (ed.) Wheat: Chemistry and technology. Vol. 2. 3rd ed. AACC, St. Paul, MN.

Dimberg, L.H., O. Theander, and H. Lingnert. 1993. Avenanthramides—a group of phenolic antioxidants in oats. Cer. Chem. 70:637–641.

Dolezel, J., S. Lucretti, and I. Schubert. 1994. Plant chromosome analysis and sorting by flow cytometry. Crit. Rev. Plant Sci. 13:275–309.

Doll, H. 1983. Barley seed proteins and possibilities for their improvement. p. 207–223. *In* W. Gottschalk and H.P. Muller (ed.) Seed proteins: Biochemistry, genetics, nutritive value. Kluwer Acad. Publ., The Netherlands.

Doll, H., and B. Koie. 1978. Influence of the high-lysine gene from barley mutant 1508 on grain, carbohydrate and protein yield. p. 107–114. *In* Proc. Seed Protein Improve. by Nucl. Tech., Vienna, Austria. 26–30 Sept. 1977. IAEA,

D'Ovidio, R. 1991. Molecular characterization of bread wheat mutants lacking the entire cluster of chromosome 6A controlled gliadin components. J. Cer. Sci. 14:125–129.

D'Ovidio, R. 1993. Single-seed PCR of LMW glutenin genes to distinguish between durum wheat cultivars with good and poor technological properties. Plant. Mol. Biol. 22:1173– 1176.

Doyle, C.J., and J. Valentine. 1988. Naked oats: an assessment of the economic potential for livestock feed in the United Kingdom. Plant Var. Seeds 1:99–108.

Duvick, D.N. 1996. Plant breeding, an evolutionary concept. Crop Sci. 36:539–548.

Dweikat, I., S. Mackenzie, M. Levi, and H. Ohm. 1993. Pedigree assessment using RAPD-DGGE in cereal crop species. Theor. Appl. Genet. 85:497–505.

Eggum, B. 1978. Protein quality of induced high lysine mutants in barley. p. 317–341. *In* M. Friedman (ed.) Nutritional improvement of food and feed proteins. Adv. Exp. Med. Vol. 105. Plenum Press, New York.

Eliasson, A.-C. 1989. Characterization of wheat starch and gluten as related to end-use properties. p. 177–192. *In* Y. Pomeranz (ed.) Wheat is unique. Structure, composition, processing, end-use properties, and products. AACC, St Paul, MN.

Erdal, K. 1986. Proanthocyanidin-free barley—malting and brewing. J. Inst. Brew. 92:220–224.

Everson, E.H. 1987. Wheat quality: A breeders point of view. Cer. Foods World 32:306–308.

Everson, E.H., and R.B. Hart. 1961. Varietal variation for dormancy in mature wheat. p. 820–829. Michigan Agric. Exp. Stn. Bull. Vol. 43, no. 4.

Fabijanski, S., and I. Altosaar. 1985. Evidence for translational control of storage protein biosynthesis during embryogenesis of *Avena sativa* L. (oat germplasm). Plant Mol. Biol. 4:211–218.

Fastnaught, C.E., P.T. Berglund, E.T. Holm, and G.J. Fox. 1996. Genetic and environmental variation in β-glucan content and quality parameters of barley for food. Crop Sci. 36:941–946.

Fedak, G. 1985. Wide crosses in *Hordeum*. p. 156–186. *In* D.C. Rasmusson (ed.) Barley. Agron. Monogr. 26. ASA, Madison, WI.

Fedak, G., and I. de la Roche. 1977. Lipid and fatty acid composition of barley kernels. Can. J. Plant Sci. 57:257–260.

Fincher, G.B., and B.A. Stone. 1993. Physiology and biochemistry of germination in barley. p. 247–295. *In* A.W. MacGregor and R.S. Bhatty (ed.) Barley: Chemistry and technology. AACC, St. Paul, MN.

Fisher, S.G., and L.S. Lerman. 1983. DNA fragments differing by single base-pair substitutions are separated in denaturing-gradient gels: Correspondence with melting theory. Proc. Natl. Acad. Sci. 80:1579–1583.

Fominaya, A., G. Hueros, Y. Loarce, and E. Ferrer. 1995. Chromosomal distribution of a DNA sequence from C-genome heterochromatin and the identification of a new ribosomal DNA locus in the *Avena* genus. Genome 38:548–557.

Forsberg, R.A., and D.L. Reeves. 1992. Breeding oat cultivars for improved grain quality. p. 751–775. *In* H.G. Marshall and M.E. Sorrells (ed.) Oat science and technology. ASA and CSSA, Madison, WI.

Forsberg, R.A., and H.L. Shands. 1989. Oat breeding. p. 167–207. *In* J. Janick (ed.) Plant breeding reviews. Vol. 6. Timberland Press, Portland, OR.

Foster, E., and N. Prentice. 1987. p. 337–396. Barley. *In* R.A. Olsen and K.J. Frey (ed.) Nutritional quality of cereal grains: Genetic and agronomic improvement. ASA, Madison, WI.

Freed, R.D., E.H. Everson, K. Ringlund, and M. Gullord. 1976. Seedcoat color in wheat and the relationship to seed dormancy at maturity. Cer. Res. Commun. 4:147–149.

Freiberg, B. 1991. Kansas value-added wheat program looks promising. Seed Industry. March, Vol. 6, p. 22.

Frey, K.J. 1977. Protein of oats. Z. Pflanzenzucht. 78:185–215.

Frey, K.J., and E.G. Hammond. 1975. Genetics, characteristics, and utilization of oil in caryopses of oat species. J. Am. Oil Chem. Soc. 52:358–362.

Friebe, B., J. Jiang, D.R. Knott, and B.S. Gill. 1994. Compensation indices of radiation-induced wheat-*Agropyron elongatum* translocations conferring resistance to leaf rust and stem rust. Crop Sci. 34:400–404.

Friebe, B., Y. Mukai, H.S. Dhaliwal, T.J. Martin, and B.S. Gill. 1991. Identification of alien chromatin specifying resistance to wheat streak mosaic and greenbug in wheat germ plasm by C-banding and in situ hybridization. Theor. Appl. Genet. 81:381–389.

Friedman, M., and D. Atsmon. 1988. Comparison of grain composition and nutritional quality in wild barley (Hordeum spontaneum) and in a standard cultivar. J. Agric. Food Chem. 36:1167–1172.

Fujigaki, J. 1995. Genetic relationship between barley and rye from morphological and classical cytogenetic viewpoints. p. 140–147. *In* Proc. U.S.-Jpn. Symp., Rep. 95-352-D. 21–23 March. Kansas Agric. Exp. Stn., Manhattan, KS.

Galili, G., and M. Feldman. 1984. Intergenomic suppression of endosperm protein genes in common wheat. Can. J. Genet. Cytol. 26:651–656.

Galili, G., A.A. Levy, and M. Feldman. 1986. Gene-dosage compensation of endosperm proteins in hexaploid wheat *Triticum aestivum*. Proc. Natl. Acad. Sci. 83:6524–6528.

Gill, B.S., and W.J. Raupp. 1987. Direct transfers from *Aegilops squarrosa* L. to hexaploid wheat. Crop Sci. 27:445–450.

Glover, D.V., and E.T. Mertz. 1987. Corn. p. 183–336. *In* R.A. Olsen and K.J. Frey (ed.) Nutritional quality of cereal grains: Genetic and agronomic Improvement. ASA and CSSA, Madison, WI.

Goblirsch, C.A., R.D. Horsley, and P.B. Schwarz. 1996. A strategy to breed low-protein barley with acceptable kernel color. Crop Sci. 36:41–44.

Goering, K.J., and R.F. Eslick. 1976. Barley starch. VI: A self-liquefying waxy barley starch. Cer. Chem. 53:174–180.

Goffreda, J.C., W.B. Burnquist, S.C. Beer, S.D. Tanksley, and M.E. Sorrells. 1992. Application of molecular markers to assess genetic relationships among accessions of wild oat, *Avena sterilis*. Theor. Appl. Genet. 85:146–151.

Graner, A., A. Jahoor, J. Schondelmaier, H. Siedler, K. Pillen, G. Fischbeck, G. Wenzel, and R. G. Herrmann. 1991. Construction of an RFLP map of barley. Theor. Appl. Genet. 83:250–256.

Graner, A., H. Siedler, A. Jahoor, R.G. Herrmann, and G. Wenzel. 1990. Assessment of the degree and type of restriction fragment length polymorphism in barley (*Hordeum vulgare*). Theor. Appl. Genet. 80:826–832.

Graybosch, R.A., C.J. Peterson, L.E. Hansen, and P.J. Mattern. 1990. Relationships between protein solubility characteristics, 1BL/1RS, high molecular weight glutenin composition, and end-use quality in winter wheat germ plasm. Cer. Chem. 67:342–349.

Graybosch, R.A., C.J. Peterson, J.H. Lee, and D.R. Shelton. 1994. Effects of glutenin protein polymorphisms on breadmaking quality of winter wheats. Crop Sci. 34:628–635.

Graybosch, R.A., C.J. Peterson, D.R. Shelton, and P.S. Baenziger. 1996. Genotypic and environmental modification of wheat flour protein composition in relation to end-use quality. Crop Sci. 36:296–300.

Greenberg, D.C. 1977. A diallel cross analysis of gum content in barley (*Hordeum vulgare*). Theor. Appl. Genet. 50:41–46.

Greene, F.C. 1981. In vitro synthesis of wheat (*Triticum aestivum* L.) storage proteins. Plant Physiol. 68:778–783.

Greenwell, P., and J.D. Schofield. 1986. A starch granule protein associated with endosperm softness in wheat. Cer. Chem. 63:379–380.

Gupta, P.K. 1991. Cytogenetics of wheat and its close wild relatives—*Triticum* and *Aegilops*. p. 243–262. *In* P.K. Gupta and T. Tsuchiya (eds.) Chromosome engineering in plants: Genetics, breeding, evolution. Part A. Elsevier, New York.

Hagberg, A. 1978. The Svalov cereal protein quality breeding programme. p. 91–105. *In* Proc. Seed Protein Improve. Nucl. Tech., Vienna, Austria. 26–30 Sept. 1977. IAEA.

Hagberg, A. 1986. Induced structural rearrangements. p. 17–36. *In* W. Horn et al. (ed.) Genetic manipulations in plant breeding. Walter de Gruyter & Co., New York.

Hagberg, A., and P. Hagberg. 1991. Production and analysis of chromosome duplications in barley. p. 401–410. *In* P.K. Gupta and T. Tsuchiya (ed.) Chromosome engineering in plants: Genetics, breeding, evolution (Part A). Elsevier, New York.

Hardin, B. 1990. Making oats more healthful. Agric. Res. 38:12–13.

Harlan, J.R. 1976. Genetic resources in wild relatives of crops. Crop Sci. 16:329–333.

Harlan, J.R., and J.M.J. de Wet. 1971. Toward a rational classification of cultivated plants. Taxon 20:509–517.

Hayes, P.M., B.H. Liu, S.J. Knapp, F. Chen, B. Jones, T. Blake, J. Franckowiak, D. Rasmusson, M. Sorrells, S.E. Ullrich, D. Wesenberg, and A. Kleinhofs. 1993. Quantitative trait locus effects and environmental interaction in a sample of North American barley germ plasm. Theor. Appl. Genet. 87:392–401.

Hayter, A.M., and T.J. Riggs. 1978. The inheritance of diastatic power and alpha-amylase contents in spring barley. Theor. Appl. Genet. 52:251–256.

He, S., H.W. Ohm, and S. Mackenzie. 1992. Detection of DNA sequence polymorphisms among wheat varieties. Theor. Appl. Genet. 84:573–578.

Helentjaris, T. 1992. RFLP analyses for manipulating agronomic traits in plants. p. 357–372. *In* H.T. Stalker and J.P. Murphy (ed.) Plant breeding in the 1990s. Redwood Press Ltd., Melksham, UK.

Helentjaris, T., G. King, M. Slocum, C. Siedenstrang, and S. Wegman. 1985. Restriction fragment polymorphisms as probes for plant diversity and their development as tools for applied plant breeding. Plant Mol. Biol. 5:109–118.

Helm, J.H., R.J. Metzer, and W.E. Kronstadt. 1974. Inheritance of high lysine in hiproly barley and its association with the hiproly endosperm gene. Crop Sci. 14:637–640.

Henry, R.J., and A.H.D. Brown. 1987. Variation in the carbohydrate composition of wild barley (*Hordeum spontaneum*) grain. Plant Breed. 98:97–103.

Heun, M., J.P. Murphy, and T.D. Phillips. 1994. A comparison of RAPD and isozyme analyses for determining the genetic relationships among *Avena sterilis* L. accessions. Theor. Appl. Genet. 87:689–696.

Heyne, E.G., and C.O. Johnson. 1954. Inheritance of leaf rust reaction and other characters in crosses among Timstein, Pawnee, and Redchief wheats. Agron. J. 46:81–85.

Ho, T.H.D., and S.C. Shih. 1980. Screening barley mutants with altered hormone sensitivity in their aleurone layers. Plant Physiol. 66:153–157.

Holthaus, J.F., J.B. Holland, P.J. White, and K.J. Frey. 1996. Inheritance of β-glucan content of oat grain. Crop Sci. 36:567–572.

Horsley, R.D., P.B. Schwarz, and A.E. Foster. 1991. Effects of gene *ant13* on agronomic and malt quality traits of barley. Crop Sci. 31:593–598.

Horsley, R.D., P.B. Schwarz, and J.J. Hammond. 1995. Genetic diveristy in malt quality of North American six-rowed spring barley. Crop Sci. 35:113–118.

Hoseney, R.C. 1987. Wheat hardness. Cer. Foods World 32:320–322.

Hou, G., H. Yamamoto, and P.K.W. Ng. 1996a. Relationships of quantity of gliadin subgroups of selected U.S. soft wheat flours to rheological and baking properties. Cer. Chem. 73:352–357.

Hou, G., H. Yamamoto, and P.K.W. Ng. 1996b. Relationships of quantity of glutenin subunits of selected U.S. soft wheat flours to rheological and baking properties. Cer. Chem. 73:358–363.

Howes, N.K., M.I. Kovaks, and E.R. Kerber. 1990. D-genome suppression of gliadin electrophoretic band 45 in hexaploid wheat and the effect upon quality. Genome 33:395–399.

Huebner, F.R., T.C. Nelson, and J.A. Bietz. 1995. Differences among gliadins from spring and winter wheat cultivars. Cer. Chem. 72:341–343.

Humphreys, D.G., and D.E. Mather. 1996. Heritability of β-glucan, groat percentage, and crown rust resistance in two oat crosses. Euphytica 91:359–364.

Hutchinson, J., and J. Postoyko. 1986. C-banding of *Avena* species. p. 157–159. *In* W. Horn et al. (ed.) Genetic manipulation in plant breeding. Proc. Int. Symp. EUCARPIA, Berlin. 8–13 Sept. 1985. Walter de Gruyter, Berlin.

Inglett, G.E., and S.B. Grisamore. 1991. Maltodextrin fat substitute lowers cholesterol. Food Tech. 45:104.

Jahoor, A., G. Backes, A. Graner, R.G. Herrmann, and G. Fischbeck. 1991. Development of RFLP markers for barley. Plant Breed. 107:73–76.

Jain S.K., C.O. Qualset, G.M. Bhatt, and K.K. Wu. 1975. Geographical patterns of phenotypic diversity in a world collection of durum wheats. Crop Sci.15:700–704.

Jan, C.C., J. Dvorak, C.O. Qualset, and K.M. Soliman. 1981. Selection and identification of a spontaneous alien chromosome translocation in wheat. Genetics 98:389–398.

Jaradat, A.A. 1991. Grain protein variability among populations of wild barley (*Hordeum spontaneum* C. Koch.) from Jordan. Theor. Appl. Genet. 83:164–168.

Jensen, J. 1984. Coordinators report. Chromosome 5. Barley Genet. Newsl. 14:62–65.

Johansson, H. 1978. Improvement of nutritional quality in barley and spring wheat: An FAO/SIDA/SAREC Proj. p. 85–90. *In* Proc. Seed Protein Improve. Nucl. Tech., Vienna, Austria. 26–30 Sept. 1977. IAEA.

Johnson, S.S., R.L. Phillips, and H.W. Rines. 1987. Possible role of heterochromatin in chromosome breakage induced by tissue culture in oats (*Avena sativa* L.). Genome 29:439–446.

Jolly, C.J., S. Rahman, A.A. Kortt, and T.J.V. Higgins. 1993. Characterisation of the wheat Mr 15000 "grain-softness protein" and analysis of the relationship between its accumulation in the whole seed and grain softness. Theor. Appl. Genet. 86:589–597.

Jones, J.L. 1992. Genetic engineering of crops: Its relevance to the food industry. Trends Sci. Technol. 3:54–59.

Joppa, L.R., K. Khan, and N.D. Williams. 1983. Chromosomal location of genes for gliadin polypeptides in durum wheat Triticum turgidum L. Theor. Appl. Genet. 64:289–293.

Joppa, L.R., and N.D. Williams. 1988. Genetics and breeding of durum wheat in the United States. p. 47–68. *In* G. Fabriani and C. Lintas (ed.) Durum chemistry and technology. AACC, St. Paul, MN.

Kahlon, T.S. 1989. Nutritional implications and uses of wheat and oat kernel oil. Cer. Foods World 34:872–875.

Kasha, K.J., D.E. Falk, and A. Ziauddin. 1993. Potential improvement of barley quality through genetic engineering. p. 419–435. *In* A.W. MacGregor and R.S. Bhatty (ed.) Barley: Chemistry and technology. AACC, St. Paul, MN.

Kasha, K.J., and E. Reinbergs. 1981. Recent developments in the production and utilization of haploids in barley. p. 655–665. *In* Barley breeding. Proc. 4th Int. Barley Genet. Symp., Vol. 5, Edinburgh, UK. 22–29 July. Redwood Burn Ltd., UK.

Kerber, E.R., and K.H. Tipples. 1969. Effects of the D genome on milling and baking properties of wheat. Can. J. Genet. Cytol. 49:255–263.

Kleine, M., W. Michalek, A. Graner, R.G. Herrmann, and C. Jung. 1993. Construction of a barley (*Hordeum vulgare* L.) YAC library and isolation of a Hor1-specific clone. Mol. Gen. Genet. 240:265–272.

Klopfenstein, C.F. 1988. The role of cereal β-glucans in nutrition and health. Cer. Foods World 33:865–866.

Knackstedt, M.A., R.G. Sears, D.E. Rogers, and G.L. Lookhart. 1994. Effects of T2BS.2RL wheat-rye translocation on breadmaking quality in wheat. Crop Sci. 34:1066–1070.

Knott, D.R. 1980. Mutation of a gene for yellow pigment linked to *Lr19* in wheat. Can. J. Genet. Cytol. 22:651–654.

Koebner, R.M.D., and K.W. Shepherd. 1988. Isolation and agronomic assessment of allosyndetic recombinants derived from wheat/rye translocation 1DL.1RS, carrying reduced amounts of rye chromatin. p.343–348. *In* Proc. 7th Int. Wheat Genet. Symp. Cambridge, UK. Univ. Cambridge, UK.

Kratochvil, R.J., and D.J. Sammons. 1991. Dormancy screening for soft red winter wheat cultivars. p. 102. *In* Agronomy abstracts. ASA, Madison, WI.

Kreis, M., and H. Doll. 1980. Starch and prolamin level in single and double high-lysine barley mutants. Physiol. Plant. 48:139–143.

Kreis, M., M.S. Williamson, J. Forde, D. Schmutz, J. Clark, B. Buxton, J. Pywell, C. Marris, J. Henderson, N. Harriswon, P.R. Shewry, B.G. Forde, and B.J. Miflin. 1986. Differential gene expression in the developing barley endosperm. Philos. Trans. R. Soc. London B 314:355–365.

Ladizinski, G., and R. Fainstein. 1977. Domestication of the protein-rich tetraploid wild oats *Avena magna* and *A. murphyi*. Euphytica 26:221–223.

Lagudah, E.S., F. MaCritchie, and G.M. Halloran. 1987. The influence of high-molecular-weight subunits of glutenin from *Triticum tauschii* on flour quality of synthetic hexaploid wheat. J. Cer. Sci. 5:129–138.

Larkin, P.J., S.A. Ryan, R.I.S. Brettell, and W.R. Scowcroft. 1984. Heritable somaclonal variation in wheat. Theor. Appl. Genet. 67:443–455.

Larkin, P.J., and W.R. Scowcroft. 1981. Somaclonal variation—a novel source of variability from cell cultures for plant improvement. Theor. Appl. Genet. 60:197–214.

Laurie, D.A., and M.D. Bennett. 1988. The production of haploid wheat plants from wheat x maize crosses. Theor. Appl. Genet. 76:393–397.

Law, C.N., and P.I. Payne. 1983. Genetical aspects of breeding for improved grain protein content and type in wheat. J. Cer. Sci. 1:79–93.

Leah, R., and J. Mundy. 1989. The bifunctional α-amylase/subtilisin inhibitor of barley: Nucleotide sequence and pattern of seed specific expression. Plant Mol. Biol. 12:673–682.

Lee, J.H., R.A Graybosch, and C.J. Peterson. 1995. Quality and biochemical effects of a 1BL/1RS wheat/rye translocation in wheat. Theor. Appl. Genet. 90:105–112.

Lee, B.T., K. Murdoch, J. Topping, M.T.J. de Both, Q.S. Wu, A. Karp, S. Steele, C. Symonds, M. Kreis, and M.G.K. Jones. 1988. Isolation, culture and morphogenesis from wheat protoplasts, and study of expression of DNA constructs by direct gene transfer. Plant Cell Tiss. Org. Cult. 12:223–226.

Lee, B.T., K. Murdoch, J. Topping, M. Kreis, and M.G.K. Jones. 1989. Transient gene expression in aleurone protoplasts isolated from developing caryopses of barley and wheat. Plant Mol. Biol.13:21–29.

Leuck, D.J. 1989. Economic and technical constraints in implementing industrial utilization of wheat in North America and Europe. p. 1–19. *In* Y. Pomeranz (ed.) Wheat is unique. Structure, composition, processing, end-use properties, and products. AACC, St. Paul, MN.

Lim, H.S., P.J. White, and K.J. Frey. 1992. Genotypic effects on β-glucan content of oat lines grown in two consecutive years. Cer. Chem. 69:262–265.

Linde-Laursen, I. 1988. Giemsa C-banding of barley chromosomes. V. Localization of break points in 70 reciprocal translocations. Hereditas 108:65–76.

Linde-Laursen, I., and R. Von Blothmer. 1986. Preferential loss and gain of specific *Hordeum vulgare* chromosomes in hybrids with three alien species? p. 179–182. *In* W. Horn et al. (ed.) Genetic manipulations in plant breeding. Walter de Gruyter & Co., New York.

Lineback, D.R., and V.F. Rasper. 1988. Wheat carbohydrates. p. 277–372. *In* Y. Pomeranz (ed.) Wheat: Chemistry and technology. Vol. 1. 3rd. ed. AACC, St. Paul, MN.

Liu, C.-Y., K.W. Shepherd, and P.W. Gras. 1994. Grain yield and quality characteristics of chromosome 1D and 1B substitution lines in durum wheat and their F2 progeny lines. I. Comparisons among the tetraploid phenotypes. J. Cer. Sci. 20:23–32.

Liu, C.-Y., K.W. Shepherd, and A.J. Rathjen. 1996. Improvement of durum wheat pastamaking and breadmaking qualities. Cer. Chem. 73:155–166.

Loi, L., B. Ahluwalia, and G.B. Fincher. 1988. Chromosomal location of genes encoding barley (1-3, 1-4)-β-glucanohydrolases. Plant Physiol. 87:300–302.

Lockhart, H.B., and H.D. Hurt. 1986. Nutrition of oats. p. 297–308. *In* F.H. Webster (ed.) Oats: Chemistry and technology. AACC, St. Paul, MN.

Lookhart, G.L., K. Hagman, and D.D. Kasarda. 1993. High-molecular-weight glutenin subunits of the most commonly grown wheat cultivars in the U.S. in 1984. Plant Breed. 110:48–62.

Lorenz, K. 1984. Cereal and dental caries. p. 83–137. *In* Y. Pomeranz (ed.) Advances in cereal science and technology. Vol. 6. AACC, St. Paul, MN.

Lorz, H., C.T. Harms, and I. Potrykus. 1976. Regeneration of plants from callus in *Avena sativa* L. Z. Pflanzenzucht. 77:257–259.

Lukaszewski, A.J., and C.A. Curtis. 1992. Transfer of the Glu-D1 gene from chromosome 1D of breadwheat to chromosome 1R in hexaploid triticale. Plant Breed. 109:203–210.

Lynch, P.J., and K.J. Frey. 1993. Genetic improvement in agronomic and physiological traits of oat since 1914. Crop Sci. 33:984–988.

MacArthur, L.A., and B.L. D'Appolonia. 1979. Composition of oat and wheat carbohydrates. II. Starch. Cer. Chem. 56:458–461.

MacGregor, A.W., and G.B. Fincher. 1993. Carbohydrates of the barley grain. p. 73–130. *In* A.W. MacGregor and R.S. Bhatty (ed.) Barley: Chemistry and technology. AACC, St. Paul, MN.

Madsen, K.O. 1981. The anticaries potential of seeds. Cer. Foods World 26:19–25.

Magnuson, K.M. 1985. Uses and functionality of vital wheat gluten. Cer. Foods World 30:179–181.

Maheshwari, N., K. Rajyalakshmi, K. Bawela, S.K. Dhir, C.N. Chowdhry, and S.C. Maheshwari. 1995. In vitro culture of wheat and genetic transformation—retrospect and prospect. Crit. Rev. Plant Sci. 14:149–178.

Mannonen, L., V. Kauppinen, and T.-M. Enari. 1994. Recent developments in the genetic engineering of barley. Crit. Rev. Biotech. 14:287–310.

Mares, D.J. 1987. Comparisons of tolerance to pre-harvest sprouting within *Triticum* species, triticale, and barley. p. 148–157. *In* D.J. Mares (ed.) Proc. 4th Int. Symp. Pre-harvest Sprout. Cer. Westview Press, Boulder, CO.

Mariani, B.M., M.G. D'egidio, and P. Novaro. 1995. Durum wheat quality evaluation: Influence of genotype and environment. Cer. Chem. 72:194–197.

Marshall, H.G., and F.L. Kolb. 1986. Relationships among grain quality indicators in oats. Crop Sci. 26:800–804.

Marshall, H.G., and G.E. Shaner. 1992. Genetics and inheritance in oat. p. 509–571. *In* H.G. Marshall and M.E. Sorrells (ed.) Oat science and technology. ASA and CSSA, Madison, WI.

Matzk, F. 1996. Hybrids of crosses between oat and Andropogoneae or Paniceae species. Crop Sci. 36:17–21.

McIntyre, C.L., B.C. Clarke, and R. Appels. 1988. Amplification and dispersion of repeated DNA sequences in the Triticeae. Plant Syst. Evol. 160:39–59.

McKechnie, R. 1983. Oat products in bakery foods. Cer. Foods World 28:635–637.

McKinnon, G.E., and R.J. Henry. 1995. Control of gene expression for the genetic engineering of cereal quality. J. Cer. Sci. 22:203–210.

Melchinger, A.E., A. Graner, M. Singh, and M.M. Messmer. 1994. Relationships among European barley germplasm: I. Genetic diversity among winter and spring cultivars revealed by RFLPs. Crop Sci. 34:1191–1199.

Michelmore, R.W., I. Paran, and R.V. Kesseli. 1991. Identification of markers linked to disease-resistance genes by bulked segregant analysis: A rapid method to detect markers in specific genomic regions by using segregating populations. Proc. Natl. Acad. Sci. 88:9828–9832.

Miller, T.E., S.M. Reader, K.A. Purdie, S. Abbo, R.P. Dunford, and I.P. King. 1995. Fluorescent in situ hybridization as an aid to introducing alien genetic variation into wheat. Euphytica 85:275–279.

Miskelly, D.M. 1984. Flour components affecting paste and noodle colour. J. Sci. Food Agric. 35:463–471.

Miura, H., S. Tanii, T. Nakamura, and N. Watanabe. 1994. Genetic control of amylose content in wheat endosperm starch and differential effects of three Wx genes. Theor. Appl. Genet. 89: 276–280.

Molina-Cano, J.L., and J. Conde. 1982. Genetic and environmental variation of gum content in barley. J. Inst. Brew. 88:30–33.

Molnar, S.J., and A. McKay. 1991. Restriction fragment analysis of ribosomal and hordein genes in eastern Canadian two-rowed barleys. Genome 34:298–302.

Moore, G., M.D. Gale, N. Kurata, and R.B. Flavell. 1993. Molecular analysis of small grain cereal genomes: Current status and prospects. Bio/Technology 11:584–589.

Morgan, A.G., A.A. Gill, and D.B. Smith. 1983a. Some barley grain and green malt properties and their influence on malt hot-water extract. I. β-Glucan, β-glucan solubilase and endo-β-glucanase. J. Inst. Brew. 89:283–291.

Morgan, A.G., A.A. Gill, and D.B. Smith. 1983b. Some barley grain and green malt properties and their influence on malt hot-water extract. II. Protein, proteinase and moisture. J. Inst. Brew. 89:292–298.

Morris, C.F., and G.M. Paulsen. 1987. Development of preharvest sprouting-resistant germplasm from 'Clark's Cream' hard white winter wheat. Cer. Res. Commun. 15:229–235.

Morrison, W.R. 1978. Cereal lipids. p. 221–348. *In* Y. Pomeranz (ed.) Advances in cereal science and technology. Vol. 2. AACC, St. Paul, MN.

Morrison, W.R. 1993. Barley lipids. p. 199–246. *In* A.W. MacGregor and R.S. Bhatty (ed.) Barley: Chemistry and technology. AACC, St. Paul, MN.

Moseman, J.G., and D.H. Smith, Jr. 1985. Germplasm resources. p. 57–72. *In* D.C. Rasmusson (ed.) Barley. Agron. Monogr. 26. ASA, Madison, WI.

Moser, H., and M. Lee. 1994. RFLP variation and genealogical distance, multivariate distance, heterosis, and genetic variance in oats. Theor. Appl. Genet. 87:947–956.

Munck, L., K. Bang-Olsen, and B. Stilling. 1986. Genome adjustment by breeding to balance yield defects in high-lysine mutants of barley. p. 49–60. *In* W. Horn et al. (ed.) Genetic manipulations in plant breeding. Walter de Gruyter & Co., New York.

Muntz, K., K. Hammer, C. Lehmann, A. Meister, A. Rudolph, and F. Scholz. 1979. Variability of protein and lysine content in barley and wheat specimens from the world collection of cultivated plants at Gatersleben. p. 183–200. *In* Proc. Seed Improve. Nucl. Tech., Neuherberg, Germany. 4–8 Sept. 1978. IAEA.

Murphy, J.P., and L. . Hoffman. 1992. The origin, history, and production of oat. p. 1–28. *In* H.G. Marshall and M.E. Sorrells (ed.) Oat science and technology. ASA and CSSA, Madison, WI.

Nagao, S. 1995. Wheat products in East Asia. Cer. Foods World 40:482–487.

Nakamura, T., M. Yamamori, H. Hirano, and S. Hidaka. 1993. Identification of three wx proteins in wheat. Biochem. Genet. 31:75–86.

Nakamura, T., M. Yamamori, H. Hirano, S. Hidaka, and T. Nagamine. 1995. Production of waxy (amylose-free) wheats. Mol. Gen. Genet. 248:253–259.

Nelson, K.A., and S.J. Eilertson. 1996. Cereal grain specialty products: A brief review. Cer. Foods World 41:383–385.

Nemeth, L.J., P.C. Williams, and W. Bushuk. 1994. A comparative study of the quality of Soft wheats from Canada, Australia, and the United States. Cer. Foods Word 39:691–700.

Nevo, E., E. Golenberg, A. Beiles, A.H.D. Brown, and D. Zohary. 1982. Genetic diversity and environmental associations of wild wheat, *Triticum dicoccoides*, in Israel. Theor. Appl. Genet. 62:241–254.

Newman, C.W., and C.F. McGuire. 1985. Nutritional quality of barley. p. 403–456. *In* D.C. Rasmusson (ed.) Barley. Agron. Monogr. 26. ASA, Madison, WI.

Newman, C.W., M. Overland, R.K. Newman, K. Bang-Olsen, and B. Pedersen. 1990. Protein quality of a new high-lysine barley derived from Riso 1508. Can. J. Anim. Sci. 70:279–285.

Nieto-Taladriz, M.T., G. Branlard, and M. Dardevet. 1994. Polymorphism of omega-gliadins in durum wheat as revealed by the two-step APAGE/SDS-PAGE technique. Theor. Appl. Genet. 87:1001–1005.

Oda, M., Y. Yasuda, S. Okazaki, Y. Yamauchi, and Y. Yokoyama. 1980. A method of flour quality assessment for Japanese noodles. Cer. Chem. 57:253–254.

O'Donoughue, L.S., S.F. Kianian, P.J. Rayapati, G.A. Penner, M.E. Sorrells, S.D. Tanksley, R.L. Phillips, H.W. Rines, M. Lee, G. Fedak, S.J. Molner, D. Hoffman, T. Salas, B. Wu, E. Autrique, and A. Van Deynze. 1995. A molecular linkage map of cultivated oat. Genome 38:368–380.

O'Donoughue, L.S., Z. Wang, M. Roder, B. Kneen, M. Leggett, M.E. Sorrells, and S.D. Tanksley. 1992. An RFLP-based linkage map of oats based on a cross between two diploid taxa (*Avena atlantica* x *A. hirtula*). Genome 35:765–771.

Olson, M., L. Hood, C. Cantor, and D. Dotstein. 1989. A common language for physical mapping of the human genome. Science (Washington, DC) 254:1434–1435.

Orth, R.A., and J.A. Shellenberger. 1988. Origin, production and utilization of wheat. p. 1–14. *In* Y. Pomeranz (ed.) Wheat: Chemistry and technology. Vol. 1. 3rd ed. AACC, St. Paul, MN.

Ostrowski, H.T. 1977. Analysis for availability of amino acid supplements in foods and feeds: Biochemical and nutritional implications. p. 497–547. *In* M. Friedman (ed.) Nutritional improvement of food and feed proteins. Adv. Exp. Med. Vol. 105. Plenum Press, New York.

Parsons, J.G., and P.B. Price. 1974. Search for barley (*Hordeum vulgare* L.) with higher lipid content. Lipids 9:804–808.

Patterson, F.L., and R.E. Allan. 1981. Soft wheat breeding in the United States. p. 33–98. *In* W.T. Yamazaki and C.T. Greenwood. (ed.) Soft wheat: Production, breeding, milling, and uses. AACC, St. Paul, MN.

Paton, D. 1977. Oat starch. I. Extraction, purification and pasting properties. Starke 29:184–187.

Payne, P.I., and G.J. Lawrence. 1983. Catalogue of alleles for the complex gene loci, Glu-A1, Glu-B1, and Glu-D1 which code for high-molecular-weight subunits of glutenin in hexaploid wheat. Cer. Res. Commun. 11:29–35.

Payne, P.I., M.A. Nightingale, A.F. Krattiger, and L.M. Holt. 1987. The relationship between HMW glutenin subunit composition and the bread-making quality of British-grown wheat varieties. J. Sci. Food Agric. 40:51–65.

Peltonen-Sainio, P. 1994. Productivity of oats: Genetic gains and associated physiological changes. p. 69–94. *In* G.A. Slafer (ed.) Genetic improvement of field crops. Marcel Dekker, Inc., New York.

Pena, R.J., J. Zarco-Hernandez, A. Amaya-Celis, and A. Mujeeb-Kazi. 1994. Relationships between chromosome 1B-encoded glutenin subunit compositions and bread-making quality characteristics of some durum wheat (*Triticum turgidum*) cultivars. J. Cer. Sci. 19:243–249.

Pena, R.J., J. Zarco-Hernandes, and A. Mujeeb-Kazi. 1995. Glutenin subunit and bread-making quality characteristics of synthetic hexaploid wheats derived from *Triticum turgidum* x *Triticum tauschii* (coss.) Schmal crosses. J. Cer. Sci. 21:15–23.

Peterson, D.M. 1991. Genotype and environmental effects on oat β-glucan concentration. Crop Sci. 31:1517–1520.

Peterson, D.M. 1992. Physiology and development of the oat plant. p. 77–114. *In* H.G. Marshall and M.E. Sorrells (ed.) Oat science and technology. ASA and CSSA, Madison, WI.

Peterson, D.M., and A.C. Brinegar. 1986. Oat storage proteins. p. 153–203. *In* F.H. Webster (ed.) Oats: Chemistry and technology. AACC, St. Paul, MN.

Peterson, D.M., D.M. Wesenberg, and D.E. Burrup. 1995. β-Glucan content and its relationship to agronomic characteristics in elite oat germplasm. Crop Sci. 33:965–970.

Peterson, G.A., and A.E. Foster. 1973. Malting barley in the United States. Adv. Agron. 25:327–378.

Phillips, R.L., H.W. Rines, E.N. Jellen, W.L. Rooney, B.X. Wu, S. Kianian, and O. Riera-Lizarazu. 1995. Oat genome analysis via molecular markers and oat x corn crosses. p. 49–56. *In* W.J. Raupp and B.S. Gill (ed.) Classical and molecular cytogenetic analysis. Proc. U.S.-Jpn. Symp., Manhattan, KS. 21–23 March 1994. Kansas Agric. Exp. Stn., Manhattan, KS.

Pickering, R.A., and P. Devaux. 1992. Haploid production: Approaches and use in plant breeding. Biotech. Agric. 5:519–547.

Poehlman, J.M. 1985. Adaptation and distribution. p. 1–17. *In* D.C. Rasmusson (ed.) Barley. Agron. Monogr. 26. ASA, Madison, WI.

Pomeranz, Y. 1975. Proteins and amino acids of barley, oats and buckwheat. p. 13–78. *In* M. Friedman (ed.) Protein nutritional quality of foods and feeds. Vol. 1. Part 2. Mercel Dekker, Inc., New York.

Pomeranz, Y., G.D. Davis, J.L. Stoops, and F.S. Lai. 1979. Test weight and groat-to-hull ratio in oats quality. Cer. Foods World 24:600–602.

Pomeranz, Y., C.J. Peterson, and P.J. Mattern. 1985. Hardness of winter wheats grown under widely different climatic conditions. Cer. Chem. 62:463–467.

Pomeranz, Y., V.L. Youngs, and G.S. Robbins. 1973. Protein content and amino acid composition of oat species and tissues. Cer. Chem. 50:702–707.

Prentice, N., S. Babler, and S. Faber. 1980. Enzymatic analysis of β-D-glucans in cereal grains. Cer. Chem. 57:198–202.

Pritchard, P.E., and C.J. Brock. 1994. The glutenin fraction of wheat protein: The importance of genetic background on its quantity and quality. J. Sci. Food. Agric. 65:401–406.

Qi, X., P. Stam, and P. Lindhout. 1996. Comparison and integration of four barley genetic maps. Genome 39:379–394.

Ragot, M., and D.A. Hoisington. 1993. Molecular markers for plant breeding: Comparisons of RFLP and RAPD genotyping costs. Theor. Appl. Genet. 86:975–984.

Ramage, R.T. 1991. Chromosome manipulations in barley breeding. p. 385–400. *In* P.K. Gupta and T. Tsuchiya (ed.) Chromosome engineering in plants: Genetics, breeding, evolution. Part A. Elsevier Publ., New York.

Rasper, V.F. 1979. Chemical and physical characteristics of dietary cereal fiber. p. 93–115. *In* G. E. Inglett and S.I. Falkehag (ed.) Dietary fibers: Chemistry and nutrition. Acad. Press, New York.

Reich, J.M., and M.A. Brinkman. 1984. Inheritance of groat protein percentage in *Avena sativa* L. x *A. fatua* L. crosses. Euphytica 33:907–913.

Rines, H.W. 1983. Oat anther culture: Genotype effects on callus initiation and the production of a haploid plant. Crop Sci. 23:268–272.

Rines, H.W., and L.S. Dahleen. 1990. Haploid oat plants produced by application of maize pollen to emasculated oat florets. Crop Sci. 30:1073–1078.

Rines, H.W., R.L. Phillips, and D.A. Somers. 1992. Applications of tissue culture to oat Improvement. p. 777–791. *In* H.G. Marshall and M.E. Sorrells (ed.) Oat science and technology. ASA and CSSA, Madison, WI.

Rines, H.W., D.D. Stuthman, L.W. Briggle, V.L. Youngs, H. Jedlinski, D.H. Smith, J.A. Webster, and P.G. Rothman. 1980. Collection and evaluation of *Avena fatua* for use in oat improvement. Crop Sci. 20:63–68.

Robbins, G.S., Y. Pomeranz, and L.W. Briggle. 1971. Amino acid composition of oat groats. J. Agric. Food Chem. 10:536–539.

Rogers, W.J., P.I. Payne, and K. Harinder. 1989. The HMW glutenin subunit and gliadin compositions of German-grown wheat varieties and their relationship with bread-making quality. Plant Breed. 103:89–100.

Rogers, W.J., J.M. Rickatson, E.J. Sayers and C.N. Law. 1990. Dosage effects of homoeologous groups 1 and 6 upon bread-making quality in hexaploid wheat. Theor. Appl. Genet. 80:281–287.

Rohde, W., D. Becker, and F. Salamini. 1988. Structural analysis of the waxy locus from *Hordeum vulgare*. Nucl. Acids Res. 16:7185–7186.

Rosenberg, H., J. Obrist, and S.J. Stohs. 1978. Production of fungal protein from oat hulls. Econ. Bot. 32:413–417.

Rousset, M., J.M. Carrillo, C.O. Qualset, and D.D. Kasarda. 1992. Use of recombinant inbred lines of wheat for study of associations of high-molecular-weight glutenin subunit alleles to quantitative traits. 2. Milling and bread-baking quality. Theor. Appl. Genet. 83:403–412.

Saastamoinen, M., J. Kumpulainen, and S. Nummela. 1989. Genetic and environmental variation in oil content and fatty acid composition of oats. Cer. Chem. 66:296–300.

Saghai-Maroof, M.A., Q. Zhang, and R. Biyashev. 1995. Comparison of restriction fragment length polymorphisms in wild and cultivated barley. Genome 38:298–306.

Saiki, R.K., D.H. Gelfand, S. Stoffel, S. Scharf, R.H. Higuchi, G.T. Horn, K.B. Mullis, and H.A. Erlich. 1988. Primer-directed enzymatic amplification of DNA with a thermostable DNA polymerase. Science (Washington, DC) 239:487–491.

Schipper, H., and K.J. Frey. 1992. Agronomic and bioenergetic consequences of selection for high groat-oil content and high protein yield in oat. Plant Breed. 108:241–249.

Schondelmaier, J., R. Martin, A. Jahoor, A. Houben, A. Graner, H.-U. Koop, R.G. Herrmann, and C. Jung. 1993. Microdissection and microcloning of the barley (*Hordeum vulgare* L.) chromosome 1HS. Theor. Appl. Genet. 86:629–636.

Schrickel, D.J., V.D. Burrows, and J.A. Ingemansen. 1992. Harvesting, storing, and feeding of oat. p. 223–245. *In* H.G. Marshall and M.E. Sorrells (ed.) Oat science and technology. ASA and CSSA, Madison, WI.

Sharma, H.C. 1995. How wide can a wide cross be? Euphytica 82:43–64.

Sharma, H.C., J. Varnum, S. Sato, S. Baenziger, and S. Metz. 1992. Analysis of plants derived from wheat tissue culture. Cer. Res. Commun. 20:75–79.

Sharma, H.C., J.G. Waines, and K.W. Foster. 1981. Variability in primitive and wild wheats for useful genetic characters. Crop Sci. 21:555–559.

Sharp, P.J., S. Chao, S. Desai, and M.D. Gale. 1989. The isolation, characterization and application in the Triticeae of a set of wheat RFLP probes identifying each homoeologous chromosome arm. Theor. Appl. Genet. 78:342–348.

Sherman, J.D., A.L. Fenwick, D.M. Namuth, and N.L.V. Lapitan. 1995. A barley RFLP map: Alignment of three barley maps and comparisons to Gramineae species. Theor. Appl. Genet. 91:681–690.

Shewry, P.R., and M. Kreis. 1987. Symposium on: Application of biotechnology to livestock production. Proc. Nutr. Soc. 46:379–385.

Shinnick, F.L., R. Mathews, and S. Ink. 1991. Serum cholesterol reduction by oats and other fiber sources. Cer. Foods World 36:815–821.

Shotwell, M.A., C. Afonso, E. Davies, R.S. Chesnut, and B.A. Larkins. 1988. Molecular characterization of oat seed globulins. Plant Physiol. 87:698–704.

Shotwell, M.A., S.K. Boyer, R.S. Chesnut, and B.A. Larkins. 1990. Analysis of seed storage protein genes of oat. J. Biol. Chem. 265:9652–9658.

Shuey, W.C. 1976. Influence of wheat cultivars and environment on agtron values and flour ash. Cer. Chem. 53:429–437.

Smith, R.L., M.E. Schweder, and R.D. Barnett. 1994. Identification of glutenin alleles in wheat and triticale using PCR-generated DNA markers. Crop Sci. 34:1373–1378.

Snape, J.W., L.A. Sitch, E. Simpson, and B.B. Parker. 1988. Tests for the presence of gametoclonal variation in barley and wheat doubled haploids produced using the *Hordeum bulbosum* system. Theor. Appl. Genet. 75:509–513.

Soliman, K.M., J.E. Bernardin, and C.O. Qualset. 1980. Effects of *Agropyron* chromosome on endosperm proteins in common wheat (*Triticum aestivum* L.). Biochem. Genet. 18:465–482.

Somers, D.A., H.W. Rines, W. Gu, H.F. Kaeppler, and W.R. Bushnell. 1992. Fertile, transgenic oat plants. Bio/Technology 10:1589–1594.

Soper, J.F., R.G. Cantrell, and J.W. Dick. 1989. Sprouting damage and kernel color relationships in durum wheat. Crop Sci. 29:895–898.

Sorenson, M.B. 1989. Mapping of the *Hor 2* locus in barley by pulsed field gel electrophoresis. Carlsberg Res. Commun. 54:109–120.

Sorrells, M.E. 1991. New technologies and their application to wheat breeding in warmer areas. p. 113–124. *In* D.A. Saunders (ed.) Wheat for the nontraditional, warm areas. Proc. Int. Conf., Foz do Iguacu, Brazil. 29 July–3 Aug. 1990. CIMMYT, Mexico.

Sorrells, M.E., A.H. Paterson, and P.L. Finney. 1989. Milling and baking quality of soft white wheat genotypes subjected to preharvest sprouting. Cer. Chem. 66:407–410.

Sorrells, M.E., A. Van Deynze, J.C. Nelson, and S.R. McCouch. 1995. Comparative mapping in the Gramineae. p. 76–84. *In* Classical and molecular cytogenetic analysis. Proc. U.S.-Jpn. Symp., Manhattan, KS. 21–23 March. Kansas Agric. Exp. Stn., Manhattan, KS.

Soulaka, A.B., and W.R. Morrison. 1985. The amylose and lipid contents, dimensions, and gelatinisation characteristics of some wheat starches and their A- and B-granule fractions. J. Sci. Food Agric. 36:709–718.

Souza, E., J.M. Tyler, K.D. Kephart, and M. Kruk. 1993. Genetic improvement in milling and baking quality of hard red spring wheat cultivars. Cer. Chem. 70:280–285.

Srivastava, J.P., A.B. Damania, and L. Pecetti. 1988. Landraces, primitive forms and wild progenitors of macaroni wheat, *Triticum durum*: Their use in dryland agriculture. p. 153–158. *In* Proc. 7th Int. Wheat Genet. Symp., Cambridge, UK. Univ. Cambridge, UK.

Steiger, D.K., E.M. Elias, and R.G. Cantrell. 1996. Evaluation of lines derived from wild emmer chromosome substitutions: I. Quality traits. Crop Sci. 36:223–227.

Stolen, O. 1979. Breeding for high protein production in barley. p. 315–325. *In* Proc. Seed Improve. Nucl. Tech., Neuherberg, Germany. 4–8 Sept. 1978. IAEA.

Stuart, I.M., L. Loi, and G.B. Fincher. 1987. Immunological comparison of (1-3), (1-4)-β-glucan in oats and barley. J. Cer. Sci. 6:45–52.

Stuber, C.W. 1992. Biochemical and molecular markers in plant breeding. *In* J. Janick (ed.) Plant breeding reviews. Vol. 9. John Wiley & Sons, Inc., New York.

Subrahmanyam, N.C., and K.J. Kasha. 1973. Selective chromosomal elimination during haploid formation in barley following interspecific hybridization. Chromosoma 42:111–125.

Svensson, B. 1994. Protein engineering in the α-amylase family: catalytic mechanism, substrate specificity, and stability. Plant Mol. Biol. 25:141–157.

Swanston, J.S. 1987. The consequences, for malting quality, of *Hordeum laevigatum* as a source of mildew resistance in barley breeding. Ann. Appl. Biol. 110:351–355.

Swanston, J.S. 1991. The assessment of two barley starch mutants as potential parents in a malting quality breeding programme. Ann. Appl. Biol. 118:417–422.

Symes, K.J. 1965. The inheritance of grain hardness in wheat as measured by the particle size index. Aust. J. Agric. Res. 16:113–123.

Symes, K.J. 1969. Influence of a gene causing hardness on the milling and baking quality of two wheats. Aust. J. Agric. Res. 20:971–979.

Taha, S.A., and F. Sagi. 1987. Quality of durum wheat (*Triticum durum* Desf.): Grouping of varieties according to their gluten strength, cooking behavior, and gliadin composition. Cer. Res. Commun. 15:281–287.

Takeda, K., and K.J. Frey. 1979. Protein yield and its relationship to other traits in backcross populations from an *Avena sativa* x *A. sterilis*. Crop Sci. 19:623–628.

Talbert, L.E. 1993. Molecular biology and wheat improvement. p. 235–250. *In* J. Janick (ed.) Plant breeding reviews. Vol. 11. John Wiley & Sons, Inc., New York.

Tanchak, M.A., M. Giband, B. Potier, J.P. Schernthaner, S. Dukiandjiev, and I. Altosaar. 1995. Genomic clones encoding 11S globulins in oats. Genome 38:627–634.

Tanksley, S.D., N.D. Young, A.H. Paterson, and M.W. Bonierbale. 1989. RFLP mapping in plant breeding: New tools for an old science. Biotech 7:257–264.

Thomas, H. 1992. Cytogenetics of *Avena*. p. 473–508. *In* H.G. Marshall and M.E. Sorrells (ed.) Oat science and technology. ASA and CSSA, Madison, WI.

Thomas, H., J.M. Haki, and S. Arangzeb. 1980. The introgression of characters of the wild oat *Avena magna* (2n=4x=28) into the cultivated oat *A. sativa* (2n=6x=42). Euphytica 29:391–399.

Tingey, S.V., J.A. Rafalski, and J.G.K. Williams. 1992. Genetic analysis with RAPD markers. p. 3–8. *In* Applications of RAPD technology to plant breeding. Proc. Symp. 1 November. AHS, AGA, and CSSA, Madison, WI.

Torp, J. 1980. Variation in the concentration of major carbohydrates in the grain of some spring barleys. J. Sci. Food Agric. 31:1354–1360.

Tragoonrung, S., V. Kanazin, P.M. Hayes, and T.K. Blake. 1992. Sequence-tagged-site-facilitated PCR for barley genome mapping. Theor. Appl. Genet. 84:1002–1008.

Tsuchiya, T. 1991. Chromosome mapping by means of aneuploid analysis in barley. p. 361–384. *In* P.K. Gupta and T. Tsuchiya (ed.) Chromosome engineering in plants: Genetics, breeding, evolution. Part A. Elsevier Publ., New York.

Ullrich, S.E., C.N. Coon, and J.M. Sever. 1981. Relationships of nutritional and malting quality traits of barley. p. 225–233. *In* Barley breeding. Vol. 5. Proc. 4th Int. Barley Genet. Symp., Edinburgh, UK. 22–29 July. Redwood Burn Ltd., UK. Redwood Burn Ltd. Publ., UK.

Ullrich, S.E., J.M. Edmiston, A. Kleinhofs, D.A. Kudrna, and M.E.H. Maatougui. 1991. Evaluation of somaclonal variation in barley. Cer. Res. Commun. 19:245–260.

Ullrich, S.E., and R.F. Eslick. 1978. Lysine and protein characterization of spontaneous shrunken endosperm mutants of barley. Crop Sci. 18:809–812.

Van Deynze, A.E., J. Dubcovsky, K.S. Gill, J.C. Nelson, M.E. Sorrells, J. Dvorak, B.S. Gill, E.S. Lagudah, S.R. McCouch, and R. Appels. 1995. Molecular-genetic maps for group 1 chromosomes of Triticeae species and their relation to chromosomes in rice and oat. Genome 38:45–59.

Van Hintum, T.J.L. 1994. Drowning in the genepool: Managing genetic diversity in genebank collections. Centre Genet. Resour., Wageningen, The Netherlands.

Vasil, V., S.M. R.D. Brown, M.E. Fromm, and I.K. Vasil. 1991. Stably transformed callus lines from microprojectile bombardment of cell suspension cultures of wheat. Biotechnology 9:743–747.

Von Bothmer, R., and N. Jacobsen. 1991. Interspecific hybrids within the genus *Hordeum*. p. 411–431. *In* P.K. Gupta and T. Tsuchiya (ed.) Chromosome engineering in plants: Genetics, breeding, evolution. Part A. Elsevier, New York.

Vos, P., R. Hogers, M. Bleeker, M. Reijans, T. Van De Lee, M. Hornes, A. Frijters, J. Pot, J. Peleman, M. Kuiper, and M. Zabeau. 1995. AFLP: A new technique for DNA fingerprinting. Nucl. Acids Res. 23:4407–4414.

Waines, J.G., B. Ehdaie, and D. Barnhart. 1987. Variability in *Triticum* and *Aegilops* species for seed characteristics. Genome 29:41–46.

Walker, J.T., and N.R. Merritt. 1969. Genetic control of abnormal starch granules and high amylose content in mutant Glacier barley. Nature (London) 221:482–483.

Wall, J.S. 1990. The structural and genetic basis for the unique properties of cereal proteins. Cer. Foods World 35:715–724.

Walther, H., and K.H. Seibold. 1979. Improved protein mutants selected from barley after multiple EMS and x-ray treatments. p. 327–343. *In* Proc. Seed Improve. Nucl. Tech., Neuherberg, Germany. 4–8 Sept. 1978. IAEA.

Wang, M.L., A.R. Leitch, T. Schwarzacher, J.S. Heslop-Harrison, and G. Moore. 1992. Construction of a chromosome-enriched *HpaII* library from flow-sorted wheat chromosomes. Nucl. Acids Res. 20:1897–1901.

Webster, F.H. 1986. Oat utilization: Past, present, and future. p. 413–426. *In* F.H. Webster (ed.) Oats: Chemistry and technology. AACC, St. Paul, MN.

Weining, S., L. Ko, and R.J. Henry. 1994. Polymorphisms in the *α-amy1* gene of wild and cultivated barley revealed by the polymerase chain reaction. Theor. Appl. Genet. 89:509–513.

Wesenberg, D.M., L.W. Briggle, and D.H. Smith. 1992. Germplasm collection, preservation, and utilization. p. 793–820. *In* H.G. Marshall and M.E. Sorrells (ed.) Oat science and technology. ASA and CSSA, Madison, WI.

Wieser, H., W. Seilmeier, and H.-D. Belitz. 1994. Quantitative determination of gliadin subgroups from different wheat cultivars. J. Cer. Sci. 19:149–155.

Wilhelm, E, W. Kempf, A. Lehmussaari, and A. Caransa. 1989. Process technology for the production of starch, protein, and fiber from oats. Starke 10:372–376.

William, M.D.H.M., R.J. Pena, and A. Mujeeb-Kazi. 1993. Seed protein and isozyme variations in *Triticum tauschii* (*Aegilops squarrosa*). Theor. Appl. Genet. 87:257–263.

Williams, J.G.K., A.R. Kubelik, K.J. Livak, J.A. Rafalski, and S.V. Tingey. 1990. DNA polymorphisms amplified by arbitrary primers are useful as genetic markers. Nucl. Acids Res. 18:6531–6535.

Williams, P.C. 1988. The influence of chromosome number and species on wheat hardness. Cer. Chem. 63:56–57.

Wood, P.J. 1986. Oat β-glucan: Structure, location, and properties. p. 121–152. *In* F.H. Webster (ed.) Oats: Chemistry and technology. AACC, St. Paul, MN.

Wood, P.J., J.W. Anderson, J.T. Braaten, N.A. Cave, F.W, Scott, and C. Vachon. 1989. Physiological effects of β-D-glucan rich fractions from oats. Cer. Foods World 34:878–882.

Woolston, C.J., R. Barker, H. Gunn, M.I Boulton, and P.M. Mullineaux. 1988. Agroinfection and nucleotide sequence of cloned wheat dwarf virus DNA. Plant Mol. Biol. 11:35–43.

Wrigley, C.W., and G.J. McMaster. 1989. Wheat: Right for bread, but for more too. p. 21–39. *In* Y. Pomeranz (ed.) Wheat is unique: Structure, composition, processing, end-use properties, and products. AACC, St. Paul, MN.

Wu, Y.V. 1990. Recovery of protein-rich byproducts from oat stillage after alcohol distillation. J. Agric. Food Chem. 38:588–592.

Wu, Y.V., A.C. Stringfellow, and J.A. Bietz. 1990. Relation of wheat hardness to air-classification yields and flour particle size distribution. Cer. Chem. 67:421–427.

Wych, R.D., and D.C. Rasmusson. 1983. Genetic improvement in malting barley cultivars since 1920. Crop Sci. 23:1037–1044.

Wych, R.D., and D.D. Stuthman. 1983. Genetic improvement in Minnesota-adapted oat cultivars released since 1923. Crop Sci. 23:879–881.

Yamamori, M., T. Nakamura, and A. Kuroda. 1992. Variations in the content of starch-granule bound protein among several Japanese cultivars of common wheat (*Triticum aestivum* L.). Euphytica 64:215–219.

Yamazaki, W.T., and L.C. Andrews. 1982. Experimental milling of soft wheat cultivars and breeding lines. Cer. Chem. 59:41–45.

Yamazaki, W.T., and J.R. Donelson. 1983. Kernel hardness in some U.S. Wheats. Cer. Chem. 60:344–350.

Yamazaki, W.T., M. Ford, K.W. Kingswood, and C.T. Greenwood. 1981. Soft wheat production. p. 1–32. *In* W.T. Yamazaki and C.T. Greenwood (ed.) Soft wheat: Production, breeding, milling and uses. Vol. 6. AACC, St. Paul, MN.

Yasui, T., J. Matsuki, T. Sasaki, and M. Yamamori. 1996. Amylose and lipid contents, amylopectin structure, and gelatinisation properties of waxy wheat (*Triticum aestivum*) starch. J. Cer. Sci. 24:131–137.

Youngs, V.L. 1974. Extraction of a high-protein layer from oat groat bran and flour. J. Food Sci. 39:1045–1046.

Youngs, V.L., and R.A. Forsberg. 1979. Protein-oil relationships in oats (genotypes). Crop Sci. 19:798–802.

Youngs, V.L., and R.A. Forsberg. 1987. Oat. p. 457–499. *In* R.A. Olsen and K.J. Frey (ed.) Nutritional quality of cereal grains: Genetic and agronomic improvement. ASA and CSSA, Madison, WI.

Youngs, V.L, P. Püskülcü, and R.R. Smith. 1977. Oat lipids. I. Composition and distribution of lipid components in two oat cultivars. Cer. Chem. 54:803–812.

Zeller, F.J., and S.L.K. Hsam. 1983. Broadening the genetic variability of cultivated wheat by utilizing rye chromatin. p. 161–173. *In* Proc. 6th Int. Wheat Genet. Symp., Kyoto, Japan. Univ. Kyoto, Jpn.

7 Genetic Potential

DAVID M. PETERSON

USDA-ARS
University of Wisconsin
Madison, Wisconsin

L. CURTIS HANNAH

University of Florida
Gainesville, Florida

WILLIAM D. HITZ and ANTHONY J. KINNEY

DuPont Experimental Station
Wilmington, Delaware

WILLIAM R. MEREDITH, JR.

USDA-ARS
Stoneville, Mississippi

MARY ANN L. SMITH

University of Illinois
Urbana, Illinois

PAMELA J. WHITE

Iowa State University
Ames, Iowa

Designing crops to achieve specific goals for value-added traits involves a complex interaction of plant genetics with environment, economic forces, cultural needs and desires, projected uses, and available technology. In this chapter, we primarily consider the genetic potential inherent within plant genomes to achieve specific quantitative or qualitative traits that may add value to the harvested crop. However, it is first necessary to consider briefly other factors that impinge on the ability and desirability of research to impose such changes in plant characteristics.

One of the first considerations is whether genetic alteration of a crop species is the best method to achieve a desired goal. Alternative methods, such as processing or substitution by other commodities must be considered with respect to cost and convenience. This can be illustrated with an example on the formulation of animal feedstuffs. Is it better to supplement an energy source, such as corn (*Zea mays* L.), with a protein source, such as soybean (*Glycine max* L. Merr.) meal, or to develop high-protein corn hybrids that can be fed without supplementation? Similarly, there is a choice between breeding cultivars with better balanced protein and supplementing feed with amino acids from microbial sources. The answer may differ among geographic regions. In the USA, protein sources are relatively cheap, whereas in other locations, such as northern Europe, imported protein meal may be more expensive. For human food, the practice of adding vitamins and minerals to cereal products is well established, whereas this may not be feasible in foods that are often marketed unprocessed, such as fruits and vegetables.

Another consideration is the ability of the market to maintain cultivar identity. The development of value-added cultivars will necessitate their segregation throughout the marketing and processing chain to the consumer. This will be essential to preserve their value and because the new cultivars may be designed for special purposes. This practice is well established for some commodities, such as malting barley (*Hordeum vulgare* L.), where cultivar identity is considered essential by the maltsters because of different optimum malting conditions.

Stability of the value-added trait across growing environments is extremely important. This, of course, has always been true for plant improvement through breeding.

Another consideration is changing needs or requirements. Our dietary needs and prohibitions are continually being updated by advances in medical and nutritional science. We must have confidence that the trait we are attempting to introduce today will still be considered important a decade or more in the future when the improved cultivars are available for use. In other words, we need to distinguish real needs from food fads.

Economic return to the grower must always be considered. Value-added traits must either not decrease yield per hectare significantly, or if they do, the increased value must offset the lowered yield. Price volatility makes it extremely difficult for the plant breeder to select appropriate goals for crop improvement that will still be economically viable in future years. For the grower, the option of selecting seed types and management practices based on predictable market prices would be desirable.

These considerations notwithstanding, the remainder of this chapter will consider some goals for adding valuable traits to crop cultivars through genetic manipulation. Examples of prior successes will be discussed as models for future research. Possible routes to attain the goals will be suggested. The chapter is subdivided by constituents, under which particular crops are considered as appropriate. The protein section (Peterson, 2000, see Chapter 7A) considers progress and prospects for increasing protein quantity or improving protein quality, primarily in the grain crops. The oil section (Kinney & Hitz, 2000, see Chapter 7B) examines altering fatty acid content in oilseed crops for edible and industrial oils. The starch section (Hannah, 2000, see Chapter 7C) takes a different approach, capitalizing on the wealth of knowledge that has been obtained from the genetic and biochemical examination of corn mutants to elucidate pathways of starch biosynthesis. The commercial usefulness of many of these mutants is pointed out. Research to increase dietary fiber in oat (*Avena sativa* L.), barley, rye (*Secale cereale* L.), mustard

(*Brassica nigra* L. Koch), and flaxseed (*Linum usitatissimum* L.) in described in White (2000, see Chapter 7D). Cotton (*Gossypium hirsutum* L.) is detailed as the primary example of an industrial fiber crop in Meredith Jr. (2000, see Chapter 7E). The last section on functional foods (Smith, 2000, see Chapter 7F) describes the multitude of phytochemicals present in vegetables and fruits that could be enhanced to improve the health benefits of food.

REFERENCES

Hannah, L.C. 2000. Starch biosynthesis and genetic potential. p.181–199. *In* C.F. Murphy and D.M. Peterson (ed.) Designing crops for added value. Agron. Monogr. 40. ASA, CSSA, and SSSA, Madison, WI.

Kinney, A.J., and W.D. Hitz. 2000. Oils. p. 167–180. *In* C.F. Murphy and D.M. Peterson (ed.) Designing crops for added value. Agron. Monogr. 40. ASA, CSSA, and SSSA, Madison, WI.

Meredith, Jr., W.R. 2000. Designing cotton for added value. p. 215–227. *In* C.F. Murphy and D.M. Peterson (ed.) Designing crops for added value. Agron. Monogr. 40. ASA, CSSA, and SSSA, Madison, WI.

Peterson, D.M. 2000. Protein. p. 151–166. *In* C.F. Murphy and D.M. Peterson (ed.) Designing crops for added value. Agron. Monogr. 40. ASA, CSSA, and SSSA, Madison, WI.

Smith, M.A.L. 2000. Functional food value of designer crops. p. 229–237. *In* C.F. Murphy and D.M. Peterson (ed.) Designing crops for added value. Agron. Monogr. 40. ASA, CSSA, and SSSA, Madison, WI.

White, P.J. 2000. Dietary fiber. p. 201–213. *In* C.F. Murphy and D.M. Peterson (ed.) Designing crops for added value. Agron. Monogr. 40. ASA, CSSA, and SSSA, Madison, WI.

7A Protein

DAVID M. PETERSON

USDA-ARS
University of Wisconsin
Madison, Wisconsin

Crops used for human food and animal feed provide energy, protein, vitamins, minerals and fiber, all essential for the diet. In grain crops, the protein component is often lower than the optimum for meeting dietary requirements, and the balance of essential amino acids is less than optimum for their most efficient utilization by humans and nonruminants. Legumes, while high in protein, are often deficient in methionine, an essential amino acid for humans. For these reasons, there has been considerable research and breeding directed towards increasing the protein concentrations of grains and altering the amino acid compositions of grains and legumes to provide a more favorable balance of essential amino acids. For the U.S. food market, protein has not been considered particularly important, due to the abundance of well-balanced protein from animal sources. With the increasing popularity of vegetarian diets, this may change. In less developed countries, where the diet may be predominantly from plant products, benefits from higher protein and better amino acid balance could be realized. In the case of animal feed, it is largely a matter of economics whether to supplement grains with protein meal from soy (*Glycine max* L. Merr.), fish, or other high-protein sources or to breed cultivars that contain adequate protein without supplementation. Supplementation of rations with amino acids from bacterial sources also is practiced to gain more efficient protein utilization.

At a recent workshop that brought together plant breeders, food and feed technologists, and animal nutritionists, the potential value of designing more nutritious food and feed grains was explored [ARS Feed (and Food) Grain Nutr. Composit. Workshop, 1992]. The consensus of workshop participants was that increasing protein and altering amino acid composition through breeding/biotechnology would be useful in many cases.

Aside from its nutritional value, protein is crucial to wheat baking quality. Considerable advances have been made in identifying specific gluten proteins associated with baking quality. Plant breeders can now screen breeding selections for the presence of these desirable proteins. This is enabling breeders to combine more favorable gluten proteins in one genotype than could be done by evaluating for baking quality directly.

Genetic engineering has led to the concept of using crops to produce industrial or pharmaceutical proteins, as has been done with microbes for many years. Genes for specific protein products are introduced into a crop such as alfalfa (*Medicago sativa* L.), along with a powerful promoter, so that the genes are over expressed in leaf tissue producing an abundance of the desired protein.

I. ADDED NUTRITIONAL VALUE FOR FOOD AND FEED

A. Corn

World production of corn (*Zea mays* L.) totaled 533×10^6 t in the 1992–1993 crop year, second only to wheat (*Triticum aestivum* L.) (USDA, 1994). U.S. production was 241×10^6 t or almost half of the world total. Worldwide, about 27% of the production is used for human food, either directly or as processed products, and 65% for animal feed (Glover & Mertz, 1987). In the USA, the proportion used for animal feed is much higher. The average protein content of corn is 9 to 11%, and in normal corn, the protein is severely deficient in the essential amino acids, lysine and tryptophan, and, to a lesser extent, threonine (Table 7–1). Because of this amino acid imbalance, corn is inadequate as the sole protein source for growing children and monogastric animals. In certain developing countries in Central and South America, eastern and southern Africa, and Mexico, where corn constitutes a significant proportion of the diet, development of corn genotypes with a better-balanced protein could lead to a more nutritious diet. Rations for monogastric animals using corn as the primary energy source are supplemented with a high-protein source, such as soybean meal, and/or amino acids. Depending on the availability and cost of the protein and amino acid supplements, development of high-yielding corn genotypes with more and better quality protein could be economically justified for animal feed.

The protein content of corn is under both genetic and environmental control. The historic selection experiment at the Illinois Agricultural Experiment Station has produced lines with 27% protein after more than 70 generations of selection (Dudley, 1977). However, protein concentration is negatively correlated with yield, and protein quality was not improved with increased quantity. Due to the availability of relatively inexpensive soybean meal, there is little emphasis on developing high-protein corn hybrids in the USA.

The discovery that a corn mutant with opaque, rather than vitreous, kernels had an altered amino acid balance that was about 50% higher in lysine and tryptophan (Table 7–1) was a major breakthrough towards development of a nutritionally superior corn (Mertz et al., 1964). *Opaque*-2 and several other mutants have increased lysine and tryptophan concentrations because of a decreased quantity of the lysine/tryptophan-poor zein fraction (Nelson, 1980). Numerous studies with growing children, adults, and animals have demonstrated the nutritional superiority of *opaque*-2 corn over its normal counterpart (Cromwell et al., 1967a,b; Bressani, 1992). Deficiencies in the *opaque*-2 mutant, such as lower yield and reduced kernel quality have been major impediments to its adoption for commercial production.

Table 7–1. Essential amino acid composition (g kg^{-1} protein) of normal and mutant corn and barley, as compared to oat and the FAO ideal pattern.

	Corn†		Barley‡			
Essential amino acid	Normal	*02*	Normal	1508	Oat§	FAO¶
Histidine	29	32	20	31	24	
Isoleucine	45	39	40	44	42	40
Leucine	188	116	72	80	75	70
Lysine	16	37	35	57	42	55
Methionine + cysteine	38	27	ND	ND	40	35
Phenylalanine + tyrosine	118	88	74	80	80	60
Threonine	35	37	28	42	33	
Tryptophan	3	7	ND	ND	ND	10
Valine	54	53	53	62	58	50

† Nelson et al. (1965)
‡ Recalculated from Bang-Olsen et al. (1987)
§ Pomeranz et al. (1973).
¶ FAO/WHO (1973).

These deficiencies have led to the development of corn genotypes retaining the high lysine/tryptophan characteristics of *opaque*-2, but with improved yield and more vitreous kernels. These genotypes, developed primarily at CIMMYT in Mexico through backcrossing and recurrent selection, are known as quality protein maize (QPM).

Several approaches were tried to produce QPM maize, but the most successful has been the use of modifier genes to restore kernel vitreousness (Villeagas et al., 1992). Initially, partially modified ears were selected from many different *opaque*-2 lines in tropical, subtropical and highland QPM germplasm. Selections were based on kernel phenotype, protein quantity, and lysine content. Modified *opaque*-2 donor stocks were used to convert a broad range of maize germplasm to modified *opaque*-2 types, resulting in pools and populations of QPM. Pools and populations were improved and monitored for yield, and many QPM genotypes available today compare favorably in yield with the best conventional varieties in developing countries. Current emphasis at CIMMYT has switched to the development of QPM inbreds with good combining ability, agronomic traits, and pest resistance that can be used in hybrid production or to form synthetics for use as open-pollinated varieties (Villeagas et al., 1992). The CIMMYT germplasm is being used in other tropical and subtropical countries, and active hybrid breeding programs are underway in several countries including South Africa and Brazil.

In the USA, breeding programs at Purdue University, University of Illinois, and Texas A & M University are working with QPM. However, with the great emphasis on breeding higher yielding hybrids, it is inevitable that the developed QPM inbreds will lag behind the best normal inbreds in producing high-yielding hybrids.

The zein (prolamin) proteins, which constitute 50% of the total seed protein, have been classified into four groups, α, β, γ, and δ, based on structural relationships (Larkins et al., 1989). These zeins contain high amounts of proline and glutamine, whereas lysine and tryptophan are absent. The protein specified by the wild type allele of the *opaque*-2 locus contains a leucine-zipper domain characteristic of some mammalian and fungal transcription activators. In an in vitro binding assay, the opaque-2 fusion protein bound to regions on the 5′ side of the coding sequence in a zein genomic clone (Schmidt et al., 1990), suggesting its role as an activator of zein synthesis. The reduction in α-zein in *opaque*-2 is accompanied by a corresponding increase in albumins, globulins and glutelins, fractions with more abundant lysine and tryptophan. Larkins and coworkers (Larkins et al., 1992) showed that in modified QPM genotypes, the γ-zein fraction is approximately doubled. It is unclear how the increased amount of γ-zein is related to the vitreous kernel phenotype. Like α-zein, γ-zein has no lysine or tryptophan, so the elevated γ-zein levels are not contributing to the higher lysine and tryptophan levels in the total protein.

Other approaches towards improving the amino acid composition in maize include double mutant combinations of high-lysine mutants, selection for a multiple aleurone characteristic, selection for improved protein quality in normal corn, selection for increased germ size, and selection for resistance to growth inhibition by lysine plus threonine in tissue culture. These techniques, none of which has progressed as far as the modified *opaque*-2 QPM, are described in some detail by Glover and Mertz (1987).

B. Wheat

Wheat (*Triticum aestivum* L. and *T. turgidum* L.) is the most important cereal crop worldwide, with production of 561 × 10^6 t in the 1991–1992 crop year. The U.S. production for the same period was 67 × 10^6 t, second only to corn in total production (USDA, 1994). The U.S. wheats have been divided into classes based upon suitability for particular uses. The hard red wheats, both winter and spring types, are used primarily for bread, whereas the soft red wheats are more suitable for cookies and pastry. Soft white wheats are used for the same end-products as the soft red grain, and also for some types of noodles. In the USA, hard white wheats are being developed that may be used for pan bread and other yeast-leavened products. The durum wheats (*T. turgidum* var. *durum*) are used for semolina and pasta.

Considerable research progress has been made relating the protein composition of wheat to baking quality. This will be covered in "Improved Baking Quality of Wheat Flour" below. The protein content and amino acid composition of wheat also are important from a nutritional standpoint, and research to increase protein content and/or alter amino acid balance is covered in this section. The protein content of wheat is under genetic and environmental control, but a large component of the variance is environmental. Hard red spring and durum wheats are relatively high in protein (13–17%), whereas the hard red winter wheats are slightly lower (11–15%)(Johnson & Mattern, 1987). Soft white wheats average about 10.5 to 11% protein with a wide range (7–15%). As with other cereal proteins, wheat protein is deficient in lysine relative to human needs for essential amino acids.

Johnson et al. (1978) showed that lysine, as a percentage of protein, declines as protein increases up to about 15%, and thereafter is level. The genetic variation for lysine was only about 0.5 percentage point, whereas that for protein was about five percentage points, indicating that lysine content can best be increased by selection and breeding for higher protein. Despite the apparent quality advantages of high-protein wheat, Johnson et al. (1985) stated that there is little economic incentive to breed high-protein cultivars, especially if they carry agronomic deficiencies. In Australia, however, where wheat for export is priced on a sliding scale for protein content, an economic model indicated a benefit of $53 million (U.S. $42 million) for each percentage point gain in protein (Voon & Edwards, 1992). In the USA, high protein in hard wheats may receive premiums, especially when high protein wheats are in short supply.

Two biological problems have hindered progress towards breeding higher protein wheats. First is the large environmental influence on protein content (Johnson & Mattern, 1987), and second is the strong negative relationship between protein and yield (Bhatia & Rabson, 1976). Johnson and Mattern with coworkers (ARS and Univ. Nebraska) screened almost 20 000 selections from the National Small Grains Wheat Collection to identify potential sources of high-protein genes (Johnson et al., 1978). Although variation from about 7 to 21% protein was found, most of the variation was environmental and not genetic.

Nevertheless, several sources of high protein genes have been identified and used to breed higher protein cultivars. Atlas 66, developed from a cross of rust-resistant 'Frondoso' with 'Redhart-Noll', was about two percentage points higher in protein than comparable soft wheat cultivars in the southeastern USA (Middleton et al., 1954). 'Nap Hal', an Indian introduction, was found to be high in protein (Watson et al., 1966). Progenies from an Atlas 66 × Nap Hal cross exhibited transgressive segregation, indicating that each cultivar had different genes for high protein (Johnson et al., 1973). An analysis of about 1000 accessions of the wild tetraploid wheat, *T. turgidum* var. *dicoccoides* revealed protein levels of 20 to 28% (population means) and 14 to 35% (individual accessions) (Avivi, 1977). This wild wheat crosses readily with both *T. turgidum* and *T. aestivum* producing fertile and semifertile F_1 hybrids (Morris & Sears, 1967).

Genetic studies have shown that protein content is a polygenic character with mostly additive gene effects (Feldman et al., 1990). However, a cross between the high-protein hard red spring cultivar, Minnpro, and MN7570 revealed a bimodal distribution for grain protein among the progeny, indicating that Minnpro has one major gene for high grain protein (Beninati & Busch, 1992).

The cultivar Lancota was developed by the ARS/Nebraska program by transferring high-protein genes from Atlas 66 into the hard red winter background. It did not receive wide acceptance due to weak straw and lack of winter hardiness (Johnson et al., 1985). Nap Hal was even less useful than Atlas 66 as a source of high-protein genes because of several associated undesirable agronomic and quality traits. 'Plainsman V', developed by a private seed company, received genes from an unknown source (Stein et al., 1992) that increased protein by two to three percentage points. Although it was not widely adapted, it has been an important source of high protein genes.

Despite considerable efforts, there have been few high-protein wheat cultivars developed that maintain their identity in the U.S. marketplace. An exception was 'Karl', a line with Plainsman V in its pedigree, that was bought on an identity preserved basis in Kansas in the early 1990s. A premium was paid for Karl based on its superior dough strength, bake performance, and protein content. Research and breeding have produced several cultivars with marginally higher protein that have received modest acceptance. Current breeding efforts are focused on improving agronomic traits, disease resistance, and baking quality, with less attention to increasing protein content.

C. Barley

World barley (*Hordeum vulgare* L.) production in 1992–1993 totaled 165 million t, with almost half of the total being produced in Europe. North American production was 21 million t, 51% in Canada, 47% in the USA, and a small amount in Mexico (USDA, 1994). Barley is a dual-purpose grain used for malting and livestock feeding. Relatively little is used directly for human food, especially in the USA.

Barley can meet protein requirements for animals at 12% protein, but 15% is preferred (Price, 1980). Constraints on breeding high-protein barley include negative correlations between protein content and starch content, lysine content, and yield (Scholz, 1976). In addition, there is a strong environmental influence on protein content. Malting cultivars, for which low protein is desired, are often used for feed if they don't meet malting specifications, leading to lower protein on average.

In the USA and Canada, the greater emphasis among barley breeders has been improved malting cultivars (Anderson & Reinbergs, 1985). Even in northern Europe, where a greater proportion of the barley crop is used for animal feed, the breeding emphasis has been on improving malting cultivars (L. Lehmann, 1993, personal communication). The primary emphasis on improving barley food and feed quality has been on increasing the lysine content, which, as in other cereal grains, is low. Hordein, the storage prolamin of barley is very low in lysine, whereas the albumins and globulins have higher levels. The discovery of 'Hiproly' barley (*lys1*), with 19.6% protein and 4.1% lysine (as percentage of protein) (Hagberg & Karlsson, 1969, Munck et al., 1970) led to research on developing a suitable high-lysine cultivar. The high-lysine content of Hiproly is attributed to elevated levels of salt-water-soluble proteins including β-amylase, protein Z, and the chymotrypsin inhibitors (Hejgaard & Boisen, 1980; Tallberg, 1982), all of which have high lysine levels. Hiproly has insufficient yield and kernel quality to achieve commercial acceptance.

Another mutant, Risø 1508, which was obtained by mutagenesis from the cultivar Bomi, has excellent amino acid composition, lysine and threonine being increased by 42 and 36%, respectively (Ingverson et al., 1973). Shrunken kernels and low yield make it unacceptable as a commercial cultivar (Doll et al., 1974). Risø 1508 has a mutation in a regulatory gene (*lys3a*) that controls hordein synthesis, resulting in lower hordein content (Kreis et al., 1984).

Extensive research efforts at the Carlsberg Research Laboratories in Denmark have been made to breed high-lysine barley cultivars with acceptable agronomic characteristics (Bang-Olsen et al., 1987; Munck, 1992). The best results have been obtained by crossing Risø 1508 with large-seeded plump cultivars and selecting and backcrossing plump recombinants. One of these was then crossed with Triumph, a high-starch malting cultivar. A selection from this cross, Ca 700202, has stable yield that approaches that of standard feed cultivars, seed is less shriveled than Risø 1508, and the kernels are smaller. Starch content, which was a major deficiency in Risø 1508, is equivalent to that of standard feed barleys. A key factor in the success of this breeding effort was the fact that the Risø 1508 phenotype had an enlarged germ and scutellum and could therefore be recognized visually without costly and time-consuming assays of the thousands of lines evaluated (Munck, 1992).

Nutritional studies have confirmed that high-lysine barley could substitute for part of the soybean meal in swine diets (Munck, 1992). Best results were obtained when the protein content of the barley was above 11%.

In the final analysis, the value of high-lysine barleys will depend not only on their relative yield potential, compared to that of the best normal cultivars, but also on the relative price of high-protein feed supplements, such as soybean meal. When soybean meal is cheap, the main purpose of the barley is for energy. Fortunately, the digestible energy of the *lys3a* lines compares favorably with that of normal cultivars, because the mutants have 50% more oil (Munck, 1992).

D. Other Cereals

There have been efforts to add value to the other major cereals, rice (*Oryza sativa* L.), oat (*Avena sativa* L.), and sorghum [*Sorghum bicolor* (L.) Moench] by breeding for higher protein and/or improved amino acid composition. Incremental gains in protein content have been made, but there is little incentive for a concerted effort in protein improvement. Higher protein often is achieved only at the expense of lower yield, reduced percentage of lysine, or resources needed to breed for more crucial objectives.

Cultivated rice has a low concentration of protein, 8 to 9% in brown rice, and about 1% lower in milled rice at 14% moisture (Juliano, 1979). The protein concentration range among over 17 000 entries in the world collection was 4.3 to 18.2% (Coffman & Juliano, 1987), indicating sufficient genetic diversity to breed for higher protein. However, protein concentration is strongly sensitive to growing environment (Gomez, 1979). Also, breeders have experienced little success in using world collection entries as a source of high-protein genes because of associated undesirable agronomic traits (Coffman & Juliano, 1979). Although some progress was made by screening for protein among advanced breeding lines, the best lines identified and tested were not competitive for yield with the newest conventional-protein cultivars (Coffman & Juliano, 1987).

The lysine content of rice protein, 3.5 to 4.0%, is better than that of most cereals. Because of this, and due to the absence of much genetic diversity for lysine, there has been almost no effort to improve this characteristic (Coffman & Juliano, 1987). In vitro selection of anther-derived rice cells in the presence of inhibitory levels of lysine and threonine, followed by treatment with the lysine analog, *S*-(aminoethyl)cysteine, produced regenerated plants, some of which, in the second and third generation, had elevated levels of protein and lysine (Schaeffer & Sharpe, 1987). However, a broad spectrum of variant types was induced in the regenerated plants, and the high-protein, high-lysine somaclones had a soft, chalky endosperm and reduced seed size.

According to Coffman and Juliano (1987), "High protein rice would be of added value to consumers only under specific situations where protein is deficient and energy and mineral intake is adequate." Therefore, the investment of significant resources in this effort is probably not justifiable at this time.

Sorghum is used primarily as a feed grain in the USA, but in Africa and other tropical areas, it is grown for food and for alcoholic and nonalcoholic beverages. A wide variation in grain types exists, as does their preparation and use by consumers. Protein content is variable, ranging from 7 to 14% (Hulse et al., 1980), and is strongly influenced by the environment. The amino acid composition of sorghum protein is similar to that of corn; lysine is the first limiting amino acid (Wall & Paulis, 1978). Protein digestibility is adversely affected by the high polyphenol contents of some sorghums. The method of preparation of sorghum foods may affect protein utilization more than the genotype. Alkaline preparation procedures actually inhibit some of the deleterious effects of polyphenols and improve protein availability (Kirleis & Axtell, 1985).

While breeding for high-protein, high-lysine cultivars has not been a high priority, several genes affecting these characters have been identified. Ethiopian lines were discovered that contained a gene, *hl*, that increased protein concentration by three to five percentage points and doubled lysine concentration on a total weight basis (Singh & Axtell, 1973). However, associated negative effects on endosperm development could not be overcome, and the gene has not been useful commercially. Another high-lysine sorghum with opaque endosperm was obtained by chemical mutagenesis (Mohan & Axtell, 1974), but recovery of full expression in crosses has not been realized. A gene was discovered that increases grain protein almost 50% and is linked to a red-leaf character (Ejeta & Axtell, 1985). This gene did not affect endosperm type or size.

Oat protein is recognized as being of superior quality, and its concentration, expressed on a groat basis, is higher than that of other cereals. A survey of selections from the U.S. National Small Grains Collection found a range in protein concentration from 12.4 to 24.4% (Robbins et al., 1971), although most commercially grown cultivars are in the 15 to 20% range. In the 1970s and early 1980s, breeders strived to increase the protein concentration of oat cultivars, in some cases introgressing germplasm from high-protein *Avena sterilis* L. genotypes. Several high-protein cultivars were released and some achieved wide acceptance, such as ‘Dal’ (University of Wisconsin), ‘Hinoat’ (Agriculture Canada), and ‘Otee’ (University of Illinois). There has never been an attempt to maintain cultivar identity in the marketplace or offer a premium for higher protein, except for a small contract program offered by The Quaker Oats Company, where cultivars known to produce grain with good milling quality and high protein concentration were specified (B. Roskins, 1994, personal communication).

Lysine in oat protein is equal to that of rice, and better than that of other cereals (Robbins et al., 1971). This is due to the high proportion of globulin in oat protein (Peterson & Brinegar, 1986), unlike other cereals where lysine-poor prolamin predominates. Peterson (1976) analyzed a series of oat samples that varied widely in protein concentration because of genetic or environmental effects, and he found very little influence of protein concentration on amino acid composition. He concluded that there was little expectation of improving the amino acid balance through conventional breeding. Conversely, elevating protein concentration through breeding would not adversely affect lysine concentration, as is the case for some other cereals.

E. Improving the Quality of Legume Proteins by Transformation with Heterologous Genes

In addition to conventional and mutant breeding approaches to improving the quality of seed proteins, some molecular approaches are being attempted. One possibility is to alter the regulatory sequences to enhance the production of specific proteins with desirable characteristics. Alternatively, the coding sequences of specific storage proteins can be altered to change their amino acid balances in a favorable manner (Hoffman et al., 1988). Another approach is to transfer a gene coding for a protein rich in the desired amino acid(s) into the target (economic) species.

Legume seed proteins typically are deficient in methionine. A protein unusually rich in methionine is the 2S albumin of Brazil Nut (*Bertholletia excelsa* H.B.K.), in which over 18% of the amino acid residues are methionine (Sun et al., 1987b). The protein consisted of a 3-kD and 9-kD subunit, both derived from a 17-kD precursor (Sun et al., 1987a). Workers reasoned that if the genes for this protein could be cloned, inserted into target species, and expressed in seeds at sufficient levels, overall methionine levels would be significantly boosted. The gene for the 17-kD sequence was coupled to a promoter from the phaseolin gene and a selectable marker and used to transform tobacco (*Nicotiana tabacum* L.)(Altenbach et al., 1989). The tobacco seeds expressed and processed the protein into its subunits. The protein accumulated at levels as high as 8% of total protein, and methionine in the total seed protein was increased by as much as 30%. Subsequently, canola (*Brassica napus* L.) plants were transformed with the same gene, and transgenic plants accumulated the heterologous protein in seeds at levels from 1.7 to 4.0% of total protein causing methionine levels to increase up to 33% (Altenbach et al., 1992). Plants with multiple gene copies produced more protein than those with a single copy.

Brazilian workers (Aragao et al., 1992), using the particle bombardment technique, achieved transient expression of the 2S Brazil nut albumin gene in bean (*Phaseolus vulgaris* L.). Other workers demonstrated the stable expression and Mendelian inheritance of the gene in the grain legume *Vicia narbonensis* L. (Saalbach et al., 1994). However, under the influence of the CaMV promoter, the gene was expressed in both vegetative and reproductive tissues.

The Altenbach group indicated that the 2S albumin gene was transferred successfully to other commercial canola varieties and speculated that it could be introduced into other crop species. Their construct, using the phaseolin promoter, appeared to give higher levels of expression than constructs used by others. While this approach to protein quality improvement appears to be viable and may eventually produce useful, value-added cultivars, the recent discovery that the 2S Brazil nut albumin is a major allergen will limit the usefulness of this particular gene (Nordlee et al., 1996).

II. IMPROVED BAKING QUALITY OF WHEAT FLOUR

Baking quality of wheat flour, as measured by several rheological procedures, is dependent on the complex biochemistry and interactions of protein, lipid, carbohydrate, and other components. However, gluten proteins are the major determinants of baking quality, and researchers have shown that the amounts and properties of specific gluten polypeptides can significantly influence baking quality. Gluten proteins have been classified as glutenins and gliadins based on differential solubilities in alcoholic solutions. High-molecular-weight (HMW) glutenins, low-molecular-weight (LMW) glutenins, and gliadins can be separated by sodium dodecyl sulfate-polyacrylamide gel electrophoresis (SDS-PAGE). There is a voluminous literature on the genetics, biochemistry, and molecular biology of wheat gluten proteins that is beyond the scope of this chapter, but several reviews are available (Payne, 1987; Tatham et al., 1990; MacRitchie et al., 1990; Shewry et al., 1992).

Wheat quality can be defined in terms of its expected use, and this discussion will focus on the baking quality of hexaploid or bread wheats, which to a large extent is a function of dough strength. High-dough-strength flours produce breads with acceptable textures and loaf volumes, whereas low-dough-strength flours are more suitable for other purposes.

Early research indicated that dough strength is a function of the gluten proteins. The network of gluten proteins in the dough has elastic and viscous properties, and poor quality doughs lack sufficient elasticity (Shewry et al., 1994). Gliadins are single, monomeric proteins associated by nonionic forces, and they confer viscosity to the dough. Glutenins consist of HMW and LMW polypeptides assembled into large polymers by disulfide bonds, and they confer elasticity. Although all gluten proteins affect baking quality, research has focused on the HMW glutenin subunits as the most important proteins affecting baking quality. These subunits are coded by *Glu-1* loci on the long arms of chromosomes 1A, 1B, and 1D (Payne, 1987). Each locus contains two genes that encode subunits of low and high mobilities by SDS-PAGE. These are called the x and y subunits, respectively. Thus, six subunits may theoretically exist in any given genotype, but an analysis of 300 cultivars has shown that only three, four, or five are present in any one genotype (Lawrence & Shepherd, 1980; Payne et al., 1981). All cultivars contain a 1Bx, 1Dx, and 1Dy subunit, and some also contain a 1By and/or 1Ax subunit. Figure 7–1 shows the frequency of the subunit patterns.

Correlative studies have related the presence or absence of particular HMW glutenin subunits with baking quality (Payne et al., 1987). As indicated in Table 7–2, lines with 5 + 10 generally have greater dough strength than lines with 2 + 12 or 4 + 12, provided protein concentrations are equal. Similarly, lines with subunits 1 or 2* are usually of higher quality than those with a null allele on chromosome 1A. Studies of lines with glutenin protein polymorphisms have indicated that the association of baking characteristics with specific HMW glutenin polypeptides does not necessarily hold across all populations (Primard et al., 1991) or cultivars (Graybosch et al., 1994).

Genetic mutants have been used to produce near isogenic lines null for HMW glutenin subunits at one, two, or all three of the *Glu-1* loci (Lawrence et al., 1988). In this study, baking quality, as measured by dough development time, was correlated with the total quantity of HMW glutenin subunits but was independent of the particular subunit composition. Fractionation and reconstitution of flours supports the conclusion that dough strength is related to the proportion of glutenin in flour protein (MacRitchie, 1987).

(B)

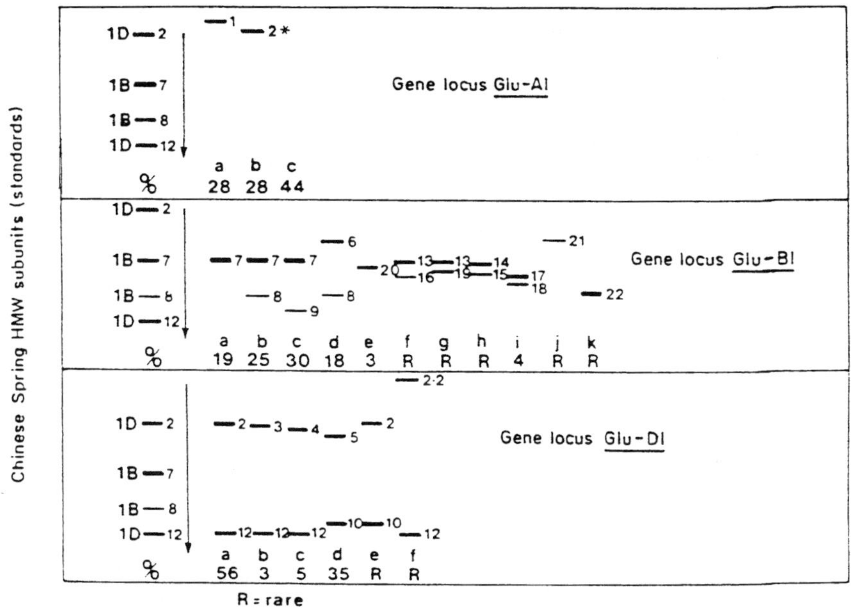

Fig. 7–1. Allelic variation of high-molecular-weight (HMW) subunits of wheat (*Triticum aestivum* L.) based on analysis of 300 cultivars. Electrophoric mobilities of subunits encoded by genes at *Glu-A1*, *Glu-B1*, and *Glu-D1* loci are shown and compared to subunits present in 'Chinese Spring' (Payne & Lawrence, 1983).

The objective of relating protein composition to quality is to enable the development, through breeding or direct gene insertion, of superior (value-added) wheat cultivars. Improving protein quality is not likely to reduce yield, in the way that increasing protein quantity does. Several approaches are being used. In conventional breeding programs, the progeny of crosses selected for agronomic characteristics are screened for baking *potential* by examining their protein compositions. This bypasses the more costly and time-consuming quality testing of early generation lines and can be done on smaller samples. Also, the protein compositions of parental lines are used to predict which crosses might successfully com-

Table 7–2. Quality scores assigned to HMW-glutenin subunits of wheat†.

	Locus		
Score‡	Glu-A1	Glu-B1	Glu-D1
4	--	--	5 + 10
3	1	17 + 18	--
3	2*	7 + 8	--
2	--	7 + 9	2 + 12
2	--	--	3 + 12
1	Null	7	4 + 12
1	--	6 + 8	--

† Payne et al. (1987).
‡ A score of 4 represents the best breadmaking quality, whereas a score of 1 represents the poorest.

bine favorable quality genes. The genes for nine HMW glutenin subunits have been cloned (see Shewry et al., 1992), indicating that transforming wheat plants with genes favorable for quality will be possible (Weeks et al., 1993). The HMW glutenin genes have been expressed in tobacco (Robert et al., 1989) and *Escherichia coli* (Galili, 1989), and the latter were tested in dough mixing experiments (Bekes et al., 1994). Results showed significant changes in functional properties when the expressed HMW glutenin subunits were incorporated into the glutenin structure. Experiments cited above indicate that the quantity of HMW glutenin subunits is correlated with baking quality and suggest that insertion of additional HMW subunit genes can improve quality (Anderson et al., 1994).

Knowledge gained about the structural differences between subunits that confer various qualities can be used to mutagenize specific genes in vitro and then reinsert the altered genes back into plants. For example, the degree of protein cross linking could be changed by inserting or replacing cysteine residues. Cysteine residues have been incorporated into storage proteins for direct evaluation of the consequences in mixing studies (Tamás et al., 1995). The length of the protein β-spiral region could be altered with insertions or deletions of one or more of the repeated domains (Shewry et al., 1992).

These approaches have not yet led to value-added wheat cultivars, but they are conceptually, if not yet technically feasible. As with the previously mentioned examples of nutritional improvement of other cereals, the problems of retaining or improving agronomic traits while simultaneously adding improved quality parameters will be a challenge for plant breeders.

III. DEVELOPMENT OF GENETICALLY ENGINEERED PLANTS AS BIOREACTORS FOR PROTEIN PRODUCTION

Genetic manipulation to enhance levels of specific proteins has occurred since the advent of plant breeding. The development of recombinant DNA technology has added a degree of control not previously available through cross-breeding and selection. Recently progress has occurred in developing plants as bioreactors to produce proteins that do not normally occur in the species. While this has been done with microbial fermentation systems for many years, a potential advantage of using plants is the economy of large-scale production.

The genetic potential for such systems is limited by the ability to transform the species with the gene of interest and achieve protein expression levels that are economical to extract and purify, without adverse effects on plant growth. Transformation systems are now in place for many species including field crops that may become useful as bioreactors.

There are many potential uses in medicine, agriculture, and industry for enzymes produced in plant bioreactors. Applications include biomass conversion for energy, biopulping, food and feed additives, cleaning products, environmental remediation, food processing, pharmaceuticals, and many others. Certain agricultural biotechnological companies are involved in expression of foreign proteins in plants, such as Agracetus (Madison, WI) and Mogen (The Netherlands), according to reports in trade journals. Agracetus has developed plant bioreactor systems based on transgenic corn and soybean that produce recombinant enzymes in seed.

A research project underway at the University of Wisconsin is producing foreign proteins in alfalfa (*Medicago sativa* L.) leaves. Alfalfa has several advantages. It is a widely grown perennial crop that is highly productive, with three or more harvests per year, and alfalfa is more environmentally benign than annual row crops. Furthermore, the residue remaining after extraction of the protein-containing juice is a valuable animal feed. This collaborative effort has involved constructing vectors, developing cell culture and transformation systems, introducing transgenes into adapted cultivars, developing methods for extraction and purification of the proteins, and calculating economic returns (S. Austin, 1994, personal communication).

Transgenic alfalfa plants that express two enzymes with potential industrial use have been produced and field tested (Mathews et al., 1993; Austin et al., 1994, 1995). One enzyme, Mn-dependent lignin peroxidase from the fungus, *Phanerochaete chrysoporium*, is useful in biopulping and biobleaching for the paper industry. Relatively high levels of expression were obtained, 0.01 to 1.0% of total soluble protein, but expression of the protein impaired plant growth and development (Austin et al., 1995). Attempts to purify the enzyme from plant tissues have been unsuccessful. The second enzyme, thermophyllic α-amylase from *Bacillus licheniformis*, is widely used in starch processing. The plasmid vector was modified to include a plant endoplasmic reticulum targeting sequence at the amino terminus and a sequence at the carboxy terminus to facilitate purification. Plant performance was unaffected, but expression in transgenic alfalfa was low, only 0.001 to 0.01% of total soluble protein.

Future plans include the transformation of alfalfa with genes for cellulases and phytase. Cellulases are widely used in biomass conversion for fuel and chemical feedstocks, and phytase is useful as a feed additive to improve the availability of P.

The successful application of this technology to use field crops to produce specific foreign proteins will add tremendous value to the crop above that obtained for traditional feed and food uses. To achieve required levels of expression, gene products may need to be targeted to be secreted into the apoplast to minimize cellular damage. Alternatively, an inducible promoter (Ward et al., 1993) could be used to limit gene expression to a period after the majority of growth has occurred.

REFERENCES

Altenbach, S.B., C.-C. Kuo, L.C. Staraci, K.W. Pearson, C. Wainwright, A. Georgescu, and J. Townsend. 1992. Accumulation of a Brazil nut albumin in seeds of transgenic canola results in enhanced levels of seed protein methionine. Plant Mol. Biol. 18:235–245.

Altenbach, S.B., K.W. Pearson, G. Meeker, L.C. Staraci, and S.S.M. Sun. 1989. Enhancement of the methionine content of seed proteins by the expression of a chimeric gene encoding a methionine-rich protein in transgenic plants. Plant Mol. Biol. 13:513–522.

Anderson, O.D., A.E. Blechl, F.C. Greene, and J.T. Weeks. 1994. Progress towards genetic engineering of wheat with improved quality. p. 87–95. *In* R.J. Henry and J.A. Ronalds (ed.) Improvement of cereal quality by genetic engineering. Plenum Press, New York.

Agagao, F.J.L., M.F. Grossi de Sa, E.R. Almeida, E.S. Gander, and E.L. Rech. 1992. Particle bombardment-mediated transient expression of a Brazil nut methionine-rich albumin in bean (*Phaseolus vulgaris* L.). Plant Mol. Biol. 20:357–359.

Agricultural Research Service. 1992. Feed (and food) grain nutrient composition workshop. In C.F. Murphy (ed.) Rep. ARS Feed (and Food) Grain Nutrient Composit. Workshop, St. Louis, MO. 4–6 May. ARS.

Austin, S., E.T. Bingham, D.E. Matthews, M.N. Shahan, J. Will, and R.R. Burgess. 1994a. Production and field performance of transgenic alfalfa (*Medicago sativa* L.) expressing alpha-amylase and manganese-dependent lignin peroxidase. Euphytica 85:381–393.

Austin, S., E.T. Bingham, R.G. Koegel, D. Matthews, M. Shahan, R.J. Straub, and R.R. Burgess. 1994b. An overview of a feasibility study for the production of industrial enzymes in transgenic alfalfa. Ann. N.Y. Acad. Sci. 721:234–244.

Avivi, L. 1977. High grain protein content in wild wheat. Can. J. Genet. Cytol. 19:569–570.

Bang-Olsen, K., B. Stilling, and L. Munck. 1987. Breeding for yield in high-lysine barley. p. 865–870. *In* S. Yasuda and T. Konishi (ed.) Barley genetics. Proc. 5th Int. Barley Genet. Symp., Vol. 5, Okayama, Japan. 6–11 Oct. 1986. Barley Genet. Symp. Organiz. Commit.

Bekes, F., O. Anderson, P.W. Gras, R.B. Gupta, A. Tam, C.W. Wrigley, and R. Appels. 1994. The contributions to mixing properties of 1D HMW glutenin subunits expressed in a bacterial system. p. 97–103. *In* R.J. Henry and J.A. Ronalds (ed.) Improvement of cereal quality by genetic engineering. Plenum Press, New York.

Beninati, N.F., and R.H. Busch. 1992. Grain protein inheritance and nitrogen uptake and redistribution in a spring wheat cross. Crop Sci. 32:1471–1475.

Bressani, R. 1992. Nutritional value of high-lysine maize in humans. p. 205–224. *In* E.T. Mertz (ed.) Quality protein maize. Am. Assoc. Cer. Chem., St. Paul, MN.

Coffman, W.R., and B.O. Juliano. 1979. Seed protein improvement in rice. p. 261–277. *In* Proc. Int. Symp. Seed Protein Improve. Cer. Grain Legumes, Vol. 2, Neuherburg, Fed. Rep. Germany. 4–8 Sept. 1978. IAEA, Vienna.

Coffman, W.R., and B.O. Juliano. 1987. Rice. p. 101–131. *In* R.A. Olsen and K.J. Frey (ed.) Nutritional quality of cereal grains: Genetic and agronomic improvement. ASA, Madison, WI.

Cromwell, G.L., R.A. Pickett, and W.M. Beeson. 1967a. Nutritional value of opaque-2 corn for swine. J. Anim. Sci. 26:1325–1331.

Cromwell, G.L., J.C. Rogler, W.R. Featherston, and R.A.Pickett. 1967b. Nutritional value of opaque-2 corn for the chick. Poult. Sci. 46:705–712.

Doll, H., B. Køie, and B.O. Eggum. 1974. Induced high lysine mutants in barley. Rad. Bot. 14:73–80.

Dudley, J.W. 1977. 76 generations of selection for oil and protein percentage in maize. p. 459–472. *In* Proc. Int. Conf. Quant. Genet. Iowa State Univ. Press, Ames.

Ejeta, G., and J.D. Axtell. 1985. Mutant gene in sorghum causing leaf reddening and increased protein concentration in the grain. J. Hered. 76:301–302.

Feldman, M., L. Avivi, A.A. Levy, M. Zaccai, Y. Avivi, and E. Millet. 1990. High protein wheat. p. 593–614. *In* Y.P.S. Bajaj (ed.) Wheat. Springer-Verlag, Berlin.

Glover, D.V., and E.T. Mertz. 1987. Corn. p. 183–336. *In* R.A. Olsen and K.J. Frey (ed.) Nutritional quality of cereal grains: Genetic and agronomic improvement. ASA, CSSA, and SSSA, Madison, WI.

Gomez, K.A. 1979. Effect of environment on protein and amylose content of rice. p. 59–68. *In* Chemical aspects of rice grain quality. IRRI, Los Baños, Philippines.

Hagberg, A., and K.-E. Karlsson. 1969. Breeding for high protein content and quality in barley. p. 17–21. *In* New approaches to breeding for improved plant protein. Proc. Panel, Rötånga, Sweden. 17–21 June 1968. IAEA, Vienna.

Hejgaard, J., and S. Boisen. 1980. High-lysine proteins in Hiproly barley breeding: Identification, nutritional significance and new screening methods. Hereditas 93:311–320.

Hoffman, L.M., D.D. Donaldson, and E. M. Herman. 1988. A modified storage protein is synthesized, processed, and degraded in the seeds of transgenic plants. Plant Mol. Biol. 11:717–729.

Hulse, J.H., E. Laing, and D.E. Pearson. 1980. Sorghum and the millets: Their composition and nutritive value. Acad. Press, New York.

Ingverson, J., B. Køie, and H. Doll. 1973. Induced seed protein mutant of barley. Experientia 29:1151–1152.

Johnson, V.A., and P.J. Mattern. 1987. Wheat, rye, and triticale. p. 133–182. *In* R.A. Olsen and K.J. Frey (ed.) Nutritional quality of cereal grains: Genetic and agronomic improvement. ASA, CSSA, and SSSA, Madison, WI.

Johnson, V.A., P.J. Mattern, and C.J. Peterson. 1985. Improvement of wheat protein by traditional breeding and genetic techniques. Cer. Chem. 62:350–355.

Johnson, V.A., P.J. Mattern, J.W. Schmidt, and J.E. Stroike. 1973. Genetic advances in wheat protein quantity and composition. p. 547–556. *In* E.R. Sears and L.M.S. Sears (ed.) Proc. 4th Int. Wheat Genet. Symp., Columbia, MO. 6–11 August. Univ. Missouri, Columbia.

Johnson, V.A., P.J. Mattern, K.D. Wilhelmi, and S.L. Kuhr. 1978. Seed protein improvement in common wheat (*Triticum aestivum* L.): Opportunities and constraints. p. 23–32. *In* Seed protein improvement by nuclear techniques. IAEA-RC-57-2. Proc. Two Res. Coord. Meet., Baden and Vienna, Austria. 28 Mar.–1 Apr. 1977 and 26–30 Sept. 1977. IAEA, Vienna, Austria.

Juliano, B.O. 1979. Chemical basis of rice grain quality. p. 69–90. *In* Chemical aspects of rice grain quality. IRRI, Los Baños, Philippines.

Kirleis, A.W., and J.D. Axtell. 1985. Chemical and physical aspects of food and nutritional quality of sorghum. p. 153–156. *In* J.F. Winn (ed.) Fighting hunger with research. INTSORMIL, U.S. Agency for Int. Develop., Washington, DC, and Lincoln, NE.

Kreis, M., P.R. Shewry, B.G. Forde, S. Rahman, M.B. Bahramian, and B.J. Miflin. 1984. Molecular analysis of the effects of the mutant *lys3a* gene on the expression of Hor loci in developing endosperms of barley (*Hordeum vulgare*). Biochem. Genet. 22:231–255.

Larkins, B.A., C.R. Lending, J.C. Wallace, G. Galili, E.E. Kawata, K.B. Geetha, A.L. Kriz, D.N. Martin, and C.E. Bracker. 1989. Zein gene expression during maize endosperm development. p. 109–120. *In* R.B. Goldberg (ed.) The molecular basis of plant development. Alan R. Liss, Inc., New York.

Larkins, B.A., M.A. Lopes, and C.R. Lending. 1992. A genetic, biochemical and ultrastructural analysis of modified *opaque*-2 maize. p. 174–204. *In* E.T. Mertz (ed.) Quality protein maize. Am. Assoc. Cer. Chem., St. Paul, MN.

Lawrence, G.J., F. MacRitchie, and C.W. Wrigley. 1988. Dough and baking quality of wheat lines deficient in glutenin subunits controlled by the *Glu-A1*, *Glu-B1* and *Glu-D1* loci. J. Cer. Sci. 7:109–112.

Lawrence, G.J., and K.W. Shepherd. 1980. Inheritance of glutenin protein subunits of wheat. Aust. J. Biol. Sci. 33:221–233.

MacRitchie, F. 1987. Evaluation of contributions from wheat protein fractions to dough mixing and breadmaking. J. Cer. Sci. 6:259–268.

MacRitchie, F., D.L. du Cros, and C.W. Wrigley. Flour polypeptides related to wheat quality. p. 79–145. *In* Y. Pomeranz (ed.) Advances cereal science technology. Vol. 10. Am. Assoc. Cer. Chem., St. Paul.

Mathews, D., S. Austin, J. Monorama, J. Will, M. Shahan, R. Amasino, and R.R. Burgess. 1993. Expression of fungal lignin degrading enzymes in plants. Plant Physiol. (Suppl.) 102: 960.

Mertz, E.T., L.S. Bates, and O.E. Nelson. 1964. Mutant gene that changes protein composition and increases lysine content of maize endosperm. Science (Washington, DC) 145:279–280.

Middleton, G.K., C.E. Bode, and B.B. Bayles. 1954. A comparison of the quantity and quality of protein in certain varieties of soft wheat. Agron. J. 46:500–502.

Mohan, D.P., and J.D. Axtell. 1974. Chemically induced hugh-lysine mutants in *Sorghum bicolor* (L.) Moench. p. 66. *In* Agronomy abstracts. ASA, Madison, WI.

Morris, C.F., and E.R. Sears. 1967. The cytogenetics of wheat and its relatives. p. 19–87. *In* K.S. Quisenberry and L.P. Reitz (ed.) Wheat and wheat improvement. ASA, Madison, WI.

Munck, L. 1992. The case of high-lysine barley breeding. p. 573–601. *In* P.R. Shewry (ed.) Barley: genetics, biochemistry, molecular biology and biotechnology. C.A.B. Int., Wallingford, UK.

Munck, L., K.-E. Karlsson, A. Hagberg, and B.O. Eggum. 1970. Gene for improved nutritional value in barley seed protein. Science (Washington, DC) 168:985–987.

Nelson, O.E. 1980. Genetic control of polysaccharide and protein synthesis in the endosperms of barley, maize, and sorghum. p. 41–71. *In* Y. Pomeranz (ed.) Advances cereal science technology. Vol. 3. Am. Assoc. Cer. Chem., St. Paul, MN.

Payne, P.I. 1987. Genetics of wheat storage proteins and the effect of allelic variation on bread-making quality. p. 141–153. *In* W.R. Briggs et al. (ed.) Annual review of plant physiology. Vol. 38. Annu. Rev. Inc., Palo Alto, CA.

Payne, P.I., L.M. Holt, and C.N. Law. 1981. Structural and genetical studies on the high-molecular-weight subunits of wheat glutenin. Part 1. Allelic variation in subunits amongst varieties of wheat. Theor. Appl. Genet. 60:229–236.

Payne, P.I., M.A. Nightingale, A.F.Krattinger, and L.M. Holt. 1987. The relationship between HMW glutenin subunit composition and the bread-making quality of British-grown wheat varieties. J. Sci. Food Agric. 40:51–65.

Peterson, D.M. 1976. Protein concentration, concentration of protein fractions, and amino acid balance in oats. Crop Sci. 16:663–666.

Peterson, D.M., and A.C. Brinegar. 1986. Oat storage proteins. p. 153–203. *In* F.H. Webster (ed.) Oats: Chemistry and technology. Am. Assoc. Cer. Chem., St. Paul, MN.

Price, P.B. 1980. Barley proteins. South Dakota State Univ. Ext. Circ. FS 759.

Robbins, G.S., Y. Pomeranz, and L.W. Briggle. 1971. Amino acid composition of oat groats. J. Agric. Food Chem. 10:536–539.

Saalbach, I., T. Pickardt, F. Machemehl, G. Saalbach, O. Schieder, and K. Müntz. 1994. A chimeric gene encoding the methionine-rich 2S albumin of the Brazil nut (*Bertholletia excelsa* H.B.K.) is stably expressed and inherited in transgenic grain legumes. Mol. Gen. Genet. 242:226–236.

Schaeffer, G.W., and F.T. Sharpe, Jr. 1987. Increased lysine and seed storage protein in rice plants recovered from calli selected with inhibitory levels of lysine plus threonine and *S*-(2-aminoethyl)cysteine. Plant Physiol. 84:509–515.

Schmidt, R.J., F.A. Burr, M.J. Aukerman, and B. Burr. 1990. Maize regulatory gene opaque-2 encodes a protein with a "leucine-zipper" motif that binds to zein DNA. Proc. Nat. Acad. Sci. (USA) 87:46–50.

Scholz, F. 1976. Problems of breeding for high protein yield in barley. p. 548–556. *In* H. Gaul (ed.) Barley genetics, Vol. 3, Proc. 3rd Int. Barley Genet. Symp. Verlag Karl Theimig, Munich.

Shewry, P.R., N.G. Halford, and A.S. Tatham. High molecular weight subunits of wheat glutenin. J. Cer. Sci. 15:105–120.

Shewry, P.R., A.S. Tatham, N.G. Halford, J. Davies, N. Harris, and M. Kreis. 1994. Improvement of barley and wheat quality by genetic engineering. p. 79–86. *In* R.J. Henry and J.A. Ronalds (ed.) Improvement of cereal quality by genetic engineering. Plenum Press, New York.

Singh, R., and J.D. Axtell. 1973. High lysine mutant gene (*hl*) that improves protein quality and biological value of grain sorghum. Crop Sci. 13:535–539.

Sun, S.S.M., S.B. Altenbach, and F.W. Leung. 1987a. Properties, biosynthesis and processing of a sulfur-rich protein in Brazil nut (*Bertholletia excelsa* H.B.K.). Eur. J. Biochem. 159:597–604.

Sun, S.S.M., F.W. Leung, and J.C. Tomic. 1987b. Brazil nut (*Bertholletia excelsa* H.B.K.) proteins: Fractionation, composition, and identification of a sulfur-rich protein. J. Agric. Food Chem. 35:232–235.

Tallberg, A. 1982. Characterization of high-lysine barley genotypes. Hereditas 96:229–245.

Tatham, A.S., P.R. Shewry, and P.S. Belton. Structural studies of cereal prolamins, including wheat gluten. p. 1–78. *In* Y. Pomeranz (ed.) Advances cereal science technology. Vol. 10. Am. Assoc. Cer. Chem., St. Paul.

U.S.Department of Agriculture. 1993. Agricultural statistics 1993. U.S. Gov. Print. Office, Washington, DC.

Villeagas, E., S.K. Vasal, and M. Bjarnason. 1992. Quality protein maize—what it is and how was it developed. p. 27–48. *In* E.T. Mertz (ed.) Quality protein maize. Am. Assoc. Cer. Chem., St. Paul, MN.

Voon, T.J., and G.W. Edwards. 1992. Research payoff from quality improvement: the case of protein in Australian wheat. Am. J. Agric. Econ. 74:564–572.

Wall, J.S., and J.W. Paulis. 1978. Corn and sorghum grain proteins. p. 135–219. *In* Y. Pomeranz (ed.) Advances in cereal science and technology. Vol. 2. Am. Assoc. Cer. Chem., St. Paul, MN.

Ward, E.R., J.A. Ryals, and B.J. Miflin. 1993. Chemical regulation of transgene expression in plants. Plant Mol. Biol. 22:361–366.

Watson, C.A., F.H. McNeal, M.A. Berg, and M. Menzel. 1966. Evaluating the quality potential of foreign wheats for breeding programs. Cer. Sci. Today 11:326–327.

Weeks, J.T., O.D. Anderson, and A.E. Blechl. 1993. Rapid production of multiple independent lines of fertile transgenic wheat. Plant Physiol. 102:1077–1089.

7B Oils

ANTHONY J. KINNEY and WILLIAM D. HITZ

Dupont Experimental Station
Wilmington, Delaware

I. INCREASING OIL CONTENT

Given that a photosynthetic plant has a limited amount of fixed C, any increase in the quantity of storage oil must be at the expense of some other biosynthetic pathway. It follows therefore, that there are two possible ways of increasing the oil content of a developing oil seed: increasing the number of oil producing cells relative to other, nonoil producing cells or increasing the flux of C through the lipid biosynthetic pathways at the expense of another pathway, such as carbohydrate biosynthesis. Although the former seems at first to be an unlikely strategy, it is in fact the basis of high oil corn (*Zea mays* L.) germplasm developed at the University of Illinois (Alexander, 1989). In corn kernels, oil is synthesized mainly in embryo cells, and starch is made predominantly in the endosperm cells. Selection of corn for a high oil content resulted in seeds that had an enlarged embryo relative to the endosperm. Thus, over 34 generations oil content increased from 6.2 to 12% and embryo size from 14 to 21.5% of the kernel.

In the past few years the genes encoding many different enzymes involved in plant metabolism have been cloned, and it is now reasonable to consider increasing the flux of C through fatty acid biosynthesis in developing oilseeds. This may be possible by increasing the activity of the flux-generating step of fatty acid biosynthesis, which is thought to be acetyl-CoA carboxylase (ACCase), increasing the supply of acetyl-CoA for lipid synthesis (by increasing the activity of enzymes such as pyruvate dehydrogenase or ATP-citrate lyase), or by diverting C away from some other pathway so that it flows into lipid biosynthesis. The obvious limitation to each of these approaches is the capacity of the enzymes of fatty acid and triacylglycerol biosynthesis to respond to the increased flux. At some stage, increasing the flux of C into fatty acid biosynthesis will result in some other fatty acid biosynthetic enzyme, such as the fatty acid condensing enzymes (see below), becoming rate-limiting. Exactly how much the rate of oil biosynthesis can be increased in any given oilseed before this new "wall" is hit is unclear and will have to be determined experimentally. However, it is likely that in oilseeds with a relatively low oil content (less than 20%) there is room for substantial improvement. For example, the relative steady state concentrations of mRNA for most fatty acid biosynthetic enzymes are somewhat comparable between canola (*Brassica napus* L.)

---

embryos and soybean (*Glycine max* L.) embryos, yet canola synthesizes triacylglycerol at about twice the rate of soybean (Kinney, 1996), and the oil content of canola seeds (40–60%) is more than twice that of soybean (18–22%). It is reasonable to postulate, therefore, that by increasing the flux of C into fatty acid biosynthesis it may be possible to increase the oil content of soybean to something close to that of canola.

II. ALTERED FATTY ACID CONTENT: EDIBLE OILS

A. Desired Traits

With a few exceptions, only five fatty acids from the several hundred found in plant storage oils are important for human food. These five fatty acids, palmitic (16:0), stearic (18:0), oleic (18:1), linoleic (18:2) and linolenic (18:3), represent the major fatty acid components of over 75% of the annual worldwide vegetable oil production of about 68 million t. The exceptions include coconut (*Cocos nucifera* L.) and palm kernel oils in which lauric acid (12:0) and medium chain fatty acids (MCFAs) of 10 C atoms or less are predominant. The latter are used in specialized diets such as premature infant formulas and athletic supplements for reasons explained below. About 75% of the world's vegetable oil is obtained from four plant species: soybean (30%), palm (17%), oilseed rape (*Brassica napus*/canola (15%) and sunflower (14%). Other sources of vegetable oil include cotton (7%), peanut (6%), coconut (5%), olive (3.5%), palm kernel (2.5%), linseed (1%) and corn (less than 1%). The relative fatty acid compositions of unmodified commodity oils from these crop species are shown in Table 7–3.

Although there is variation among different species in the relative fatty acid composition of the oil, almost all of the important commodity vegetable oils are rich in polyunsaturated fatty acids (linolenic and/or linoleic acids). Polyunsaturated fatty acids are prone to oxidation during storage (linolenic acid, for example, is about 100 times more chemically reactive than oleic acid) and chemical changes during cooking, which result in off-flavors. Oils rich in polyunsaturated and monounsaturated fatty acids are liquid at room temperatures that makes them, in their unmodified form, unsuitable for applications such as butter and lard substitutes. Thus, most edible vegetable oils used in food service applications such as commercial frying, margarines, baking shortenings and hydrophobic coatings for cookies, crackers and other snack foods currently use commodity vegetable oils that have been modified by chemical hydrogenation after refining.

Hydrogenation adds about 3¢ lb^{-1} (1.4¢ kg^{-1}) to the cost of producing oil, also it results in the formation of monounsaturated fatty acids with a double bond in the *trans* configuration. These so-called *trans* fatty acids have physical properties resembling a fully saturated fatty acid rather than like those of a monounsaturated fatty acid with a *cis* double bond, the naturally occurring form. For that reason, the consumption of *trans* fatty acids has been correlated with unfavorable changes in blood cholesterol, although additional studies are needed to fully establish the role of *trans* fatty acids in heart disease. Therefore, value could possibly be added to generic vegetable oils by eliminating the need for hydrogenation and producing *trans*-free oils with a fatty acid composition directly suited to the intended food ap-

Table 7–3. Relative fatty acid content of major vegetable oils.†

Species‡	Palmitic 16:0	Stearic 18:0	Oleic 18:1	Linoleic 18:2	Linolenic 18:3
			%		
Soybean (*Glycine max*)	11	4	22	53	8
Palm (*Elaeis guineensis*, *E. oleifera*)	40–69	4–5.5	20–44	4–12	<1
Canola (*Brassica napus*)	4	2	56	26	10
Sunflower (*Helianthus annus*)	6	5	20	69	<1
Cottonseed (*Gossypium* spp.)	27	2	18	51	Trace
Peanut (*Arachis hypogaea*)	12.5	2.5	38	41	<1
Coconut (*Cocus nucifera*)	7	2	4.5	1.5	
Olive (*Olea europa*)	10–20	2–3	55–78	7–19	<1
Palm kernel	5–14	1–3	5–38	2–33	
Linseed (*Linum usitatissimum*)	6	3	17	14	60
Corn (*Zea mays*)	13	2.5	31	52	1

† From Gunstone et al. (1995).
‡ Coconut oil is 50% lauric acid (12:0) and 22% medium chain fatty acids (MCFAs; 6:0, 8:0 and 10:0). Palm kernel oil is 15 to 56% lauric acid, 1 to 30% myristic acid (14:0) and 0.5 to 18% MCFAs.

plication. In addition, because palmitic acid consumption also has been correlated with unfavorable changes in blood cholesterol, table oils with lower levels of saturated fatty acids also may command a premium over commodity oils.

B. Fatty Acid Mutants

For more than 20 yr, plant breeders in both the public and private sectors have performed germplasm selection and mutation breeding of oilseed crops for modified fatty acid composition. Mutants with variations in oleic, linoleic and linolenic acids have been made in soybean, canola, sunflower, safflower, peanut, flax and corn (Ohlrogge et al., 1991; Kinney, 1994). Mutants of soybean, canola and sunflower with variations in palmitic and stearic acids also have been produced (Kinney, 1994). Mutation breeding has produced some agronomically and commercially successful crops, such as low linolenic acid soybean and high oleic acid sunflower. A number of different soybean lines with linolenic acid contents of around 2.5 to 3% have been independently produced (Mounts et al., 1994; Kinney, 1994), and Kraft Foods now markets a low linolenic soybean oil under the brand name *Soy-LL*. The best high oleic acid sunflower mutants have an oleic acid content in their seed oil of 80 to 90% (compared with about 20% in conventional sunflower oil).

The high-oleic acid mutation in sunflower appears to be predominantly inherited as a single, dominant gene, and the trait is expressed independently of the environment in which the plants are grown (Miller et al.,1987; Schmidt et al., 1989). Sunflower remains something of an exception, however, as the fatty acid profile of most mutated crop species can be highly variable and subject to modification by environmental factors such as temperature and day length. In other cases, plants with a stable change in fatty acid composition express the mutation in a constitutive manner leading to changes in root and leaf fatty acid composition that result in the plant itself being sensitive to temperature (Kinney, 1994). In addition to all of this, significant changes in fatty acid composition, even of one or two fatty acids, are often inherited as a number of dominant and recessive genes (Ohlrogge et al., 1991). The task, therefore, of integrating many different alleles, some recessive, into high-yielding, elite lines becomes particularly complex, especially if several different fatty acid changes are required. The situation is further complicated in seeds, such as soybean, in which the meal is commercially important in addition to the oil. In such seeds it may be desirable to combine value-added oil traits with value-added meal traits, such as improved amino acid composition or modified carbohydrates, in order to obtain maximum benefits from the expense of identity preservation. Unfortunately, this strategy greatly increases the difficulty for the plant breeder trying to combine genes into elite lines, particularly if there are no useful molecular markers for most of the traits. Thus, it has become apparent that the potential for producing further edible oil traits by mutation breeding alone, especially in important crops such as soybean, is somewhat limited.

C. Engineering Fatty Acid Metabolism

Tremendous advances in our understanding of the synthesis of the five major fatty acids have been made in the past 5 yr or so (Ohlrogge et al., 1993; Heinz, 1993). Most of the genes encoding enzymes involved in fatty acid biosynthesis have been cloned, and many of these genes have been, or are being, introduced into transgenic plants, including oilseeds such as canola and soybean (Knauf, 1993; Kinney, 1994).

To elucidate the potential for manipulating fatty acid metabolism it is first necessary to summarize how fatty acids are made and assembled into oil (triacylglycerol) molecules (Fig. 7–2). The synthesis of triacylglycerols in developing oilseeds may be divided into four stages: (i) synthesis, in the soluble phase of plastids, of 16:0-ACP (acyl-carrier protein) and 18:0-ACP by fatty acid synthetases and of 18:1-ACP by an acyl-ACP desaturase; (ii) release, by acyl-ACP thioesterases, of 16:0, 18:0, and 18:1 into the cytoplasm where they are esterified to coenzyme A (CoA); (iii) incorporation of 18:1 into membrane phosphatidylcholine (PC), where it is desaturated to 18:2-PC and 18:3-PC by microsomal membrane-associated desaturases, followed by return of these desaturated fatty acids into the acyl-CoA pool; (iv) incorporation of 16:0-CoA, 18:0-CoA, 18:1-CoA, 18:2-CoA and 18:3-CoA into triacylglycerol by acyl-CoA acyltransferases. In addition to these reactions fatty acid desaturases in the plastid membranes add double bonds to fatty acids esterified to the backbone of monogalatosyl- and digalactosyl- diacylglycerols (not shown in Fig. 7–2). In *Arabidopsis* leaves, a mutation in the structural gene of the plastid linoleoyl desaturase, *fad7,* results in a reduction of 18:3 content in both plastid and microsomal lipids (Browse & Sommerville, 1991). Based on this observation Browse and Sommerville have postulated that polyunsaturated diacylglycerols derived from the chloroplast desaturases are transported to the cytoplasm and incorporated into microsomal lipids. If this also is true in developing oilseeds then it is possible that the plastid desaturases could contribute to the polyunsaturate content of seed oil. In some cases, soybean for example, it is apparent that this is indeed the case since the phenotypes of low-18:3 mutants and of microsomal linoleoyl desaturase antisense transgenics indicate that the plastid linoleoyl desaturase contributes about 2% 18:3 to the total fatty acid content of seed oil (Kinney, 1994).

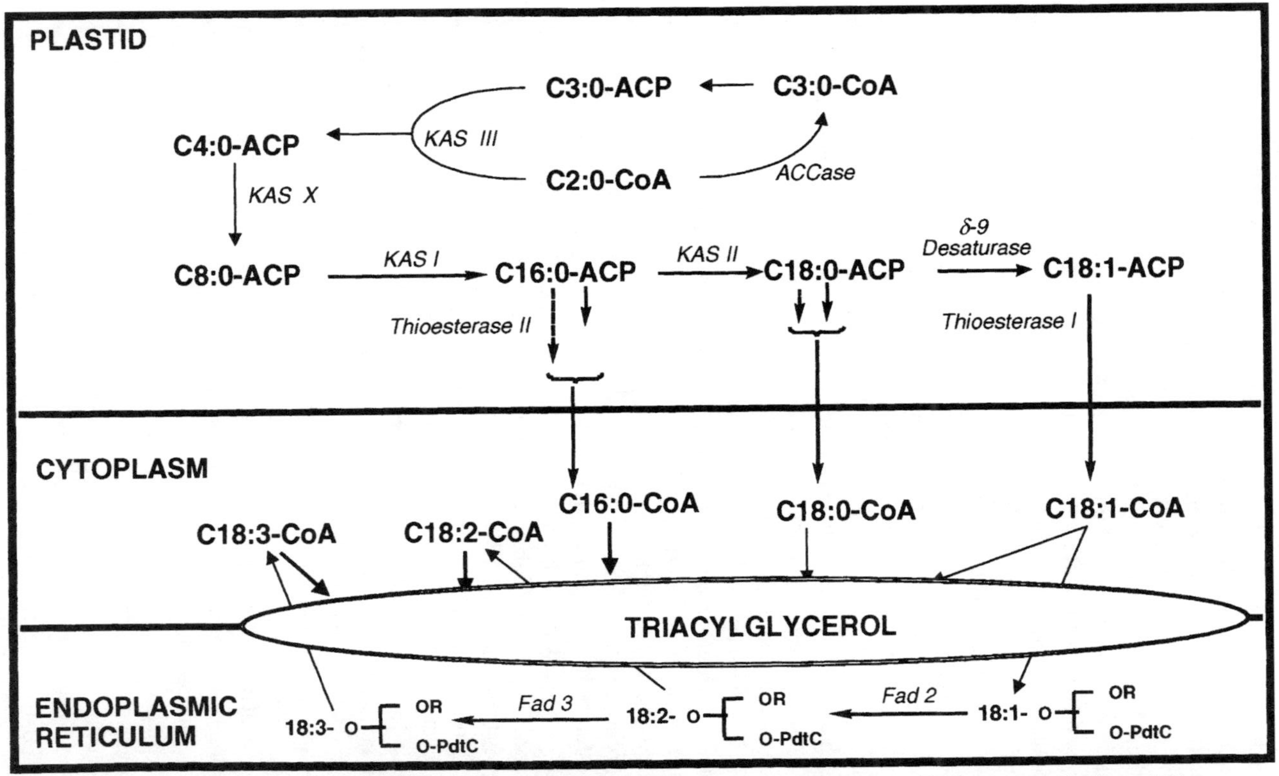

Fig. 7-2. Oil (triacylglycerol) biosynthesis in the developing oilseed. Important enzymatic steps, which are discussed in the text, are highlighted. These are: acetyl-CoA carboxylase (ACCase), ketoacyl-ACP synthetases (KAS 1, KAS 2, KAS 3), delta-9 desaturase, fatty acyl-ACP thioesterases (Fat A, Fat B), fatty acyl-phosphatidylcholine desaturases (FAD 2, FAD 3), and lysophosphatidylcholine acyltransferase (LPAAT). Other abbreviations: ACP = acyl carrier protein, CoA = coenzyme A, PtdC = phosphatidylcholine, and R = acyl chain.

By manipulating one or more of the above steps in developing seeds it is, in theory, possible to alter the relative content of the five major fatty acids at will. If this is done by using seed specific promoters to overexpress a gene or to down regulate an endogenous gene, then the growth, yield and environmental sensitivity of the plant should remain unaffected. The key to obtaining a particular phenotype is understanding which steps to change. While it may seem obvious that concentration of one fatty acid may be increased by preventing it from being converted to another, the effect of blocking a metabolic step is not always predictable. For example, while blocking the desaturation of steroyl-ACP leads to the accumulation of stearic acid (Knutzon et al., 1992; Yadav et al., 1993b), the effect of blocking oleate desaturation is less certain. Is plastid-derived 18:2, for instance, able to compensate fully or partially for the drop in 18:2-CoA generated by the oleate desaturase? The effect of increasing the concentration of an enzyme catalyzing a metabolic step is even harder to predict since different enzymes have different control strengths. Thus while a small increase in the activity of an enzyme with high control strength may cause a large change in flux through that reaction, a large increase in the activity of an enzyme with low control strength may have little effect on the flux through the reaction it catalyzes and thus on flux through the pathway.

By manipulating the concentration of fatty acid biosynthetic enzymes in transgenic soybean somatic embryos some of the important control steps in oilseed fatty acid biosynthesis were determined (Kinney, 1996). In most oilseeds all of the fatty acid C leaving the plastid is either 16:0 or 18:1. There are at least two thioesterases in soybean and other oilseeds that control the hydrolysis of 16- and 18-C acyl-ACPs. Acyl-ACP thioesterase I catalyzes the release of 18:1 at 10 times the rate of release of 16:0 or 18:0 (Yadav et al., 1993b; Loader et al., 1993). Acyl-ACP thioesterase II has equal catalytic activity towards 16:0-ACP and 18:1-ACP and little or no activity towards 18:0-ACP (Dormann et al., 1995; Jones et al., 1995). Thioesterase II has a very high control strength and any increase in thioesterase II activity leads to an increased 16:0 content of the triacylglycerol in soybean somatic embryos. Conversely, if the thioesterase II is blocked, either by an antisense construct or by sense suppression, most of the fatty acid C leaves the plastid as 18:1, resulting in a decreased saturated fat content of the oil (<4%) and an increased oleate content. In contrast, the fatty acid synthase II reaction, which catalyzes the elongation of 16:0 to 18:0, has a very low control strength in soybean. Thus increasing its activity has no effect on the final ratios of 16:0 to 18:0 and 18:1 in the soybean oil (Kinney, 1996).

Almost all of the polyunsaturated fats in soybean oil appear to be derived from the microsomal oleate desaturase reaction with little or no contribution from the plastid oleate desaturase reaction. This means that even the small quantity of 18:3 in soybean oil that is derived from the activity of the plastid linoleoyl desaturase probably originates from microsomal 18:2- CoA. Blocking the microsomal oleate desaturase reaction by antisense or sense suppression of the microsomal oleate desaturase gene results in a large increase in the 18:1 content of soybean oil (from ≈20%–>80%) and a decrease in the total polyunsaturates to less than 5%. Thus it may be possible to produce soybean oil with over 90% oleic acid by inhibiting a combination of the thioesterase and oleate desaturase reactions. What limitations are there in increasing the saturated fat content of vegetable oils? One limitation is the specificity of the acyl transferase that places fatty acids in the *sn*-2 position of glycerol phosphates. With the exception of activities from seeds that store large quantities of nonmembrane fatty acids, these enzymes have a strong tendency to transfer only unsaturated fatty acids. This in itself limits the saturate content of the triacylglyceride to about 66%. Additional restrictions may be imposed by the amount of stearate that can be tolerated by the membranes of the developing embryo or the cotyledons during seed germination. In canola sense suppression of stearoyl-ACP desaturase in combination with over expression of thioesterase I have produced oil with a saturated fat content of about 55% (W.D. Hitz, N. Yadav, R. Reiter, 1994, personal communication), and there is the possibility that the same phenotype may be achieved in soybean.

The question remains: what is the cause of the wide variation in relative fatty acid content observed among different species (Table 7–3)? For example, the kinetics and level of expression of all of the enzymes discussed above are probably similar in both soybean and canola, yet canola oil has a very different fatty acid composition from soybean oil. The answer may lie in the rate of flux through the pathway, and in particular through the thioesterase II and oleate desaturase reactions. Canola has twice the oil content of soybean, but the period of time in which that oil is made is roughly the same in both species, so the flux through the oil biosynthetic pathway is probably about twice as fast in canola as in soybean. Because the thioesterase II and oleate desaturase reactions have a high control strength and can be thought of as "rate limiting", one would predict that greater flux through the pathway would result in a lower 16:0 content and a higher oleate content, which is the case when canola is compared with soybean. This hypothesis is testable: if the flux through the fatty acid biosynthetic pathway in soybean could be increased to a rate as fast as that of canola, then the predicted phenotype in soybean would be a total oil content and a fatty acid profile similar to that of canola. To increase the flux through the pathway the activity of a flux-generating step would have to be increased, and a likely candidate for this step in oilseeds is the enzyme ACCase. It is entirely possible that soybean plants overexpressing a single enzyme, ACCase, could be produced, and their seeds would have an oil content and fatty acid composition comparable to canola.

In addition to modifying fatty acid metabolism by altering the expression of existing enzymes, it also is possible to create new edible oils by expressing lipid genes from other organisms. The desired targets may be similar to those described above, such as reducing saturates (by expressing stearoyl-CoA desaturases from yeast and rats, see Kinney, 1994) or avoiding hydrogenation. This latter alternative to increasing endogenous saturated fatty acids involves the expression in canola of a number of unique coriander enzymes—desaturases, thioesterases and synthases—to produce petroselinic acid (Ohlrogge, 1994). Petroselinic acid is an isomer of oleic acid with a single double bond at the Δ-6, rather than the Δ-9, position. This change in the bond position results in the fatty acid having a higher melting temperature, so oils containing petroselinic acid may be useful in the manufacture of edible spreads and shortenings. In addition to these well defined targets, there are potentially useful fatty acid profiles, such as oils rich in MCFAs, which can only be obtained in domesticated oilseeds by expressing heterologous genes. The MCFAs are rapidly absorbed by humans via the portal system and their uptake, unlike other fatty acids, is not carnitine dependant. This makes them useful in infant formulas, particularly premature infant formulas, since carnitine is not formed in human infants until they are about 6 wk old. In addition they have twice the calorie content of proteins and carbohydrates and yet are metabolized by the mitochondria as rapidly as glucose. Thus MCFAs are finding additional uses in athletic supplements and in long-term, intravenous feeding to avoid ketosis from excess sugar intake. Oils rich in MCFAs may be produced by expressing medium-chain acyl-ACP thioesterases from *Cuphea* species, an exotic plant with oil rich in MCFAs, in domesticated oilseed plants such as canola (Jones et al., 1994). A related acyl-ACP thioesterase from California bay laurel (*Umbellularia californica*), with specificity for lauryl-ACP, has been used to produce canola oil rich in lauric acid (Voelker et al., 1992). Lauric acid has many uses outside of the field of human nutrition, and it is in the production of oils for nonfood uses that the expression of novel lipid enzymes from exotic plant species and other organisms has the greatest potential.

III. ALTERED FATTY ACID CONTENT: INDUSTRIAL OILS

A. Choice of Targets and Approaches

The ability to introduce genes into oilseed plants for expression in a seed specific manner opens the possibility of producing fatty acids other than the five main fatty acids important in food oils. If markets can be found for novel fatty acids or even unique combinations of the more common fatty acids, new crops might be created. In thinking about what novel fatty acids or combinations of fatty acids might be desirable, potential uses or markets must be considered as well as the biochemistry and biology of producing fatty acids that are not currently produced by the traditional temperate-zone oilseed crops. Markets with unit values greater than that of commodity vegetable oils currently exist for plant-produced fatty acids that are not made in the most widely grown temperate-zone oilseeds. It may be possible to alter the metabolism of crops such as soybean to produce these fatty acids to fill these markets. The value of these types of products may be assumed to be that of the current market, although the actual market share and price obtainable are uncertain.

A longer term strategy is to use fatty acid metabolism in seeds to produce ingredients for current products whose feedstocks are not currently made in plants or even to produce feedstocks for entirely new end products. As noted above, this strategy also places products produced in plants into competition with current feedstocks, in this case usually from petrochemical sources. While this competition may currently be economically unfavorable, in the long term the ability to produce some industrial ingredients from renewable resources may make this a profitable route. The potential economic return also may play a role in determining the feasibility of transferring these genes into oilseed crops. The traits conferred by the genes must be of sufficient value to not only make the initial production of the transgenic plant attractive, it also must be of sufficient value to support the additional cost to the breeding effort required to keep the crop agronomically competitive. If the trait itself forces a yield penalty on the crop, the value of the trait must offset the economic cost of lower yields. Thus, traits conferred by a single transgene may not place much additional burden on the conventional breeding effort, whereas traits that require the transfer of several genes may pose a large and continuing breeding problem if they cannot be genetically linked.

The biological problems that must be addressed in considering the production of fatty acids that are not normally found in the common oilseed crops can be broken into two parts. The first is: how can the existing pathways be modified to produce the desired fatty acids, and do the genes required to make the changes exist? The second is: will the production of large quantities of the new fatty acid be compatible with both seed development and seed germination? Both of these processes are determinants of agronomic yield, so in a practical sense the resulting transgenic must not just survive but should fill the seed, germinate and emerge as well as elite cultivars.

A large number of fatty acids are found in the storage lipid of different plant species (Browse & Somerville, 1991). Many of these are modifications to the 18-C fatty acids, which are the main products of fatty acid biosynthesis and the initial substrates for desaturation. The synthetic steps involved in producing these fatty acids often appear to be catalyzed by single genes, so that in principle, at least, the wild species that produce these acids are a source of genes for producing unique fatty acids in existing crop plants. The multiple subunit fatty acid synthetase in plants also contributes to the ability to produce unique fatty acids. Intervention by transgenically produced enzymes has been shown to be effective in altering chain length and introducing unique double bonds. The main challenge in implementing these changes is identifying the genes responsible for fatty acid modifications in these wild species. The fact that synthesis of nonmembrane fatty acids is limited to the storage tissue makes tissue collection difficult in many cases and decreases the practicality of studying biosynthetic pathways and identifying genes through protein purification. Continuing improvement in techniques for identifying genes by homology and reverse genetics make advances in this area more rapid.

The fact that unique fatty acids make up the bulk of the fatty acids in the storage lipids of the species that produce them indicates that their existence in plant cells can be tolerated. In oilseeds such as soybean, however, it appears that the phospholipid pools, which are the intermediates of triacylglyceride synthesis, are in equilibrium with the membrane phospholipids. In plants that store large quantities of unusual fatty acids that may not be compatible with membrane function, these fatty acids are largely excluded from the bulk membrane lipid pools (Stahl et al., 1995). While many plants seem to have phospholipid lipases capable of specifically removing hydroxylated fatty acids (Banas et al., 1992), two other unusual fatty acid types, MCFAs and epoxidized fatty acids, were only effectively hydrolyzed from phosphatidyl choline by membrane preparations from species that produced large amounts of these fatty acids. Membrane preparations from other seeds, even those that produce other unusual fatty acids were not effective. Whether or not preferential hydrolysis from membrane lipids is the primary or even an essential part of excluding disruptive fatty acids from membranes remains to be seen. It does, however, indicate that seeds that produce these fatty acids may have alterations in fatty acid metabolism in addition to those required for synthesis of the nonmembrane fatty acids.

B. Alternative Production Systems for Industrial Fatty Acids

There are several examples of fatty acids that, when present in high quantity in seed oils, make those oils of higher value than commodity soybean oil and whose synthesis might require comparatively little metabolic modification. Drying oils are typified by linseed oil and are used either as the triacylglyceride or as the fatty acids derived from the triacylglyceride in coating applications because of their tendency to oxidatively crosslink to form solid films. In the case of linseed oil this property is due to the very high (60% of total fatty acids) linolenic acid content of its triacylglyceride. While soy and canola oils have linolenic acid contents that are too high for use in commercial food frying operations, their linolenate content is not high enough to allow their extensive use as drying oils. Since the activity of the ω-3 desaturase in both of these species is limiting for the production of linolenic acid, it is likely that overexpression of this activity would result in oils with linolenic acid contents similar to that of linseed oil.

Tung (*Aleurite fordii* Hemsl., *A. montana* Willd.) oil has similar uses to linseed oil but is more desirable and commands a higher unit price. Its advantage over linseed oil comes from the configuration of the double bonds in the 18:3 fatty acid of tung oil. In contrast to the methylene-interrupted double bond series found in linolenic acid (18:3, Δ9, 12, 15), tung oil is composed of about 70% 18:3 (Δ9, 11, 13 in the *cis, trans*, *trans* configuration, respectively). This conjugated double bond system is more prone to oxidative cross linking that the nonconjugated system. The nature of the desaturase responsible for its production is not known. Both of these fatty acids might be produced with single-gene modifications because their physical structure is either the same as membrane fatty acids or differs only slightly from the fatty acids normally found in the membranes of temperate plants.

Hydroxylated and epoxidized fatty acids are the major fatty acids in the storage triacylglyceride of certain plant species. These presumably result from single gene-encoded modifications of oleic acid. The hydrolxylated fatty acid used in commerce today is ricinolic acid (18:1, 12-OH, produced by the castor bean. The industrial uses of castor oil are quite varied, and it is often postulated that further chemical modifications relying on the hydroxyl group at position 12 of the hydrocarbon could provide new uses if a large, reliable and competitively priced supply of ricinolate were available. Naturally occurring epoxidized fatty acids might become effective replacements for chemically epoxidized oils in drying oil applications, if they could be economically produced in crop plants.

The chemistry of the introduction of the hydroxyl group into oleic acid as it is esterified to PC in the endoplasmic reticulum is very similar to the chemistry of double bond introduction by the membrane-bound fatty acid desaturases. Seed-specific messages for sequences very similar to the cytoplasmic ω-3 desaturase are expressed in developing castor bean seeds during the portion of seed fill in which ricinolate is synthesized (Lightner & Okuley, 1994) and may encode such an hydroxylating activity. As mentioned above, hydroxlyated fatty-acid specific phospholipases, which could form part of the mechanism required for excluding ricinolate from the membranes, appear to be ubiquitous in plants. Production of ricinolate outside of castor seeds may, therefore, be possible provided that very high quantities of it do not interfere with the energy storage function of seed triacylglyceride during germination.

A different class of fatty acids that may require exclusion from membranes but which could be of value if produced in temperate crops are the MCFAs. Fatty acids in this class are used in the manufacture of many different surfactants and have other specialty uses as well. The current production systems for fatty acids with chain lengths less than 16 C atoms are the tropical oils (palm, palm kernel and coconut). Several trees and weed species also produce triacylglycerides in their seeds that are comprised largely of fatty acids of 12 C and less.

Specific acyl-acyl carrier protein thioesterases modify usual fatty acid biosynthesis pathways to produce these fatty acids (Jones et al., 1995). These thioesterases effectively compete with the normal fatty acid elongation enzymes to remove fatty acids prior to their reaching 16 or 18 C atoms in length. Incorporation of these MCFAs into the triacylglyceride of transgenic canola plants has been demonstrated (Jones et al., 1995), although incorporation to the levels found in seeds that naturally produce them requires one or more acyl-CoA acyl transferases with unique specificity for the MCFAs (Stymne, 1993). As with the epoxidized fatty acids, exclusion from membranes could pose a problem when the MCFAs are produced at high concentrations; however, in practice laurate concentration at 40% of the triacylglyceride has been obtained in canola with retention of seed viability (Volker et al., 1992).

C. Creating New Feedstocks from Fatty Acids

Most of the above examples have as their initial target markets applications that are currently filled by exotic plant oils. The possible production of the Δ-6 monounsaturated fatty acid petroselenate in oilseed crop plants is an example of the use of a novel fatty acid to supply an industrial feedstock currently obtained from petrochemicals. As described in the section covering food uses of fatty acids, the transfer of two or three genes from coriander under the control of seed-specific promoters to soybean could introduce the synthesis of petroselenate into soybean seeds. If purified petroselinic acid could be oxidatively split at the double bond on a commercial scale, as is done on a laboratory scale by ozonolysis, the resulting products would be the 6-C diacid, adipic acid and the 12-C monoacid, lauric acid. This distribution of products is obtained due to the placement of the double bond at the 6 position instead of the 9 position found in the most common monounsaturated fatty acid, oleic acid. Adipic acid is an intermediate in the production of nylon 6,6, and a very large market exists for it at a weight unit price about twice that of soybean oil. Since the yield of adipic acid from the conversion is less than one-half of the petroselinic acid starting material, the market for lauric acid and glycerol byproducts would be important economic determinants of the overall process.

One of the ways in which soybean oil is used in drying and coating applications is by activating some of the double bonds present in the fatty acids of the triacylglycerol to allow them to rapidly form intermolecular cross links. This activation is achieved by chemical formation of the cyclic epoxide of the existing double bonds. In some species, epoxy fatty acids are naturally synthesized in the storage tissues and assembled into triacylglycerides with a high epoxy fatty acid content. For some applications, oils made from these sources may be more suitable than epoxidized soybean oil.

The biochemistry of fatty acid epoxidation has been studied in oilseeds rich in epoxy fatty acids (Bafor et al., 1993), and the formation of these modified fatty acids appears to take place while fatty acids are esterified to PC in a manner analogous to fatty acid desaturation. The enzyme that catalyzes the reaction converting oleic acid to its C12 epoxide in the seeds of *Crepis* spp. and *Vernonia* spp. has been shown to be closely related to plant omega-12 desaturases (Shanklin & Cahoon, 1998). Further, it has been shown to be possible to direct soybean seeds to synthesize triacylglycerides containing epoxy fatty acids by introducing a gene encoding the Vernonia epoxidase into soybean (Kinney, 1999). As in the previous section, proper sequestration of epoxidized fatty acids in the storage lipid and out of membrane lipid is a requirement that is not be met in seeds such as soybean and suggests that modification of metabolism is necessary.

The possibility of driving the oleate content of soybean oil to 90% of fatty acids by modification of the expression of two of the genes in fatty acid biosynthesis was discussed above. The production of a very homogenous oil also may provide applications as chemical feedstocks. One of the attributes of a useful feedstock is homogeneity or at least an inexpensive route to homogeneity. While there is no current large scale industrial use for pure oleate, erucic acid (*cis*Δ13-C22:1) in a pure state is used in lubricant applications and could be split at the double bond to produce monomers in a manner similar to that described for petroselinic acid. High erucic rapeseed, which is the source of erucic acid, produces triacylglyceride that is about 60% erucate. The specificity of the acyl transferase, which places fatty acids at the *sn*-2 position of PC in the assembly pathway for triacylglyceride, shows a strong preference against fatty acids other than 18-C unsaturated acids in this species. This specificity may be the limiting factor in the production of higher levels of erucic acid. Transfer of a lysophosphatidic acid acyl transferase with altered specificity to allow transfer of erucate to the *sn*-2 position (Hanke et al., 1995) into high erucic rapeseed might, either alone or in combination with the cytoplasmic fatty acid elongase responsible for erucate synthesis (James et al., 1995), allow the production of essentially pure erucic acid.

IV. CONCLUSIONS

These examples of modifications to fatty acid and triacylglycerol synthesis to produce new products in traditional crops are biased towards products that might be produced in the near term. As the limits of the capability of the biological system to produce novel products are tested, and the economic feasibility of producing these transgenic crops is proven, other technically feasible possibilities may become evident. The ability to transform economic plant species routinely and reproducibly will speed development and allow more complex engineering of metabolic pathways to become a reality. More sophisticated seed production systems, such as hybridization to produce seeds that express a trait but need not produce the next generation of plants, may extend the range of molecules that can be produced so that plants can in fact be used as renewable factories to produce complex food and industrial intermediates.

REFERENCES

Alexander, D.E. 1989. Maize. p. 431–437. *In* G. Robbelen et al. (ed.) Oil crops of the world. McGraw-Hill, New York.

Bafor, M., M.A. Smith, L. Jonsson, A.K. Stobart, and S. Stymne. 1993. Biosynthesis of vernoleate (*cis*-12-epoxyoctadeca-*cis*-9-enoate) in microsomal preparations from developing endosperm of *Euphorbia lagascae.* Arch. Biochem. Biophys. 303:145–151.

Banas, A., I. Johansson, and S. Stymne. 1992. Plant microsomal phospholipases exhibit preference for phosphatidylcholine with oxygenated acyl groups. Plant Sci. 84:137–144.

Browse, J., and C. Sommerville. 1991. Glycerolipid synthesis: biochemistry and regulation. Ann. Rev. Plant Physiol. Plant Mol. Biol. 42:467–506.

Dormann, P., T.A. Voelker, and J.B. Ohlrogge. 1995. Cloning and expression in *E. coli* of a novel thioesterase from *Arabidopsis thaliana* specific for long chain acyl-acyl carrier proteins. Biochem. Biophys. Res. Comm. 316:612–618.

Hanke, C., G. Peterek, F.P. Wolter, and M. Frentzen. 1995. cDNA clones from *Limnanthes douglasii* encoding an erucoyl-CoA specific 1-acylglycerol-3-phosphate acyltransferase. p. 531–533. *In* J.C. Kader and P. Mazliak (ed.) Plant lipid metabolism Kluwer Acad. Publ., Netherlands.

Heinz, E. 1993. Biosynthesis of polyunsaturated fatty acids. p.36–90. *In* T.S. Moore (ed.) Lipid metabolism in plants. CRC Press, Boca Raton, FL.

James, D.W., E.L. Lim, J. Keller, I. Plooy, E. Ralston, and H.K. Dooner. 1995. Directed tagging of the Arabidopsis *Fatty Acid Elongation 1 (Fae 1)* gene with the maize transposon activator. Plant Cell 7:309–319.

Jones, A., H.M. Davies, and T.A. Voelker. 1995. Palmitoyl-acyl carrier protein (ACP) thioesterase and the evolutionary origin of plant acyl-ACP thioesterases. Plant Cell 7:359–371.

Kinney, A.J. 1994. Genetic modification of the storage lipids of plants. Curr. Opin. Biotechnol. 5:144–151.

Kinney, A.J. 1996. Development of genetically engineered soybean oils for food application. J. Food Lipids 3:273–292.

Kinney, A.J. 1999. New and improved oils from genetically modified seeds. Lipid Technol. Newsl. 5:36–39.

Knauf, V.C. 1993. Progress in the cloning of genes for plant storage lipid biosynthesis. p. 149–164. *In* J.K. Setlow (ed.) Genetic engineering. Vol. 15. Plenum Press, New York.

Knutzon, D.S., G.A. Thompson, S.E. Radke, W.B. Johnson, V.C. Knauf, and J.C. Kridl. 1992. Modification of brassica seed oil by antisense expression of a stearoyl-ACP desaturase gene. Proc. Natl. Acad. Sci. (USA) 89:2624–2628.

Lightner, J.E., and J.J. Oukely. 1992. Genes for fatty acid desaturase enzymes. PCT Patent Application WO 9411516.

Loader, N.M., E.M. Woolner, A. Hellyer, A.R. Slabas, and R. Safford. 1993. Isolation and characterization of two *Brassica napus* embryo acyl-ACP thioesterase cDNA clones. Plant Mol. Biol. 23:769–788.

Miller, J. F., D.C. Zimmerman, and B.A. Vick. 1987. Genetic control of high oleic acid content in sunflower oil. Crop Sci. 27:923–926.

Mounts, T.L., K. Warner, G.R. List, W.E. Neff, and R.F. Wilson. 1994. Low-linolenic acid soybean oils: alternatives to frying oils. J. Am. Oil Chem. Soc. 71:495–499.

Ohlrogge, J.B., J. Browse, and C.R. Somerville. 1991. The genetics of plant lipids. Biochim. Biophys. Acta 1082:1–26.

Ohlrogge, J.B., J.G. Jaworski, and D. Post-Beittenmiller. 1993. *De novo* fatty acid biosynthesis. p.3–32. *In* T.S. Moore (ed.) Lipid metabolism in plants. CRC Press, Boca Raton, FL.

Ohlrogge, J.B. 1994. Design of new plant products: Engineering of fatty acid metabolism. Plant Physiol. 104:821–826.

Stahl, U., A. Banas, and S. Stymne. 1995. Plant microsomal phospholipid acyl hydrolases have selectivities for uncommon fatty acids. Plant Physiol. 107:953–962.

Schmidt, L., R. Marquard, and W. Friedt. 1989. Status and perspectives of breeding high-oleic sunflower for central Europe. Fett. Wiss. Technol. 91:346–349.

Shanklin, J., and E.B. Cahoon. 1999. Desaturation and related modifications of fatty acids. Annu. Rev. Plant Physiol. Plant Mol. Biol. 49:611–641.

Stymne, S. 1993. Biosynthesis of 'uncommon' fatty acids and their incorporation into triacylglycerols. p. 60–66. *In* N. Murata and C.R. Somerville (ed.) Biochemistry and molecular biology of membranes and storage lipids of plants. Am. Soc. Plant Physiol., Rockville MD.

Voelker, T.A., A.M. Worrell, L. Anderson, J. Bleibaum, C. Fran, D.H. Hawkins, S.E. Radke, and H.M. Davies. 1992. Fatty acid biosynthesis redirected to medium chains in transgenic oilseed plants. Science (Washington, DC) 257:72–74.

Yadav, N., A. Wierzbicki, S. Knowlton, J. Pierce, K. Ripp, W. Hitz, M. Aegerter, and J. Browse. 1993. Genetic manipulation to alter fatty acid profiles of oilseed crops. p.60–66. *In* N. Murata and C.R. Somervile (ed.) Biochemistry and molecular biology of membranes and storage lipids of plants. Am. Soc. Plant Physiol., Rockville, MD.

Yadav, N., A. Wierzbicki, M. Aegerter, C.S. Caster, L. Perez-Grau, A.J. Kinney, W.D. Hitz, R.B. Booth, B. Schweiger, K.L. Stecca, and other authorities. 1993. Cloning of higher plant omega-3 desaturases. Plant Physiol. 103:467–476.

7C Starch Biosynthesis and Genetic Potential

L. CURTIS HANNAH

University of Florida
Gainesville, Florida

A succinct but compelling case can be made that no molecule is more important to modern agriculture than starch. Cereals account for the vast majority of the important products of plant agriculture, and within the seed of these cereals, starch is the major component. Furthermore, tubers are important in the daily caloric intake of many people, and the major component of this foodstuff is starch. Given the importance of starch, it is amazing and indeed perplexing that it has received relatively little attention in plant biology. We do, however, understand many important steps in its biosynthesis, and these are emphasized in this short review. Speculations concerning the exploitation of this knowledge base in future uses of starch and sugars, and how actual genetic alterations in starch synthesis are presently being used are presented as well. Some of these topics have been covered in an earlier review (Hannah et al., 1993). A broader coverage of genes affecting not only starch but other kernel components was recently published (Boyer & Hannah, 1994).

At first approximation, starch is a simple molecule. It is composed of two components, amylose and amylopectin. Both are polymers of the sugar glucose held together by an α-1,4 linkage between the neighboring residues. Amylopectin is more complex than amylose; it contains branches of glucose polymers linked by α-1,6 linkages. Variation of this central theme exists: e.g., a more highly branched polymer, termed phytoglycogen and operationally defined as water-soluble polysaccharide, can be found in conventional sweet corns (*Zea mays* L.) containing a mutant allele of the *sugary1* (*su1*) locus of maize. Quantitative differences also exist: e.g., the average size of the glucose polymers and phosphate content distinguish potato (*Solanum tuberosum* L.) and maize starch. These differences have important ramifications in the industrial uses of various plant materials.

Starch is an end-product of photosynthesis. In nonphotosynthetic tissues (seeds, roots, tubers, etc.—plant organs important for agriculture), the glucose residues of starch arise from sugar exported from leaf tissue. In most but not all plants the transported sugar is sucrose. The life span of starch synthesized in these nonphotosynthetic tissues varies depending on the tissue and plant but is usually measured in days to months and sometimes in years. On the other hand, starch in leaves turns over much more rapidly, usually daily. Except for these differences in life span and the fact that sucrose does not serve as the primary substrate for starch synthesis in the leaf, the actual enzymological steps in starch synthesis in photosynthetic and nonphotosynthetic tissues appear identical.

I. AN OVERVIEW AND PHILOSOPHY

Most studies concerning starch synthesis in nonphotosynthetic tissues focus on the path from sucrose to starch. Several approaches are used. Of these, the classic "biochemical genetics" approach has provided the most insight. Mutants are compared to their wild-type counterpart in terms of intermediates in the starch biosynthetic pathway, and various candidate enzymes are then assayed. Meaningful insight is achieved when it is found that one candidate enzyme is greatly reduced in the mutant. If the enzymological lesion is the primary effect of the mutation, then one can safely conclude that the enzyme is important in starch synthesis and that the C flow from sucrose to starch must pass through the particular enzyme. Historically, this approach could only be used in plants containing starch mutants; however, the advent of antisense technology and plant transformation have opened the door to studies in other plants not amenable to the mutational approach.

Other more indirect approaches have been used for studies of starch biosynthesis as well. For example, in many studies a stress is imposed during development of the plant tissue (heat, drought, low levels of CO_2, etc.), and the effects on starch levels, enzyme levels and the flux of intermediates through the pathway are monitored. This approach suffers from the fact that not all of the effects of the imposed stress are known. Thus, cause and effect relationships become problematic. A further complication arises from inferences concerning the reduction in an enzymic activity and the resulting physiological effect.

Do small reductions in enzymic activity have an effect or are most biological systems buffered such that small changes in enzymic activity have little to no physiological consequence? Analysis of mutants provides some insight here. Most enzymic activities show a linear dosage effect when the number of functional or wild-type alleles is altered, yet in almost all cases, the nonfunctional or mutant alleles are recessive. For example, and as reviewed again below in a different context, the *waxy* (*wx*) gene of maize is the structural gene for the starch-bound glucosyl transferase involved in amylose synthesis. Measurements of enzymic activity in endosperms containing zero, one, two, three, four, or six doses or copies of the wild type allele showed that the enzyme increased linearly with the number of wild type alleles. However, the level of amylose produced does not show this pattern (Nelson & Rines, 1962). Kernels with two, three, four, or six doses of *Wx* have identical amounts of amylose, and kernels with only one dose of the wild type gene have almost as much amylose. Clearly, only two doses of the functional gene are needed to obtain the wild-type phenotype, and one dose nearly conditions it. Therefore, in this particular case, the system is buffered in such a way that reductions of enzymic activity of at least one-third and perhaps more than that of a wild-type seed can be tolerated. The remaining activity is sufficient for normal function. The fact that most enzymes show a linear dosage response with the number of structural genes and that virtually all mutants are recessive indicate that the *Wx* paradigm is true in the vast majority of cases. Hence, it may well be that small reductions in the enzymic activity noted in many experiments are physiologically insignificant.

In another approach, particular enzymes in the pathway are studied in detail and inferences are drawn from the kinetic and allosteric properties of the enzyme. These insights are indirect since rarely are the cellular or subcellular levels of the important reactants known, and in some cases the actual placement of the enzyme within the cell is suspect. These studies ignore the effects of unknown effector molecules and other proteins. Also, rarely is the possibility addressed that the allosteric and kinetic properties of the observed enzyme are more of a reflection of the evolutionary history of the genes encoding the particular enzyme than the physiological relevance of the allosteric properties in the tissue under study.

While the mutational approach is most insightful, it has its limitations. Some are obvious, some not. These are addressed below. While the finding that the loss of a particular enzyme is the primary lesion of a particular mutant shows that the enzyme is important, the lack of a mutant doesn't mean that an enzyme is unimportant. From biochemical investigations, some enzymes in the starch biosynthetic pathway must be important, yet we lack mutants for them. Two classic explanations are given for this: (i) genetic redundancy—the active enzyme is encoded by two or more duplicate genes, and (ii) lethality—mutations in the gene encoding the enzyme are lethal to the organism. While this point may appear obvious, it seems to be ignored in discussions of single gene vs. quantitative traits. There is not a single trait in any organism that is under the control of a single gene, and the single gene vs. polygene distinction refers only to the number of heteroallelic genes segregating in the population under test. Classical Mendelian segregation data tell us nothing about the underlying biochemistry of the trait but rather point to the number of physiologically significant heteroallelic genes in the population. There is an assumption in the literature that placement of a gene from one plant into another will condition the phenotype seen in the former plant. It may or may not, depending on the presence of other genes and gene products in the transformed plant that are needed for the trait.

A second point concerns the relationship between the mutant gene and the affected enzyme. Before valid conclusions can be drawn, it is critically important to know that the primary effect of the mutant gene is on the particular enzyme under investigation. An example of this can be found in the starch literature (reviewed in Nelson & Burr, 1973). The *shrunken4* (*sh4*) mutation of maize reduces starch content in the seed and also reduces phosphorylase activity. These observations led to the early conclusion that phosphorylase was involved in starch synthesis. Later work (Burr & Nelson, 1973) showed that *sh4* affected the synthesis of pyridoxal phosphate, a cofactor for phosphorylase. Since pyridoxal phosphate also is a cofactor for many enzymes, including transaminases, the relationship between starch synthesis and phosphorylase activity had to be re-evaluated.

In this review, the mutational approach to the understanding of starch biosynthesis will be highlighted. Cases in which the mutant gene has been shown to encode the affected enzyme will be emphasized. Fortunately for these studies, in every case where definitive data have been generated, the gene has been found to encode the affected enzyme. One surprising feature of these studies is that bona fide regulatory genes, as found in other pathways in plants, have not been uncovered. The significance of this awaits further investigation.

II. BIOCHEMICAL GENETICS OF STARCH BIOSYNTHESIS

A. An Overview

From analysis of mutants, it is now quite clear that the catabolism of sucrose by sucrose synthase to yield UDP-glucose and fructose is important to starch synthesis. Furthermore, invertase, which cleaves sucrose to glucose and fructose also may be important although definitive data remain lacking. The importance of these parallel steps becomes explainable if sucrose is degraded and resynthesized during its journey from the chloroplast to the amyloplast. The preliminary data suggest that invertase may be important in cleaving sucrose as it enters the seed. Sucrose is then resynthesized in the endosperm. The breakdown of endosperm sucrose by sucrose synthase occurs in the cells actually synthesizing starch. On the other hand, there is evidence that not all C entering cells synthesizing starch follows this path of sucrose catabolism, anabolism and catabolism and that reducing sugars have the ability to enter the growing endosperm (Cobb & Hannah, 1988, and references cited therein).

The next step shown by genetic data to be important is the synthesis of ADP-glucose from glucose-1-phosphate and ATP. Mutants affecting the flow of C from UDP-glucose or fructose to glucose-1-phosphate have not been reported. The lack of mutants may not be surprising if glucose-1-phosphate can be synthesized from UDP-glucose and from fructose by two separate and independent pathways. Many if not all of the necessary enzymes for these syntheses have been reported in nonphotosynthetic tissues synthesizing starch, and abundant levels of UDP-glucose pyrophosphorylase [UDP-glucose + PPi (PP = pyrophosphate) ⇄ glucose-1-phosphate + UTP] are common in these tissues.

It is clear that the glucose residues of ADP-glucose form the backbone of both amylose and amylopectin. Mutants are known that affect the synthesis of ADP-glucose, amylose and amylopectin. The mutations lie in the structural genes for ADP-glucose pyrophosphorylase, starch-bound ADP-glucose glucosyl transferase, and starch branching enzyme activity, respectively. Mutants of the first class reduce starch levels, mutants of the second class affect amylose only while mutants of the last class primarily affect amylopectin.

A detailed discussion of the pathway is given below. We will start with the end-product, starch, and work backwards to sucrose that enters the nonphotosynthetic tissue. The approach taken in this discussion is genetic. We will focus on enzymic activities for which there are definitive data that the structural gene(s) have been identified. A second criterion for selection is the identification and characterization of loss-of-function mutations in these genes. In virtually all cases, clones isolated from one plant (usually maize) have been used to isolate homologous genes in other species. Because of space constraints, such derived genes/clones will not be discussed unless they provide additional important insight into the biochemistry of starch synthesis.

B. The *Waxy* (*Wx*) Gene And Amylose Synthesis

One of the earliest genes characterized in any organism is the *waxy* locus of maize (reviewed in Nelson & Rines, 1962). Loss of *Wx* function conditions an almost opaque phenotype in the endosperm that can be seen by the trained eye. Early in the studies of *Wx* it was found that wild-type and mutant tissue could easily be distinguished by iodine staining. [It should be noted that *Wx* was one of the three genes Barbara McClintock (1953) followed that allowed her to decipher the movement and therefore the existence of transposable elements.]

The nature of the differential iodine staining became apparent when it was discovered that *waxy* starch lacks amylose. The starch of this mutant is exclusively amylopectin. This was an unexpected result because molecular structures would suggest that amylose, the straight-chained glucose polymer, would simply be branched to form amylopectin. This simple precursor-product relationship does not occur. Early in the history of studies of starch synthesis, a number of alternative hypotheses were considered. In general, these fell into two classes: (i) amylopectin serves as the precursor of amylose and the role of *Wx* is to debranch amylopectin to amylose, (ii) amylose and amylopectin are synthesized by parallel pathways, and *Wx* affects only the amylose path.

Fundamental insight into these alternative hypotheses came when Nelson and Rines (1962) showed that starch-bound ADP-glucose glucosyl transferase (granule-bound starch synthase) was greatly reduced in a series of independent *wx* mutants. That *wx* affected a starch synthetic enzyme and not a debranching enzyme provided strong evidence against *Wx* being involved in amylopectin degradation. As reviewed earlier (Hannah et al., 1993), one hypothesis put forward was that the lack of starch granule-bound starch synthase was the result of a lack of amylose. If the enzyme binds to amylose and *wx* mutants by unknown means abolish amylose synthesis, then the loss of starch synthase could be an indirect effect of the mutation. This alternative hypothesis was ruled out in the following way. Starch-bound starch synthase activity was measured in kernels in which the dosage of the functional *Wx* allele and the recessive *wx* allele were varied. It was observed that whereas enzymic activity increased in a linear manner with the number of functional *Wx* alleles, amylose levels were indistinguishable in kernels with two, three, four, or six doses of the wild-type allele. This observation rules out the possibility that starch granule-bound glucosyl transferase activity is dependent on amylose presence since if it were, the enzyme activity should show the same response to gene dosage as that exhibited by amylose content.

The binding of the *Wx*-controlled starch synthase to starch granules precluded early investigation at the enzymological level. The binding, however, was exploited in purifying a starch-bound, 58- to 60-kD protein that was missing in some *wx* mutations. The cloning and sequencing of the *Wx* gene were done independently in the laboratories of Heinz Saedler and N. Fedoroff and were reviewed previously (Hannah et al., 1993).

There is ample evidence showing that the *Wx* locus encodes the starch granule-bound glucosyl transferase in most if not all plants. First, while the original *wx* mutants analyzed lacked a detectable protein, other *wx* mutants were shown to contain an altered protein or transcript. In some mutants, the levels of protein or transcript were equal to that found in the wild type (Echt & Schwartz, 1981). This observation rules out the possibility that *Wx* operates at a level upstream from the structural gene for the enzyme (perhaps as a regulatory gene) or in some indirect manner (perhaps as the primer for amylose synthesis) to affect starch synthase activity. Second, sequencing of the *Wx* gene revealed significant sequence similarity to *Escherichia coli* glycogen synthase (reviewed in Sivak et al., 1993). Finally Macdonald and Preiss (1983) reported the solubilization of an enzymically active protein from maize starch granules of a molecular mass of 60 kD—the size of the *Wx* protein.

Recently, however, Smith (1990) proposed that the *Wx* protein is not the major starch-bound glucosyl transferase in pea seeds. She solubilized the proteins from the pea (*Pisum stivum* L.) starch, separated the proteins, and synthesized antibodies to various fractions. Antibodies against a 77-kD protein inhibited glucosyl transferase activity. Furthermore, a protein of this size was reported to have enzymic activity. In attempts to continue these investigations, Sivak et al. (1993) solubilized proteins from pea starch granules and separated them by size. In contrast to the results of Smith, Sivak and coworkers found that the major if not exclusive form of glucosyl transferase was 60 kD in size. Furthermore, this fraction cross reacted with antibody synthesized against the maize *Wx* protein. Thus, it appears from the recent results of the Preiss laboratory that a gene homologous to the maize *Wx* gene encodes the starch-bound glucosyl transferase in pea as well. The nature of the 77-kD protein and the resolution of possible artifacts, however, requires further investigation.

A point of emphasis concerning the *wx* mutations is the insight given into the biosynthetic relationship between amylose and amylopectin. The mutation and resulting physiological and molecular investigations clearly show that amylose does not serve as the precursor of amylopectin. While this fact has been known for many years, it is rarely mentioned in text books and in talks and papers given by present-day investigators. A remaining unresolved issue is the true substrate for amylopectin synthesis. One would expect that its structure must be similar to amylose, and the possibility that there is a form of compartmentalization of amylose and amylopectin should be considered.

That amylose and amylopectin are synthesized by separate pathways is consistent with the finding of multiple starch synthases. Both insoluble or bound activity and a soluble form of the enzyme have been reported and, as expected, the soluble activity is not reduced in a *wx* mutant. While multiple forms of soluble starch synthase have been reported, mutants in the various structural genes have not been found. This activity is discussed again in the *sugary1* section below.

Starch composed exclusively of amylopectin is advantageous for several commercial purposes. Since amylose is not synthesized, separation steps for removing amylose from the amylopectin fraction are not needed, thereby reducing processing costs. This has made *waxy* maize in the USA a viable industry. In Europe, potato represents a major source of starch, and investigators in The Netherlands (see Visser et al., 1991 for a review) have focused on the development of an amylose-free potato. This has been accomplished in two ways. In one case, conventional mutagenesis in potato tissue cultures of reduced ploidy levels and staining with iodine uncovered mutants lacking amylose. The *waxy* gene from potato was cloned and inserted in an inverted orientation of the gene into potato via transformation. Expression of the reverse orientation (antisense) of this gene, as in many cases of genetic transformation, blocks expression of the endogenous functional gene by an incompletely understood mechanism. Such transgenic antisense plants lacking starch-bound starch synthase were devoid of amylose.

C. Mendel's *wrinkled pea*, Maize *amylose-extender* and Starch Branching Enzymes

One interesting aside of starch synthetic genes is the role they have played in our understanding of genetics. Cryptic reference was made above to the role played by *Wx* in McClintock's unraveling of maize transposable elements. Recent evidence from Cathie Martin's laboratory (Bhattacharyya et al., 1990) showed that the very first gene used by Mendel to decipher the inheritance of genes turns out to be a starch synthetic gene. Mendel described a mutant termed *wrinkled pea,* which aptly describes the phenotype. The Martin laboratory first found that this mutant was low in starch branching activity (formation of α-1,6 bonds) and used conventional technology to clone the gene. Ironically, Mendel's mutant appears to be due to a transposable element.

Multiple forms of starch branching activity have been described. This is the case in maize, and the mutant *amylose-extender* (*ae*), which decreases amylopectin content and therefore increases the ratio of amylose to amylopectin (hence the name), affects one of the major branching activities. Several plant starch branching enzymes have been cloned, including *Ae* (see Fisher et al., 1993).

The fact that mutations at *Ae* decrease branching activity and decrease amylopectin content leads to the logical conclusion that branching activity is important in amylopectin synthesis. Furthermore, *ae* mutations affect only one of several branching enzymes and do not totally abolish amylopectin content. Also, the remaining amylopectin in an *ae* mutant is less highly branched as one might expect with a reduction in branching enzyme activity. These observations provide strong evidence that several branching enzymes are important in amylopectin synthesis.

Like *wx, ae* mutants are important in commerce. The significant increase in percentage of amylose allows for the manufacture of different products as well as economically significant changes in the way starch is processed. A mutant conditioning 100% amylose would be advantageous for many purposes, and conventional breeding and selection have generated heritable alterations in amylose content in *ae* materials ranging from 50 to 80% (B. Bear, 1982, personal communication). Based on our understanding of the biochemistry and genetic control of starch branching activity (several forms, and only one controlled by *Ae)*, it is not surprising that a single mutant totally lacking amylopectin has not been found. Several approaches can be envisaged to circumvent this problem. If there exists sufficient sequence similarity among the transcripts for the various branching enzymes, then antisense technology with a single antisense gene might suffice to remove all activity.

Another approach has recently become available. Taking advantage of the high rates of mutability of Robertson's Mutator in maize, S. Briggs (1992, personal communication) at Pioneer Hi-bred International has synthesized sufficient populations of maize to allow the insertion of a Mutator element into at least one copy of each maize gene. By pooling DNA preparations, it is feasible to screen the entire population by PCR technology. Thus in those cases, such as starch branching activity in maize, in which a gene has been cloned but no mutant is presently available, the Briggs population can be scanned via use of a Mutator-based primer and a primer from the cloned gene. Individual families from pooled samples giving positive signals can then be analyzed further. This approach would be ideal for starch branching activity since knock-out mutants for the other branching enzymes may not have a discernable phenotype in the presence of a functional *Ae* gene. However, such candidate mutants, either singly or in multiple combinations, should have greatly reduced branching activity and total loss of amylopectin in a mutant *ae* background. Given a PCR tag, appropriate mutant materials could be synthesized and then analyzed.

D. The *sugary1 (su1)* Mutation and Starch Debranching Activity

Starch debranching activity is important to normal starch synthesis. This totally unexpected conclusion was deduced from analysis of the *sugary1 (su1)* mutation of maize. The mutant of this locus termed *su1-R* is best known for its role in conventional sweet corns. This allele conditions a marked increase in the levels of phytoglycogen and sucrose. The increased sugar was exploited in early generation sweet corn to condition the sweetness; however more recent sweet corn use mutants of *sh2, bt1* or *bt2* (described below) or *su1* in combination with *sugary-enhancer.*

That the primary lesion of the *su1* locus involves a debranching activity was suggested by the work of Pan and Nelson (1985). They showed that debranching activity is greatly reduced in the *su1-R* mutant allele. Subsequently, the *sugary1* locus was cloned via transposon tagging (James & Myers, 1994), and sequence analysis of the gene revealed significant homology to a family of genes involved in cleavage of the α-1,6 bond of starch. Apparently, the absence of a specific debranching activity in a *su1* mutant allows the build up of the highly branched polysaccharide, phytoglycogen. It is suggested that the balance between branching and debranching activity in nonmutant maize results in the formation of amylopectin.

How does the absence of debranching activity lead to an increased level of sucrose? In other words, why is the traditional sweet corn sweet? It is embarrassing that we do not really know the answer to this. Several observations are relevant. First, not all mutant alleles of *su1* enhance sugar levels. As reviewed elsewhere (Hannah et al., 1993), some *su1* alleles increase phytoglycogen content without increasing sucrose levels. Thus it is possible to mutationally separate these two traits. Second, soluble starch synthase is low in a line homozygous for the *su1-R* allele (G. Singletary and P. Keeling, 1993, personal communication) and, from molecular analysis of the soluble starch synthase, it appears that a structural gene for this enzyme maps to the region of chromosome 4 defined by the *Su* locus. These observations raise the interesting possibility that *su1-R* is, in fact, a double mutant for both debranching activity and soluble starch synthase activity. A small deletion in two closely linked genes is a distinct possibility; however, Southern blots of *su1-R* probed with the debranching enzyme clone exhibit a band of hybridization (M.

James, 1993, personal communication). The nature of the relationship, if any, between elevated phytoglycogen and sucrose remains to be elucidated.

E. The *Brittle1* (*Bt1*) Gene and the Adenylate Translocator

Three genes, when mutant, condition a severe block in starch synthesis and a concomitant increase in sucrose. The *Bt1* locus is one of them. Sullivan et al.(1991) cloned the *Bt1* locus via tagging with the transposable element *suppressor/mutator* (*Spm*). Subsequent sequencing and sequence analysis revealed significant sequence similarity to a membrane-bound adenylate translocator as well as to other metabolite transporters in several plant and nonplant systems. As expected, the BT1 protein is found in the membrane fraction of maize endosperm amyloplasts (Sullivan & Kaneko, 1995), and the BT1 protein can be imported into pea chloroplasts (Li et al., 1992) Thus, the molecular genetic analysis and the subsequent biochemical/organelle analysis clearly show that the BT1 protein is plastid-membrane bound.

What is the nature of the molecule transferred by the BT1 protein? Clearly loss of BT1 function dramatically decreases starch content, and therefore, the transported molecule must be important to starch synthesis. Shannon and colleagues (Liu et al., 1992) noted that levels of ADP-glucose, the substrate for starch synthesis, are greatly elevated in developing endosperms of the *bt1* mutant. This was a surprising result, as detailed below. In several systems, ADP-glucose pyrophosphorylase, the enzyme important in the synthesis of ADP-glucose, has been reported to be in the amyloplasts. From investigations of transgenic potato (Stark et al., 1992), placement of a leader sequence onto the bacterial gene for ADP-glucose pyrophosphorylase leading to localization in the amyloplasts was necessary to obtain a significant alteration in starch content.

Is the physiologically important molecule that crosses into the amyloplasts for starch synthesis ADP-glucose? Interestingly, several studies have shown that plastids (both chloroplasts and amyloplasts) can take up ADP-glucose and incorporate the glucose into starch. Akazawa and colleagues (Pozueto-Romero et al., 1991a,b) first demonstrated this in sycamore (*Ficus sycamorus* L.) cells, and subsequent studies have shown this to be true in other plants, including amyloplasts of developing wheat (*Triticum aestivum* L.) endosperms (Tetlow et al., 1994). Shannon and colleagues (Liu et al., 1992) showed that this was true in amyloplasts from young maize seeds and, most significantly, this activity is greatly reduced in *bt1* amyloplasts.

In many studies, for example Tetlow et al. (1994), uptake of ADP-glucose is considered nonsignificant since ADP-glucose pyrophosphorylase is thought to be located in the amyloplasts in all plant tissues. However, there are two lines of evidence that suggest that this may not be the case. First, no evidence for a transit peptide could be found in a comparison of in vitro- and in vivo-synthesized maize endosperm ADP-glucose pyrophosphorylase subunits (Giroux & Hannah, 1994). Similar observations have been made for the barley (*Hordeum vulgare* L.) endosperm pyrophosphorylase (Villand & Kleczkowski, 1993). Second, the universality of the amyloplast placement must be questioned on theoretical grounds. Comparison of the N-terminal region of the pyrophosphorylase subunits reveals little to no sequence similarity among the small subunits and among the large subunits (L.C. Hannah, 1991, unpublished data). A compilation of these sequences can be found in the report by Smith-White and Preiss (1992). Since the transit peptide resides in the N-terminal region and no homology exists among comparable proteins in each subunit class, there is no logical reason, on biochemical grounds, to argue that cell localization should be universal. Fortunately, the appropriate mutants are known in maize to determine whether the cytoplasmic buildup of ADP-glucose in a *bt1* mutation is dependent upon the major ADP-glucose pyrophosphorylase found in the endosperm. J. Shannon and colleagues (1993, personal communication) have genetically removed the major ADP-glucose pyrophosphorylase activity and found that this abolished the buildup of ADP-glucose in a *bt1* background. These data point to a cytoplasmic localization of ADP-glucose pyrophosphorylase, at least in the maize endosperm.

F. The Genes *Shrunken2 (Sh2) and Brittle2 (Bt2)* and Adenosine 5′-Diphosphate-Glucose Pyrophosphorylase

Of all the starch synthetic enzymes, the one which has received the most attention from biochemists is ADP-glucose pyrophosphorylase (AGP). This enzyme catalyzes the formation of ADP-glucose and pyrophosphate from glucose-1-phosphate and ATP. In the maize endosperm, two structural genes, *shrunken2 (Sh2)* and *brittle2 (Bt2)* encode the two different subunits for AGP (Hannah & Nelson, 1976). Both genes have been cloned and sequenced (Bae et al., 1990; Bhave et al., 1990). In plants, AGP is a tetramer composed of two different subunits, whereas in bacteria, AGP is a homotetramer. Early reports that potato tubers contained a homotetrameric enzyme have been corrected (Smith-White & Preiss 1992). The ADP-glucose is the first unique intermediate in the synthesis of starch, and therefore, its synthesis would represent a logical step for control. As reviewed extensively elsewhere (e.g., Preiss & Levi, 1980; Preiss, 1982; Preiss et al., 1991), AGP is an allosteric enzyme in almost every organism and tissue analyzed to date. The individual molecules serving as activators in bacteria differ. In every case however, the most effective activating molecule is derived from the sugar preferred by the bacterium for growth. In chloroplasts, which synthesize starch during the sunlight and break it down at night, 3-phosphoglyceric acid (3-PGA) is the most potent activator of AGP. Phosphate inhibits this enzyme. These properties of AGP are viewed as control mechanisms since 3-PGA builds up during photosynthesis while phosphate increases during nonphotosynthetic periods.

Surprisingly, AGP from nonphotosynthetic plant tissues also is activated by 3-PGA (reviewed by Preiss, 1982). If AGP is localized in amyloplasts and if the C entering the amyloplasts for starch synthesis is a C-6 sugar and not 3-PGA or a close analogue, as the present data from a number of studies suggest, then the physiological significance of activation of this enzyme by 3-PGA is not obvious. Perhaps, this activation is not physiologically relevant but rather represents a remnant of the evolution of this enzyme. Indeed, the recent report that wheat endosperm AGP is not activated by 3-PGA (Kleczkowski et al., 1993) is consistent with this idea. Several reports (Dickinson & Preiss, 1969; Hannah & Nelson, 1975) have shown that the maize endosperm enzyme is much less sensitive to 3-PGA than the chloroplast enzyme. Plaxton and Preiss (1987) questioned this result and reported data showing that blockage from proteolytic cleavage maintains an enzyme quite sensitive to 3-PGA. We (Giroux & Hannah, 1994) and others have attempted to repeat the results of Plaxton and Preiss and find that while the subunits of this enzyme can be specifically cleaved by endogenous proteolytic attack, this cleavage has little effect on the activation properties of the enzyme. Activation by 3-PGA of the uncleaved maize endosperm enzyme is not comparable to that of the chloroplast AGP.

Regardless of the specific interactions with various putative effector molecules, ADP-glucose pyrophosphorylase is a rate-limiting step in starch biosynthesis in the potato tuber. Investigators at Monsanto (Stark et al., 1992) placed a bacterial-derived AGP into potato and expressed it in the tuber. Starch content increased 30%. Clearly, in the case of the wild-type potato tuber, this enzymic reaction represents a rate limiting step. We (Hannah and Giroux, in preparation) have created genetic variation within *Sh2*-encoded subunit of AGP by transposon mutagenesis of the endogenous maize gene expressed in the endosperm. One variant increases seed weight 15%. In collaboration with Greg Cobb and with others, we detect no increase in percentage of starch. Thus it appears that alteration of this enzyme leads not only to an increase in the amount of total starch but also to an increase in other seed components. In effect, the alteration in starch synthesis creates a stronger sink for the flow of C into the maize seed. While this variant produces an AGP with altered kinetic parameters, 3-PGA activation is unaffected.

G. Sucrose Synthase, *Shrunken1 (Sh1)* and *Sucrose synthase1 (Sus1)*

While all of the enzymic steps discussed above come relatively late in the path of starch synthesis, mutants are known to affect earlier steps as well. Mutations at the *shrunken1* (*Sh1*) gene of maize were described early in the history of genetics. Drew Schwartz found that a highly abundant endosperm protein was missing in some mutant alleles of this locus and, as reviewed previously (Hannah et al., 1993), investigations in Oliver Nelson's laboratory showed that the biochemical function associated with this protein is sucrose synthase. Enzymological investigations of complementing *sh1* mutants (Chourey & Nelson, 1979) as well as more direct feeding experiments (Cobb & Hannah, 1988) show that the physiological role of this enzyme for starch synthesis is actually the backwards reaction, namely: sucrose + UDP $\rightleftarrows$ UDP-glucose and fructose. One or both of these products then serves as the C source for starch synthesis.

Mutations at *sh1* give rise to a mild phenotype relative to that conditioned by mutations at *sh2, bt1* or *bt2*. Burr and Burr (1981) showed by Southern analysis that the mutant allele in the McClintock *sh1 bzm-4* totally lacked *Sh1* locus, yet still conditioned the mild phenotype. Hence, the mild phenotype is not due to partial expression of the examined *sh1* mutant alleles. Subsequent analysis revealed a low level of sucrose synthase activity in the McClintock mutant (Chourey et al., 1986). The finding of low levels of sucrose synthase activity in the *sh1* deletion mutant clearly showed that not all sucrose synthase activity is under the control of the allele at the *sh1* locus and provided an explanation for the fact that mutations at *sh1* did not condition the severe phenotype noted with other starch mutants.

Subsequently, we (McCarty et al., 1986) cloned the locus for the second sucrose synthase and determined its structure (Shaw et al., 1994). That the function of this second gene, now termed *sucrose synthase1* or *Sus1,* may not account for the underlying starch synthesis found in the *sh1* deletion mutant has recently been suggested by Chourey and Talierco (1994). They reported a mutant of *sus1* that lacks gene function, yet the mutation is associated with no visual effect on the endosperm or any other tissue of the maize plant. Double mutants for *sus1* and *sh1* are reported to have a kernel phenotype no more severe than that found in *sh1* single mutants. If so, then another enzymatic activity, perhaps invertase, accounts for the low level of starch found in *sh1* mutant endosperms.

If mutation in *Sus1* alleles conditions no effect on the maize plant, the gene should be dispensable. We recently characterized the *Sus1* gene in order to test this hypothesis (Shaw et al., 1994). The structures of *Sh1* and *Sus1* are identical except for one intron found only in *Sh1*. The two genes share 14 introns in identical positions. The most parsimonious explanation for this is that the genes arose from a duplication and these introns were present in the progenitor gene. A comparison of the sequences of the corresponding exons and introns of these two genes reveals that whereas the exons are quite similar, the introns show no obvious similarity in size or sequence. Since many mutations within introns can be tolerated, whereas most exon mutations lack this neutrality, Shaw et al. (1994) suggested that the differential rate of mutation most likely is explained by selection. This result is not what one would expect if loss of *Sus1* function conditions no physiological change in the maize plant. Furthermore, all monocots that have been investigated in sufficient detail have both the *Sus1* type and the *Sh1* type (termed *Sus2* when referring to the gene in all plants, Hannah et al., 1994) of sucrose synthase genes, as predicted if both genes are important to the respective plant. Clearly additional work is necessary to resolve the importance of *Sus1*-encoded sucrose synthase activity to starch biosynthesis, particularly in characterizing the *sus1* mutant.

H. Invertase Activity and the *miniature* Mutant of Maize

Analysis of the *miniature (mn1)* mutant may reveal a further complexity in the flow of C from sucrose to starch in this biosynthetic pathway. As described above, sucrose synthase, acting to hydrolyze sucrose, is clearly important to starch synthesis. Recent evidence suggests that invertase, which also cleaves sucrose but produces glucose and fructose, also may be important in the path of starch synthesis. Miller and Chourey (1992) showed that the *mn1* mutant, which as the name implies produces a seed much reduced in size, has greatly reduced invertase activity in the tissue residing at the base of the seed. The mutant also reduces invertase activity in the maternal tissue adjacent to the mutant seed. The relationship between the gene and the enzyme is not clear. Several invertases (acid and neutral, soluble and bound) are known and it is clear that there are several structural genes for these isoenzymes (Xu et al., 1993; K. Koch, 1992, personal communication). How *mn1* affects all of these activities and the enzymes in both the seed and the maternal tissue are not clear. The *mn1* mutant exhibits simple Mendelian inheritance and seed resulting from the pollination of a mutant female by a wild type male are nonmutant. In this cross, the cob tissue underlying the normal seed is genetically mutant, *mn1*. Hence, if this invertase activity were encoded by *Mn1* and this invertase were important in the maternal tissue, one would expect such hybrid seed to be mutant. This prediction is not borne out by experimental data. The complexity of the relationship between the *mn1* allelic state and invertase activity leaves open the possibility that the primary lesion involved in this mutation lies elsewhere. If so, the effect of *mn1* on invertase may simply be a secondary effect and the loss of wild-type levels of invertase may simply be a consequence of altered seed development.

While the primary lesion of *Mn* is clearly unknown at this time, there are independent observations that suggest that sucrose actually does appear twice in the flow of C from the leaf to the kernel. Hence, the finding that mutations occur in the two separate sucrose degrading enzymes (sucrose synthase and invertase) would make sense. Pulse-chase experiments of Shannon (reviewed in Cobb & Hannah, 1986; and Hannah et al., 1993) in which $^{14}CO_2$ was fed to maize leaves revealed peaks in sucrose label at two different times following the chase period. Cobb and Hannah (1986) showed that kernels grown in petri dishes in a medium containing only reducing sugars contained high levels of sucrose. Clearly then, maize seeds (or the small amount of cob tissue attached to them in these experiments) have the ability to synthesize sucrose; starch synthetic schemes containing both sucrose synthase and invertase may be correct. However, it should be noted that when kernels are grown in petri dishes in the presence of high levels of fructose or glucose, the added sugar accumulates to high levels in the developing kernels. Furthermore, such seeds grown on either reducing sugar develop normally. Clearly, sugars other than sucrose can be taken up by the developing kernel. It also should be noted that when labeled sucrose is fed to kernels in the presence of high levels of glucose or fructose, the hexose residues retain their original specific activities after passage from the petri dish to the developing kernel. This result suggests that the sucrose had not been degraded and resynthesized.

I. Antisense Mutants of Uridine 5′-Diphosphate-Glucose Pyrophosphorylase

One of the most active enzymes in the maize endosperm is UDP-glucose pyrophosphorylase. The enzyme is thought to convert UDP-glucose arising from the action of sucrose synthase to UTP and glucose-1-phosphate. The latter product is then used by ADP-glucose pyrophosphorylase for the synthesis of ADP-glucose, the actual substrate for starch synthesis. The UDP-glucose pyrophosphorylase also is thought to be important in cell wall synthesis since UDP-glucose is important in that biosynthetic pathway as well. No spontaneous mutants are known for UDP-glucose pyrophosphorylase; however, Zrenner et al. (1993) recently synthesized transgenic potato containing an antisense gene for this enzyme. The gene was expressed in all tissues by use of the 35S promoter, and endogenous activity was decreased 95 to 96%. Surprisingly, this gave rise to no noticeable change in the growth and development of the plant nor to starch metabolism in the tuber. Perhaps only a small percentage of wild-type activity is necessary and sufficient for wild-type starch synthesis, as discussed in general terms above, or perhaps these data point to the unimportance of this enzymic activity. Resolution of this important question awaits further investigation.

III. AN OVERVIEW

In the discussions above, emphasis was placed on those cases where the primary lesion of the mutant gene is known. This insight reduces the complexity of deciphering complex cause and effect relationships, as may be occurring in the case of the *mn1* mutant. No gene operates independent of other genes, and secondary effects of mutations are common. For example, we (Giroux et al., 1994) recently showed that transcript levels from nonmutant starch-synthetic genes are elevated when starch synthesis is genetically blocked. The elevation occurs at the transcriptional level, becomes amplified during development, is correlated with the severity of the imposed block, and is associated with the increase in sugar rather than the reduction in starch content. Surprisingly, this networking of gene expression includes genes involved in synthesis of the storage proteins as well. The network does not include genes functioning in other pathways or genes expressed in the adjacent embryo. Given this type of complexity, it is imperative that the primary lesion of a mutant is known before meaningful insight into the pathway can be gleaned.

A. Rate Limiting Steps, Dominant Functional Alleles, Myths and Realities

From an historical perspective of biochemistry, the holy grail of pathway decipherers is the identification of the rate-limiting step in the pathway under study. In the case of starch biosynthesis, the recent demonstration that the addition of a bacterial-derived AGP gene to potato with a concomitant increase in starch content as well as the transposable element-derived alteration in the maize endosperm AGP resulting in increased seed weight clearly point to a rate limiting role for AGP.

Are there other "rate-limiting" steps in the starch biosynthetic pathway? Do we think of a pathway as a chain with one weak link or are all of the links of the same strength? It has been argued that over evolutionary time, selection pressure has favored cases in which substrate levels are approximately equal to the K_m values of the cognate enzymes. This allows the system to be responsive to changes in substrate concentrations without a large waste in the production of unused enzyme molecules. In such a system, alterations in any one of the steps would affect the amount of the endproduct. This can best be understood by an admittedly oversimplified example. Suppose we introduce an altered form of the enzyme catalyzing the second step in a hypothetical pathway. Suppose the altered enzyme has a K_m for the physiologically relevant substrate some 10-fold lower than that of the wild-type enzyme. We should then expect that the amount of product produced by this enzyme to be approximately doubled since the former enzyme was operating at K_m substrate levels. Given more product from the second enzyme, then the rate of the reaction by the third enzyme would increase since it was previously operating at one-half its maximum velocity. In this simple case there is nothing special about the second enzyme, and alterations at any of the steps should give rise to an increase in product formation.

With the advent of gene cloning and transformation, the scenario described above can be tested by altering each of the known steps in the starch pathway and testing the effects of each added enzyme on starch content. This approach can be taken once an enzyme with a lowered K_m has been identified and cloned. Even when such altered enzymes are known, this approach is labor intensive, expensive, and problems of cosuppression and position effects still make interpretation of resulting data problematic.

A much less expensive, although somewhat less powerful approach, has been taken recently by G. Singletary and P. Keeling (in preparation). Rather than adding enzymic activity and monitoring its effect, these investigators measured the effect on starch levels of small reductions in the activities of various enzymes known, by mutant analysis, to be important in starch synthesis. Reductions in enzymic activity were made by producing kernels containing a reduced number of functional alleles of particular enzymes. This approach takes advantage of the fact that one usually finds a simple dosage effect between the number of functional alleles of a structural gene and the amount of its cognate enzymic activity. For example, developing endosperms containing approximately two-thirds or one-third wild type levels of AGP can be synthesized by simply crossing a *bt2* mutant reciprocally with wild type. Since the endosperm is triploid and receives two of the three genomes from the female parent, reciprocal crosses between the mutant and wild type yield the endosperms with the levels of AGP mentioned above.

In the case of AGP, Singeletary and Keeling found that starch content increased in a linear manner with the number of functional *Bt2* alleles, a result expected if AGP is a rate-limiting step in starch biosynthesis. Significantly, while kernels with one or three doses of the functional *Bt2* allele are indistinguishable to the human eye, they differ in starch content. This, I believe, is a key observation, of relevance for a number of reasons. First, it suggests that other rate-limiting steps may be identified simply by looking at starch content in mature seeds that differ in genotype at one particular locus. It also suggests that searches for dominant or semi-dominant starch mutants (as classified by the eye) might be a productive exercise in identifying important steps in this pathway. Finally it points out that had the early maize investigators used starch content rather than visual kernel phenotype in their analysis of the *Bt2* locus, we would today not think of mutants of this locus as being recessive (or, more importantly, as wild-type allele being dominant) but rather would have classified the early mutants as being semi-dominant.

While the approach above of exploiting dosage effects and starch content to identify important steps is a powerful one, it does have limitations. First, while one might find that reducing enzymic activity by one-third reduces starch content, there is no guarantee that increasing the amount of activity above that found in wild type will increase starch content. Likewise the lack of a dosage effect does not automatically negate the possibility that the addition of a kinetically different isoenzyme might increase starch content. Notwithstanding these criticisms, this approach is powerful and it is surprising that it has not been exploited in previous studies.

V. THE FUTURE OF STARCH MODIFICATION IN PLANTS

Perhaps no better examples can be found in the use of single-gene alterations for improvement than the cases of modification of starch quality and quantity in maize. The incorporation of the mutant *su* allele for sweet corn, the use of *ae* and of *wx* for starch quality in field corn, and the recent use of the mutants *sh2, bt2, bt1* and *sugary-enhancer (se)* in sweet corn are prime examples. In an earlier review (Hannah et al., 1993), the synthesis of a construct containing the *Sh2* coding region and a promoter expressed relatively late in endosperm development was discussed. Placement of this construct into transgenic *sh2* maize should give rise to seeds that are high in sugar early in development (at the eating stage) but high in starch content in the mature seed. This may circumvent the seed and seedling vigor problems associated with the high-sugar, super-sweet corn. This work is in progress.

Evidence was reviewed above showing that the bacterial AGP or the altered form of the maize endosperm AGP are advantageous in increasing starch content in potato and seed weight in maize. Undoubtedly, these genes will be used in future plant improvement programs.

From the standpoint of conventional wet corn millers, several traits are deemed desirable. These processors would much prefer starting materials composed entirely of amylopectin or of amylose. These polymers could then be mixed in the desired ratio. Today, millers must first process conventional plant materials to separate amylose from amylopectin before the appropriate mixing can be performed. Starch composed entirely of amylopectin is produced by *wx* in maize. In potato, amylose-free starch has been produced by conventional mutagenesis and by the insertion of an antisense *Wx*-homologue (Visser et al., 1991). Starch composed entirely of amylose has not yet been produced, presumably because of the presence of multiple, genetically independent branching enzymes. Approaches to silence all of these genes were mentioned above.

Starch processors also would like larger starch molecules and a higher phosphate content. The size of the glucose polymer should be a product of the rate of de novo synthesis of primers for starch synthesis and the rate of elongation and branching of the molecule. If so, this could be approached via genetic and molecular techniques. Potato starch contains a higher percentage of phosphate than corn starch. In some instances, this is advantageous for manufacturing. While the genetics of this difference is not presently understood, it is clearly approachable with the proper methods.

What does the future hold? Polymers of sugars are all around us. From grocery bags to complex carbohydrates that possibly suppress the growth and internal spread of viruses affecting humankind. It is fair and perhaps wise to view plants as inexpensive, efficient factories that can be used to synthesize these polymers that are ever more important in our daily lives. Genetic blockage of the starch assembly line in maize leads to a huge build up in sucrose, this sugar accounts for 30% (Creech, 1968) of the seed dry weight in some mutants. Clearly, the placement of the appropriate genes encoding the enzymes involved in the synthesis of these other complex carbohydrates into plants and their expression in the appropriate tissues (seed, tuber, etc.) is a fruitful area for long-term research. The particular carbohydrate to synthesize is, of course, dependent on its value, its cost of synthesis or isolation via conventional means, and whether the associated enzymes involved in its synthesis are known. At the plant level, appropriate promoters giving rise to enhanced tissue expression are known in many cases, and transformation, while still difficult in some plants, is possible. Given isolation of the genes conditioning the appropriate enzymology and their insertion into plants, the next major hurdle is whether the desired carbohydrate will accumulate. This may not be a problem since the newly synthesized sugar polymer should be "foreign" to the plant and therefore one might not expect that the cognate degradative enzymes will be present.

The research scenario pictured above should not be viewed as short term. Many of the steps are challenging, yet none is insurmountable. Given the value of carbohydrates in our society, the rewards from such a research program are immense. Plants are cheap, can be replanted, and are highly efficient in synthesizing sugars. The groundwork has been laid to harness this system in ways not possible 25 yr ago. Now, it is just a matter of hard work, time, money, patience, foresight, and some luck. (Literature review for starch biosynthesis was completed in December 1994.)

VI. ACKNOWLEDGMENT

I thank Martha James, Peter Keeling, Jack Shannon, George Singletary and Tom Sullivan for sharing unpublished data and for valuable critiques of an earlier version of this review. I thank Don McCarty, Scott Nichols, Karen Koch and especially Oliver Nelson for many insightful discussions concerning starch synthesis in maize.

REFERENCES

Bae, J.M., M. Giroux, and L.C. Hannah. 1990. Cloning and characterization of the *brittle-2* gene of maize. Maydica 35:317–322.

Bhattadharyya, M.K., A.M. Smith, T.H.N. Ellis, C. Hedley, and C. Martin. 1990. The wrinkled-seed character of pea described by Mendel is caused by a transposon-like insertion in a gene encoding starch branching enzyme. Cell 60:115–122.

Bhave, M.R., S. Lawrence, C. Burton, and L.C. Hannah. 1990. Identification and molecular characterization of *shrunken-2* cDNA clones of maize. Plant Cell 2:581–588.

Boyer, C.D., and L.C. Hannah. 1994. Kernel mutants of corn. p.1–28. *In* A.R. Hallauner (ed.) Speciality corn. CRC Press, Boca Raton, FL.

Burr, B., and F.A. Burr. 1981. Controlling-element events at the shrunken locus in maize. Genetics 98:143–156.

Burr, B., and O.E. Nelson. 1973 The phosphorylases of developing maize seeds. Annu. New York Acad. Sci. 210:129–138.

Chourey, P., M. Latham, and P. Still. 1986. Expression of two sucrose synthase genes in endosperm and seedling cells of maize: Evidence of tissue specific polymerization of promoters. Mol. Gen. Genet. 203:251–255.

Chourey, P., and O.E. Nelson. 1979. Interallelic complementation at the *sh* locus in maize at the enzyme level. Genetics 91:317–325.

Chourey, P.S., and E.W. Taliercio. 1994. Epistatic interaction and functional compensation between the two tissue- and cell-specific sucrose synthase genes in maize. Proc. Natl. Acad. Sci. USA 91:7917–7921.

Clancy, M., V. Vasil, L.C. Hannah, and I.K. Vasil. 1994. Maize *shrunken-1* intron and exon regions increase gene expression in maize protoplasts. Plant Sci. 98:151–161.

Cobb, B.G., and L.C. Hannah. 1988. *Shrunken-1* encoded sucrose synthase is not required for sucrose synthesis in the maize endosperm. Plant Physiol. 88:1219–1221.

Creech, R.G. 1968. Carbohydrate synthesis in maize. Adv. Agron. 20:275.

Dickinson, D.B., and J. Preiss 1969. Presence of ADP-glucose pyrophosphorylase in shrunken-2 and brittle-2 mutants of maize endosperm. Plant Physiol. 44:1058–1062.

Echt, C., and D. Schwartz. 1981. Evidence for the inclusion of controlling elements within the structural gene at the *waxy* locus in maize. Genetics 99:275–284.

Fisher, D.K., C.D. Boyer, and L.C. Hannah. 1993. Starch branching enzyme II from maize endosperm. Plant Physiol 102:1045–1046.

Giroux, M.J., C. Boyer, G. Feix, and L.C. Hannah. 1994. Coordinated transcriptional regulation of storage product genes in the maize endosperm. Plant Physiol. 106:713–722.

Giroux, M.J., and L.C. Hannah. 1994. ADP-glucose pyrophosphorylase in *shrunken2* and *brittle2* mutants of maize. Mol. Gen. Genet. 243:400–408.

Hannah, L.C., E. Duke, K. Koch, and B.G. Cobb. 1993. Starch synthetic genes of maize. *In* V. Walbot and M. Freeling (ed.) Maize methods.

Hannah, L.C. , W. Frommer, J.-C. Su, P. Chourey, and W. Park. 1994. Sucrose synthases. Plant Mol Biol. 12:72–73.

Hannah, L.C., M. Giroux, and C.D. Boyer. 1993. Biotechnological modification for sweet corn improvement. A review. Sci. Hort. 55:177–197.

Hannah, L.C., and O.E. Nelson, Jr. 1975. Characterization of adenosine diphosphate glucose pyrophosphorylase from developing maize seeds. Plant Physiol. 55:297–302.

Hannah, L.C., and O.E. Nelson, Jr. 1976. Characterization of ADP-glucose pyrophosphorylase from *shrunken-2* and *brittle-2* mutants of maize. Biochem. Genet. 14:547–560.

James, M.G., and A. Myers. 1994. Cloning of *sugary1* by transposon tagging with *Mutator*. Maize Genet. Coop. Newsl. 68:8.

Kleczkowski, L.A., P. Villand, E. Luthi, O.-A. Olsen, and J. Preiss. 1993. Insensitivy of barley endosperm ADP-glucose pyrophosphorylase to 3-phosphoglcerate and orthophosphate regulation. Plant Physiol. 101:179–186.

Li, H.M., T.D. Sullivan, and K. Keestra. 1992. Information for targeting to the chloroplastic inner envelope membrane is contained in the mature coding region of the maize Bt1-encoded protein. J. Biol. Chem. 267:18 999–19 004.

Liu, K.C. , C.D. Boyer, and J.C. Shannon. 1992. Carbohydrate transfer into isolated maize amyloplasts. Plant Physiol 99:39.

Macdonald, F.D, and J. Preiss. 1983. Solubilization of the starch-granule-bound starch synthases of normal maize kernels. Plant Physiol 73:175–178.

McCarty, D.R., J.R. Shaw, and L.C. Hannah. 1986. The cloning, genetic mapping, and expression of the *constitutive sucrose synthase* locus of maize. Proc. Natl. Acad. Sci. USA 83:9099–9103.

McClintock, B. 1953. Induction of instability at selected loci in maize. Genetics 38:579–599.

Miller, M.E., and P.S. Chourey. 1992. The maize invertase-deficient *miniature-1* seed mutation is associated with aberrant pedicel and endosperm development. Plant Cell 4:297–302.

Nelson, O.E., Jr., and H.W. Rines. 1962. The enzymatic deficiency in the waxy mutant of maize. Biochem. Biophys. Res. Commun. 9:297–300.

Pan, D., and Nelson, O.E., Jr. 1985. The deficiency of a starch granule-bound enzyme phospho-oligosaccharide synthase in developing *bt 1* endosperms. Maize Genet. Coop. News Lett. 59:105–106.

Parera, C.A., D.J. Cantliffe, and L.C. Hannah. 1993. Improvement of vigor in *shrunken2* corn seedlings. Search. p. 15–28.

Plaxton, W.C., and J. Preiss. 1987. Purification and properties of non-proteolytically degraded ADP-glucose pyrophosphorylase from maize endosperm. Plant Physiol. 83:105–112.

Pozueto-Romero, J., F. Ardila, and T. Akazawa. 1991a. ADP-glucose transport by the chloroplast adenylate translocator is linked to starch biosynthesis. Plant Physiol. 97:1565–1672.

Pozueto-Romero, J., M. Frehner, A.M. Viole, and T. Akazawa. 1991b. Direct transport of ADP glucose by an adenylate translocator is linked to starch biosynthesis in amyloplasts. Proc. Natl. Acad. Sci. 88:5769–5773.

Preiss, J. 1982. Regulation of biosynthesis and degradation of starch. Ann. Rev. Plant Physiol. 33:431–454.

Preiss, J., K. Ball, J. Hutney, B.J. Smith-White, L. Li, and T.W. Okita. 1991. Regulatory mechanisms involved in the biosynthesis of starch. Pure Appl. Chem. 63:535–544.

Preiss, J., and C. Levi. 1980. Starch biosynthesis and degradation. p. 371–423. *In* J. Preiss (ed.) The biochemistry of plants. A comprehensive treatise. Vol. 3. Acad. Press, New York.

Prioul, J.-L., E. Jeannette, A. Reyss, N. Gregory, M. Giroux, L. C. Hannah, and M. Causse. 1994. Expression of ADP-glucose pyrophosphorylase in maize grain and source leaf during grain filling. Plant Physiol. 104:179–187.

Shaw, J.R., R.J. Ferl, J. Baier, D. St. Clair, C. Carson, D. McCarty, and L.C. Hannah. 1994. Structural features of the maize *Sus1* gene and protein. Plant Physiol. 106:1659–1665.

Sivak, M.N., M. Wagner, and J. Preiss. 1993. Biochemical evidence for the role of the *waxy* protein from pea as a granule-bound starch synthase. Plant Physiol. 103:1355–1359.

Smith, A. 1990. Evidence that the "waxy" protein of pea is not the major starch-granule-bound starch synthase. Planta 182:599–604.

Smith-White, B.J., and J. Preiss. 1992. Comparison of proteins of ADP-glucose pyrophosphorylase from diverse sources. J. Mol. Evol. 34:449–464.

Stark, D.M., K.P. Timmerman, G. Barry, J. Preiss, and G.M. Kishore. 1992. Regulation of the amount of starch in plant tissues by ADP-glucose pyrophosphorylase. Science (Washington, DC) 258:287–292.

Sullivan, T.D., and Y. Kaneko. 1995. The maize *brittle1* gene encodes amyloplast membrane polypeptides. Planta 196:477–484.

Sullivan, T.D., L.I. Strelow, C.A. Illingworth, R.L. Phillips, and O.E. Nelson, Jr. 1991. The maize *brittle-1* locus: Molecular characterization based on DNA clones isolated using the *dSpm*-tagged *brittle-1-mutable* allele. Plant Cell 3:1337–1348.

Tetlow, I.J., K.J. Blissett, and M.J. Evans. 1994. Starch synthesis and carbohydrate oxidation in amyloplasts from developing wheat endosperm. Planta 4:454–460.

Villand, P., and L.A. Kleczkowski. 1993. Is there an alternative pathway for starch biosynthesis in cereal seeds? Z. Naturforsch. 49c:215–219.

Visser, R.G., I. Somhorst, G.J. Kuipers, N.J. Ruys, W.J. Feenstra, and E. Jacobsen. 1991. Inhibition of the expression of the gene for granule-bound starch synthase in potato by antisense constructs. Mol. Gen. Genet. 225:289–296.

Xu, J., D.R. McCarty, and K. Koch. 1993. Maize invertase gene expression is carbohydrate-responsive. Plant Physiol. 102:33–44.

Zrenner, R., L. Willmitzer, and U. Sonnewald.1993. Analysis of the expression of potato uridinediphosphate-glucose pyrophosphorylase and its inhibition by antisense RNA. Planta 190:247–252.

7D Dietary Fiber

PAMELA J. WHITE

Iowa State University
Ames, Iowa

In the early 1900s a great deal of interest was exhibited in cereal fiber, or as it was commonly known, roughage. Even with this recognition, refined cereals, especially white wheat (*Triticum aestivum* L.) flour, became popular. Since the 1970s there has been a renewed interest in the importance of fiber in the human diet. Work of Dr. Denis P. Burkitt revealed a link between the health and diet of rural native Africans who ate much more fiber than did typical people in the industrialized west (Burkitt et al., 1972, 1974; Burkitt, 1977). Further studies revealed that the traditional measurement for fiber analyses, that of crude fiber, did not adequately measure and thus, account for these observations. Today, we recognize that dietary fiber is extremely important to human health and that there are several different types of fiber, each of which should be measured separately. Analytical methods have undergone extensive development since the mid-1970s, which has helped in our definitions of dietary fiber. Unfortunately, the amount of dietary fiber reported for any product is heavily dependent on the analytical method, which limits the comparison and interpretation of data. In general, fiber data should be interpreted with caution and with careful note of the analysis method.

Fibers important in the human diet are referred to as total dietary fiber (TDF), composed of soluble (SF) and insoluble (IF) dietary fiber components. The TDF can be defined as that part of the plant material in the diet that is resistant to digestion by the secretions of the gastrointestinal tract (Hall, 1989), including the phenolic polymer lignin (Theander et al., 1993). "Dietary fiber" also refers to the supporting structures of plant cell walls and the substances intimately associated with them. Cereals are one of the best sources of TDF. The SF consists of pectin, certain hemicelluloses including arabinoxylans (pentosans), β-glucan, and other vegetable gums. The SF is believed to increase intestinal transit times, delay gastric emptying, bind serum cholesterol decreasing its absorption, and enhance glucose tolerance by slowing glucose absorption (Best, 1987; Walker, 1993). The SF is almost completely fermented into short-chain fatty acids while it is present in the colon, which may inhibit hepatic cholesterol synthesis and increase the body's clearance of low-density-lipoprotein cholesterol (Walker, 1993). Fruit, oat (*Avena sativa* L.), barley (*Hordeum vulgare* L.), and legumes are the best natural food sources of SF. Many researchers have demonstrated a link between consumption of fiber from oat and barley and lowering of total plasma cholesterol in humans (Anderson et al.,

1984; Davidson et al., 1991) and in animals (Chen & Anderson, 1979; Chen et al., 1981; Klopfenstein & Hoseney, 1987; Ranhotra et al., 1987, 1991; Jennings et al., 1988; Shinnick et al., 1988; Kahlon et al., 1990, 1993; Newman et al., 1992). The hypocholesterolemic effects of barley and oat have been attributed to their β-glucan content (Chen et al., 1981; Fadel et al., 1987; Klopfenstein & Hoseney, 1987).

The IF, which should reflect the amount of cellulose, lignin, and some hemicelluloses, has been linked to the shortening of transit times in the small intestine, the softening and bulking of stool, and delaying glucose absorption and starch hydrolysis (Walker, 1993). It reduces the risk of constipation, appendicitis, irritable bowel syndrome, and possibly colon cancer (Best, 1987). Wheat, most grain products, and vegetables are particularly good sources of IF. Unfortunately, the separation of components into SF and IF is not always exact, as just noted, and some components such as β-glucan may be split between the two fiber types.

The intent of this discussion is to identify crops targeted for genetic alteration of one or more of their fiber components, to indicate strategies for attaining these goals, and to characterize progress and ongoing research. Most importantly, this section will identify reasonable goals for this value-added trait in crops as they leave the field. This discussion does not serve as a comprehensive literature review for dietary fiber in crops.

I. CROPS TARGETED FOR ALTERED FIBER COMPOSITION

A. Oat

The crop receiving the most attention in recent years for its TDF content is oat. The specific fiber fraction of interest is (1→3),(1→4)-β-D-glucan, which along with pentosans (arabinoxylans) comprise the major polymers of the oat endosperm cell walls (Henry, 1987). The β-glucan fiber is a linear, unbranched polysaccharide composed of 4-O-linked and 3-O-linked β-D-glucopyranosyl units (Peat et al., 1957; Parrish et al., 1960; Clarke & Stone, 1966; Aspinall & Carpenter, 1984). In both oat and barley, the ratio of the 4-0 linked units to the 3-O linked units is about 7:3. Oat pentosans contain primarily xylose and arabinose (Henry, 1987), and presumeably have a main chain composed of (1→4)-β-D-xylopyranosyl units with an occasional arabinose branching (Southgate et al., 1993).

The oat groat contains approximately 5% β-glucan, with slightly more of the fiber concentrated in the bran where β-glucan might be approximately 7.5% (Wood et al., 1991). The β-glucan contents of the groat are highly variable. Surveys by Lim et al. (1992) found ranges of 3.0 to 6.5% (in 1989) and 3.9 to 6.4% (in 1990) for the same 102 oat lines. Cho and White (1993) examined 243 genetically variable oat lines and found that most samples ranged between 4.5 and 5.5% β-glucan, with an overall range of 3.0 to 6.5%. In all instances, there was a normal distribution for β-glucan content. Fulcher and Miller (1993) described a range of less than 2% in some primitive landraces to almost 7% in exceptional domestic lines. In a survey of about 6000 oat genotypes from the National Small Grains Collection (Aberdeen, Idaho), a range of 2.5 to 8.5% β-glucan was found (Peterson et al., 1996). Although oat varieties typically rank in similar order between environments, they may range widely in the absolute β-glucan content from year to year or from location to location (Peterson, 1991; Lim et al., 1992; Fulcher & Miller, 1993). Thus, variations in β-glucan content of groats is a result of both genetic and environmental factors. A study by Holthaus et al. (1996) clearly showed that β-glucan content in oat is heritable. Additive gene effects were more important, and dominant gene effects were detected in only two out of five biparental populations. These results suggested that plant breeders should be successful in altering β-glucan in oat through direct selection for this genetically variable trait. The distinct genotypic differences suggest that groat β-glucan content can be selected for without accompanying undesirable changes (Welch & Lloyd, 1989; Peterson, 1991).

Furthermore, microscopic examinations show that β-glucan distribution within the groat is variable among different oat varieties (Fulcher & Miller, 1993). In general, the β-glucan is highly concentrated in the subaleurone layer; however, in high-β-glucan varieties, there is a more even distribution of the β-glucan throughout the kernel (Miller & Fulcher, 1994). Additional work has focused on the molecular structure of the β-glucan polymer. This topic was thorougly reviewed in a recent book chapter (Wood, 1993). These factors should be considered in developing oat with altered levels of β-glucan.

Clinical studies have shown that daily diets containing 100 g of oat bran were effective at significantly lowering blood cholesterol, primarily low-density-lipoprotein (LDL) or "bad" cholesterol by 13 to 36% (Gould et al., 1980; Kirby et al., 1981; Anderson et al., 1984; Anderson et al., 1991). Furthermore, a diet containing just 25 g of oat bran cereal reduced total blood cholesterol by 5.4% and LDL-cholesterol by 8.5% (Anderson et al., 1990). To obtain the same amount of β-glucan from oat groats, rather than the oat bran, 38 to 150 g, or approximately one to four 40-g servings, of oatmeal would need to be eaten (assuming that bran contains about 1.5 times as much β-glucan as the groat, as noted above). If the β-glucan level inherently present in oat were doubled, the amount of oat needed to aid the lowering of blood cholesterol would approximate one to two servings a day, which might be considered an achievable quantity. The oat might be consumed as hot breakfast cereal (oatmeal) or incorporated into other products such as ready-to-eat breakfast cereals, breads, crackers or other baked goods, or as a filler in main-dish foods.

The oat also contains some pentosans, although this component has not been studied as much as the β-glucan. In 19 diverse oat cultivars, Bhatty (1992a) found a mean pentosan level of 3.7% with a range of 2.4 to 4.5%. In two oatmeal samples, Englyst et al. (1989) found 2.1 to 2.2% pentosan.

The same fiber components that contribute to the nutritional benefits of oat also influence the physical and functional characteristics of the grain when incorporated into foods. Oat gum, composed mainly of β-glucan, is viscous even at low concentrations, indicating its potential use as a food hydrocolloid to replace substituted celluloses such as guar gum and locust bean gum (Wood, 1986; Wood et al., 1989; Jansson & Lindahl, 1991; Dawkins & Nnanna, 1993). To the food industry, the most important physical characteristic of cereal β-glucans is their viscosity (Autio et al., 1987). Whereas this viscous characteristic may be desirable at low levels as a food additive, it is possible to create a product that is too thick and pasty. In raising levels of β-glucan in oat, scientists must be careful not to develop so much fiber as to produce an end product that is too thick, sticky, and gummy. A bowl of oatmeal already is known for its gluey consistency. Likely, it will be possible to increase the SF components of oat and still produce a product that is sensorily desirable, but the quality of the altered grain should be monitored. In some instances, the SF component may create difficulties in the manufacture of a product. For example, oat and other SF-containing cereals routinely are incorporated into batters extruded to form dry breakfast cereals of various shapes and sizes. The SF interferes with the extrusion of the cooking batter so cereals with low levels of SF are desirable for these applications.

As just mentioned, oat also is consumed after being incorporated into many baked products. For example, in two studies, commercial oat bran was substituted for 10 and 15% of white wheat flour in white pan bread to determine the effects on dough properties and bread quality (D'Appolonia & Youngs, 1978; Krishnan et al., 1987). The latter study also evaluated the effects of bran particle size on dough and bread characteristics. In general, absorption requirements in oat bran doughs increased with increased bran levels and also with decreased bran particle size. Both 10 and 15% bran blends increased dough mixing time slightly over the control and made the dough sticky; however, breads made with 10% bran of medium-size particles and breads with 15% bran of large-size particles received the highest overall scores by a sensory panel. Results from these types of studies suggest an acceptance in baked products of cereals with high SF content and a market for their sales.

Although most recent discussions of TDF center around its positive effects, fiber in food also may interfere with vitamin and mineral absorption, as demonstrated by reduced bioavailability of Vitamin E in rats consuming increased amounts of wheat bran (Kahlon et al., 1986). So far, research does not point to an exact amount of fiber needed for optimal health. Furthermore, TDF will influence the physical properties of the grain and of foods into which the grain is incorporated. Obviously, in many instances, an increase in a fiber component may be beneficial; however, care must be taken not to overdo a good thing.

B. Barley

As a human food, the goals for altering the fiber content in barley are quite similar to those just mentioned for oat. However, the viscosity of barley extract as it moves through the digestive tract of animals or of the wort as it is filtered through the lauter tun is a major concern for both animal feeders and brewers (Newman et al., 1981; Ullrich et al., 1981; Morgan et al., 1983). High viscosity, primarily caused by β-glucans, reduces digestibility of the feed in the animal and reduces flow rates in the brewery (Kasha et al., 1993). High levels of endogenous soluble β-glucans in barley have been associated with reduced feed efficiency compared with other feed grains such as wheat or corn (*Zea mays* L.).

Barley contains at least as much TDF and SF as oat, largely because of its greater amount of β-glucan (Ranhotra et al., 1991). Cellulose and hemicellulose are present mainly in the hull, and β-glucans and arabinoxylans (pentosans) are present mainly in the endosperm. In fact, the endosperm cell walls are composed almost entirely of $(1\rightarrow3)(1\rightarrow4)$-β-glucans and pentosans (Fincher, 1975), as are cell walls of other barley grain tissues such as the aleurone (Bacic & Stone, 1981a,b). Barley is unique among cereals in that almost all of the β-glucan is found in the starchy endosperm, in contrast with 47 to 48% in wheat and oat and 70% in rye (*Secale cereale* L.) (Henry, 1987; Bhatty, 1993). The measured level of β-glucan in barley may vary considerably, depending upon the cultivar, method of determination, and growing conditions (Anderson et al., 1978; Henry, 1986; Åman & Graham, 1987; Lehtonen & Aikasalo, 1987a). Ranges of 1.9 to 10.7% have been reported by various researchers (Anderson et al., 1978; Aastrup, 1979a; Smith et al., 1980; Bamforth & Martin, 1981; Ahluwalia & Ellis, 1984; Åman et al., 1985; McCleary & Glennie-Holmes, 1985; Jørgensen & Aastrup, 1988b) as described elsewhere (Jørgensen & Aastrup, 1988a). Two-rowed barley varieties were somewhat greater in β-glucans than six-rowed (Lehtonen & Aikasalo, 1987b). 'Azhul' barley contains about 20% TDF, with about one-third (7%) being SF. This barley variety has a total β-glucan content of 11% (Bhatty, 1992b). Corresponding values for a sample of commercial oat are 9% TDF, 3% SF, and 4% β-glucan (Bhatty, 1992b). An amount of 54 to 57% TDF was measured as SF in roller-milled Azhul barley bran and flour, in contrast with 31 to 34% in oat bran and flour.

Some researchers have begun to investigate the genetics of β-glucan content in barley. Powell et al. (1985) indicated that β-glucan content was controlled by a simple additive genetic system, and observed significant negative genetic correlations between β-glucan content, thousand grain weight, and plant height in random inbred lines produced by doubled haploidy. In a later study of random inbred lines produced in the same fashion (Thomas et al., 1991) the *denso* dwarfing gene was shown to be associated with late heading, low thousand kernel weight, main-stem single plant yield, and high levels of β-glucan. In another study (Moline-Cano et al., 1989), where mutation breeding was used to improve the speed of germination of the barley variety Troubadour, a decrease in β-glucan was noted. The changes were explained by a single mutation event affecting the quantity of β-glucans present in the endosperm. Loi et al. (1988) reported locating one of the barley $(1{\rightarrow}3)(1{\rightarrow}4)$-β-glucanase genes on chromosome 1.

Current research at the University of Saskatchewan has produced agronomically acceptable barley lines containing 8 to 9% β-glucan (R. Bhatty & B. Rossnagel, 1995, personal communication). Most of these lines are *waxy* (low amylose) types, but a few are high amylose lines. When milled to produce 30% bran and 70% flour, the resulting bran has a β-glucan content of about 12 to 13% (R. Bhatty, 1995, personal communication). Other researchers have reported a link between the *waxy* starch gene in barley and high β-glucan content (Newman & Newman, 1991). In particular, *waxy* hulless barley was reported to have high TDF and SF contents and greater amounts of β-glucan than most covered barley and oat cultivars (Bhatty, 1987; Åman & Graham, 1987; Newman & Newman, 1991). Where hulled barley was reported to contain 3 to 7% β-glucan (Åman & Graham, 1987), hulless varieties had as much as 16% β-glucan (Newman et al., 1989).

It should be feasible to develop low β-glucan barley for feeding to monogastric animals, particularly broiler chicks (Bhatty, 1992b), considering the range of β-glucan contents now available in different barley types. Techniques that measure the viscosity of barley extracts as an indirect indication of soluble β-glucan content (Aastrup, 1979b) are routinely used in feed-barley breeding programs (Boros et al., 1993).

It also should be possible to develop lines high in β-glucan for use in human foods as a source of TDF and SF. Currently, little barley is used in human foods in the USA. One of the important considerations in genetically altering or breeding for barley and other crops containing high levels of SF is the effect on the physical characteristics of the grain. For example, arabinose, xylose, β-glucan, total pentosan, and total β-glucan contents were negatively correlated with malt extract (Henry, 1986), illustrating the problems of high TDF in barley intended for malting.

Recently, barley-rice (*Oryza sativa* L.) and barley-wheat blends were tested as extruded, ready-to-eat cereals (Berglund et al., 1994). Sensory panelists rated cereals extruded from four 50:50 barley-rice blends and a 65:35 barley-rice blend higher than the 100% rice cereal for crispness and color, and rated them similar to the rice cereal for flavor and overall acceptability. The TDF of the barley cereals ranged from 5.8 to 9.0%, with 2.0 to 4.7% being SF. The bulk densities of 100% barley-containing cereals were high, likely because fiber inhibits expansion during extrusion. Several problems resulted, including formation of an odd-shaped extruded cereal, reduced hydration capacities, and high moisture contents that would reduce shelf-life. Blending the barley with rice or wheat corrected these physical properties.

Pentosans made up 4.4 to 7.8% dry-weight basis (dwb) of 17 Australian varieties of barley (Henry, 1986), 6.7 to 9.8% dwb of two-rowed varieties, and 7.3 to 11% dwb of six-rowed varieties in Finland (Lehtonen & Aikasalo, 1987a). Values varied, depending upon the method of analysis, environment and, of course, variety. Interestingly, whereas six-rowed barley had a greater pentosan content, two-rowed barley had a greater β-glucan content (Lehtonen & Aikasalo, 1987a,b). But β-glucan and pentosan contents were not correlated with each other, indicating that the influence of these two fractions on grain quality may be independent. Thus, estimation of both β-glucan and pentosan levels may be useful in breeding and selecting quality barleys (Henry, 1985) for both increased and decreased TDF content. In contrast, wheat flour contained only 2.3 to 2.4% pentosan (dwb) (Amado & Neukom, 1984). Only 12 and 22%, respectively, of the pentosans of barley and oat are found in the endosperm, in contrast with 35% in wheat and 44% in rye (Henry, 1987). The effect of the pentosans on nutritional quality of barley is not known, but they may contribute to acid-extract viscosity (Bhatty et al., 1991). A discussion of pentosans in baked products is covered in the section on rye.

C. Rye

As with the other grains mentioned, high SF has been a major impediment to a broader acceptance of rye as an animal feed (Annison & Choct, 1991; Bedford & Classen, 1992). The main SF in rye is pentosans, which make up a major portion of the endosperm cell wall (Saini & Henry, 1989). Rye also contains some β-glucan, with 71% of this fiber present in the endosperm (Henry, 1987). Whole-grain rye was reported to contain about 8 to 10% pentosan (dwb) and about 2.5% β-glucan (Henry, 1987). Other work reported ranges in rye of 6.5 to 11.5% pentosans and 1.0 to 2.0% β-glucans (Saastamoinen et al., 1989).

Pentosans can be a desirable part of rye, barley, and other grains because they have a high water-absorbing capacity that can be important in baking (Lehtonen & Aikasalo, 1987a). Bushuk (1966) demonstrated that 23% of the water in a wheat dough is bound to pentosans, thus contributing to loaf volume, cell uniformity, crumb characteristics, and elasticity. Pentosans were found to retain gas in a wheat dough (Hoseney, 1984) and to retard the rate of staling (Kim & D'Appolonia, 1977; Jankiewicz & Michniewicz, 1987). Delcour et al. (1991) demonstrated that the optimum concentration of a rye pentosan-protein fraction was 2 to 3% for improving the volume of gluten-starch loaves. They also explained that the improvement was related to an increase in dough viscosity. The physical properties of rye dough depend mainly upon water-binding substances (Pomeranz, 1985; Holas et al., 1992), because unlike wheat, rye lacks enough of the gluten protein needed for a cohesive dough structure. Also, the pentosans interfere with the film-making ability of the proteins present in rye. The viscous properties of rye doughs contributed by the pentosans help form a sticky dough needed for baking quality (Drews & Seibel, 1976). The pentosans may be either water soluble or water insoluble, with the former fraction having a beneficial impact on the color, taste, and texture of rye baked goods and the latter fraction reducing baked product quality (Meuser & Suckow, 1986).

With the recognition that high SF in rye interferes with animal feed efficiency but may contribute to improved baking quality in breads, scientists have developed techniques to estimate the amount of SF in rye grain for the purpose of selecting lines that are high or low in the fiber component (Scoles et al., 1993). Grain extract viscosity measured with a Brookfield viscometer was used to estimate the pentosan level. Viscosity was highly correlated with the amount of high-molecular-weight soluble pentosans, rather than with the total soluble pentosans. Also, Boros et al. (1993) demonstrated that nearly all of the viscosity in rye extracts was attributable to the pentosans, with starch, protein, and β-glucan having no contribution. Because viscosity of feed also has been correlated with the viscosity of chick intestinal contents (Bedford et al., 1991) and thus feed efficiency, viscosity measurement is likely to be useful for rapidly estimating feed efficiency in rye. Similarly, barley exhibiting low extract viscosity (caused by low levels of β-glucan) has superior feeding value to high viscosity types (Campbell et al., 1989). Scoles et al. (1993) found variability in grain viscosity for 28 inbred lines of spring rye and in an F_2 population derived by crossing lines with high and low viscosity. The SF content was not related to seed characteristics such as color, plumpness, or starch content.

D. Yellow Mustard

The issues surrounding the interest in SF from yellow mustard (*Brassica hirta* Moench, also known as *Sinapis alba* L.) are quite different from those mentioned for cereals. White or yellow mustard seeds are grown primarily for their use as condiments. The SF components contribute to the consistency of prepared mustard products, such as salad dressings and food pastes (Weber et al., 1974), and although the cost might be prohibitive, some researchers have suggested that the SF from yellow mustard might be extracted and used to replace gums imported into Canada, for instance, such as guar (*Cyamopsis tetragonoloba* L. Taubert) and acacia (Woods & Downey, 1980). Recent work suggested that the SF from mustard mimicked the interfacial activity and shear thinning behavior of another widely used product, xanthan gum (Cui et al., 1993a).

The SF from yellow mustard is known as mucilage, generally defined as polysaccharides containing galactose, galacturonic acid residues, and often xylose and arabinose (Southgate et al., 1993). Structurally, mucilage resembles hemicelluloses and is water soluble, obtained as slimy, colloidal solutions. Mucilages are found in the endosperm or in special cells in the seed coat. They retain water and so protect the seed against desiccation. Several researchers have studied the chemical composition of mucilage from yellow mustard. Siddiqui et al. (1986) found galactose, glucose, and galacturonic acid to be the major sugar components and mannose, arabinose, xylose, and rhamnose to be minor components. Other researchers reported similar findings (Vose, 1974; Theander et al., 1977). By using a modified extraction procedure, Cui et al. (1993b) recovered an increased yield of mustard mucilage, containing glucose (22–35%), galactose (11–15%), mannose (6.0–6.4%), rhamnose (1.6–4.0%), arabinose (2.8–3.2%), and xylose (1.8–2.0%), which they then separated into water-soluble and -insoluble fractions. Both mucilage fractions had similar functional behavior. Additional characterization revealed that the major part of the water-soluble fraction was a mixture of a pectic material consisting primarily of galacturonic acid, galactose, and rhamnose and a 1,4-linked β-D-glucan (Cui et al., 1993a).

Although the chemical characterization of mustard mucilage has occurred more recently, over 60 yr ago the outer seed coat of mustard seed was first reported to be rich in mucilagenous material (Bailey & Norris, 1932). Breeding efforts to alter the amount of mucilage in mustard have been ongoing for 40 to 45 yr (C. Lynch, 1995, personal communication). The actual percentage of mucilage present in mustard has been debated recently. Earlier research reported that yellow mustard contained about 2% mucilage, with the level subject to environmental influence (Weber et al., 1974; Woods & Downey, 1980; Anguilar & Ziegler, 1990). But these studies all used the Weber method (Weber et al., 1974) for extraction of the mucilage, now thought to be incomplete. A more recent study reported a yield of about 5% mucilage from yellow mustard seeds of the Tilney variety (Cui et al., 1993a) by using a modified extraction procedure (Sharafabadi, 1987). Approximately 90 to 95% of the yellow mustard grown in Canada is of the Tilney variety (Chris Lynch, personal communication). Efforts to alter the mucilage in mustard may change the 5% figure. A small quantity of the registered variety, Viscount, so named because of its high viscosity, was just released in Canada by Colman's Mustard of Norwich, England. The 1994 crop has been milled and is currently being evaluated for mucilage content.

E. Flax or Linseed

Flax (*Linum usitatissimum* L.), one of the oldest crops known to humans, traditionally was cultivated for its fiber (linen) and oil (linseed oil, used in paints and as a raw material in other industrial products such as linoleum) (see Kinney & Hitz, 2000, Chapter 7B; Wannerberger et al., 1991). Recent interest in developing agricultural-based industrial products has renewed interest in flax. The oil has been altered through plant breeding efforts to make it more suitable for human consumption (low linolenic acid), and the fiber has been evaluated for its mucilage content and quality. Consumption of the mucilage as TDF is of nutritional interest for reasons discussed earlier in this chapter, and the thickening and emulsifying properties of the mucilage may provide useful functions in foods and other products. Two patents were issued for the use of hydrocolloids (mucilage) from flax for cosmetic, therapeutic, and medicinal purposes (Attström et al., 1993; O'Mullane & Hayter, 1993).

Flaxseed mucilage ranges from 3.5 to 9.4% by weight, depending upon the extraction method used (Cui et al., 1994; Fedeniuk & Biliaderis, 1994; Mazza & Biliaderis, 1989). Also, the polysaccharide content and composition are affected by genotype and environment (Schuster et al., 1978). Wannerberger et al. (1991) measured great differences, by using a Bohlin VOR rheometer (Bohlin Reologi AB, Lund, Sweden), in the viscosities of 1% solutions of mucilage from 23 linseed varieties, with values ranging from 0.02 to 0.28 Pas at a shear rate of 10 s^{-1}. From these experiments, they selected five varieties having quite different viscosities for further chemical study. The mucilage from the different varieties contained uronic acids (21–36%), xylose (19–38%), rhamnose plus fucose (11–16%), galactose (12–16%), arabinose (8–13%), glucose (4–6%) and mannose (1–2%). Oomah et al. (1995) evaluated 109 accessions of flax from 12 geographical regions for concentration and composition of carbohydrates, finding a range in water-soluble carbohydrates of 3.6 to 8.0%. The mucilages contained glucose (21–40%), xylose (11–32%), galactose (13–27%), rhamnose (8–24%), arabinose (3–15%), and fucose (2–9%). The high concentration of glucose in the latter study was attributed to the use of a phosphate buffer for extraction of the carbohydrates. Extraction with water resulted in glucose levels comparable to those previously reported (Wannerberger et al., 1991). Oomah et al. (1995) reported great differences in the monosaccharide percentages and carbohydrate yields among accessions from various parts of the world, suggesting that parental types could be selected to breed flax cultivars with desired levels and composition of mucilage.

II. FUTURE OF ALTERING FIBER IN CROPS

Only a few crops have been considered for widespread changes in TDF content; however, many more crops have been studied for the content, function, and nutritional effects of their fiber fractions. In particular, corn (*Zea mays* L.) (Ohata et al., 1994), rice (*Oryza sativa* L.) (Aoe et al., 1993), wheat (*Triticum aestivum* L.) (Ranhotra et al., 1992), and their bran portions have been thoroughly analyzed for these properties in recent years, and some publications have evaluated such crops as amaranth (Danz & Lupton, 1992), sorghum (*Sorghum bicolor* L. Moench.) (Woolard et al., 1976; Klopfenstein and Hoseney, 1987), and soybean (*Glycine max* L. Merr.) (Sasaki et al., 1985; Schweizer et al., 1983).

REFERENCES

Aastrup, S. 1979a. The effect of rain on β-glucan content in barley grains. Carlsberg Res. Commun. 44:381–393.

Aastrup, S. 1979b. The relationship between the viscosity of an acid flour of barley and its β-glucan content. Carlsberg Res. Commun. 44:289–304.

Ahluwalia, B., and E.E. Ellis. 1984. A rapid and simple method for the determination of starch and β-glucan in barley and malt. J. Inst. Brew. 90:254–259.

Amado, R., and H. Neukom. 1984. Minor constituents of wheat flour: The pentosans. p. 241–251. *In* R.D. Hill and L. Munck (ed.) New approaches to research on cereal carbohydrates. Elsevier Sci. Publ. B.V., Amsterdam.

Åman, P., and H. Graham. 1987. Analysis of total and insoluble mixed-linked (1-3,(1-4)-β-d-glucans in barley and oats. J. Agric. Food Chem. 35:704–709.

Åman, P., K. Hesselman, and A. Tilly. 1985. The variation in chemical composition of Swedish barleys. J. Cer. Sci. 3:73–77.

Anderson, M.A., J.A. Cook, and B.A. Stone. 1978. Enzymatic determination of 1,3:1,4-β-glucans in barley grain and other cereals. J. Inst. Brew. 84:233–239.

Anderson, J.W., N.H. Gilinsky, D.A. Deakins, S.F. Smith, D.S. O'Neal, D.W. Dillon, and P.R. Oeltgen. 1991. Lipid responses to oat bran versus wheat bran intake in hypercholesterolemic men. Am. J. Clin. Nutr. 54:678–683.

Anderson, J.W., D.B. Spencer, C.C. Hamilton, S.F. Smith, J. Tietyen, C.A. Bryant, and P. Oeltgen. 1990. Oat bran cereal lowers serum total and LDL cholesterol in hypercholesterolemic men. Am. J. Clin. Nutr. 54:678–683.

Anderson, J.W., L. Story, B. Sieling, W.-J.L. Chen, M.S. Petro, and J. Story. 1984. Hypocholesterolemic effects of oat-bran or bean intake for hypercholesterolemic men. Am. J. Clin. Nutr. 40:1146–1155.

Anguilar, C.A., and G.R. Ziegler. 1990. Effect of temperature and electrolytes on the viscosity of aqueous dispersions of mustard seed (*Sinapsis alba*) mucilage. Food Hydrocol. 4:161–166.

Annison, G., and M. Choct. 1991. Anti-nutritive activities of cereal non-starch polysaccharides in broiler diets and strategies minimizing their effects. World's Poult. Sci. J. 47:232–242.

Aspinall, G.O., and R.C. Carpenter. 1984. Structural investigations on the non-starchy polysaccharides of oat bran. Carbohydr. Polym. 4:271–282.

Attström, R., P.O. Glantz, H. Hakansson, and K. Larsson. 1993. Saliva substitute. U.S. Patent 5 260 282. Date issued: 9 November.

Autio, K., O. Myllymäki, and Y. Mälkki. 1987. Flow properties of solutions of oat β-glucans. J. Food Sci. 52:1364–1366.

Bacic, A., and B.A. Stone. 1981a. Isolation and ultrastructure of aleurone cell walls from wheat and barley. Aust. J. Plant Physiol. 8:453–474.

Bacic, A., and B.A. Stone. 1981b. Chemistry and organization of aleurone cell wall components from wheat and barley. Aust. J. Plant Physiol. 8:475–495.

Bailey, K., and F.W. Norris. 1932. The nature and composition of the seed of white mustard (*Brassica alba*). Biochem. J. 26:1609–1623.

Bamforth, C.W., and H.L. Martin. 1981. β-glucan and β-glucan solubilase in malting and mashing. J. Inst. Brew. 87:365–371.

Bedford, M.R., and H.L. Classen. 1992. Reduction of intestinal viscosity through manipulation of dietary rye and pentosanase concentration is effected through changes in the carbohydrate composition of the intestinal aqueous phase and results in improved growth rate and food conversion efficiency of broiler chicks. J. Nutr. 122:560–569.

Bedford, M.R., H.L. Classen, and G.L. Campbell. 1991. The effect of pelleting, salt and pentosanase on the viscosity of intestinal contents and the performance of broilers fed rye. Poult. Sci. 70:1571–1577.

Berglund, P.T., C.E. Fastnaught, and E.T. Holm. 1994. Physicochemical and sensory evaluation of extruded high-fiber barley cereals. Cer. Chem. 71:91–95.

Best, D. 1987. Building fiber into foods. Prep. Foods 156:112–115.

Bhatty, R.S. 1987. Relationship between acid extract viscosity and total soluble and insoluble β-glucan contents of hulled and hulless barley. Can. J. Plant Sci. 67:997–1008.

Bhatty, R.S. 1992. β-Glucan content and viscosities of barleys and their roller-milled flour and bran products. Cer. Chem. 69:469–471.

Bhatty, R.S. 1993. Nonmalting uses of barley. p. 355–417. *In* A.W. MacGregor and R.S. Bhatty (ed.) Barley: Chemistry and technology. Am. Assoc. Cer. Chem., Inc., St. Paul, MN.

Bhatty, R.S., A.W. MacGregor, and B.G. Rossnagel. 1991. Total and acid-soluble β-glucan content of hulless barley and its relationship to acid-extract viscosity. Cer. Chem. 68:221–227.

Boros, D., R.R. Marquardt, B.A. Slominski, and W. Guenter. 1993. Extract viscosity as an indirect assay for water-soluble pentosan content in rye. Cer. Chem. 70:575–580.

Burkitt, D.P. 1977. Food fiber, benefits from a surgeon's perspective. Cer. Foods World 22:6–9.

Burkitt, D.P., A.R.P. Walker, and N.S. Painter. 1972. Effect of dietary fibre on stools and transit-times and its role in the causation of disease. Lancet 2:1408–1411.

Burkitt, D.P., A.R.P. Walker, and N.S. Painter. 1974. Dietary fiber and disease. J. Am. Med. Assoc. 229:1068–1074.

Bushuk, W. 1966. Distribution of water in dough and bread. Bakers Digest. October Vol. 40. p. 38–40.

Campbell, G.L., B.G. Rossnagel, H.L. Classen, and P.A. Thacker. 1989. Genotypic and environmental differences in extract viscosity of barley and their relationship to its nutritive value for broiler chickens. Anim. Feed Sci. Technol. 26:221–230.

Chen, W.L., and J.W. Anderson. 1979. Effects of plant fiber in decreasing plasma total cholesterol and increasing high-density lipoprotein cholesterol. Proc. Soc. Exp. Biol. Med. 162:310–313.

Chen, W.L., J.W. Anderson, and M.R. Gould. 1981. Effects of oat bran oat gum and pectin on lipid metabolism of cholesterol-fed rats. Nutr. Rep. Int. 24:1093–1098.

Cho, K.C., and P.J. White. 1993. Enzymatic analysis of β-glucan content in different oat genotypes. Cer. Chem. 70:539–542.

Clarke, A.E., and B.A. Stone. 1966. Enzymic hydrolysis of barley and other β-glucans by a β-(1→4)-glucan hydrolase. Biochem. J. 99:582–588.

Cui, W., N.A.M. Eskin, and C.G. Biliaderis. 1993a. Chemical and physical properties of yellow mustard (*Sinapis alba* L.) mucilage. Food Chem. 46:169–176.

Cui, W., N.A.M. Eskin, and C.G. Biliaderis. 1993b. Water-soluble yellow mustard (*Sinapis alba* L.) polysaccharides: Partial characterization, molecular size distribution and rheological properties. Carbohydr. Polym. 20:215–225.

Cui, W., G. Mazza, and C.G. Biliaderis. 1994. Chemical structure, molecular size distributions, and rheological properties of flaxseed gum. J. Agric. Food Chem. 42:1891–1895.

Davidson, M.H., L.D. Dugan, J.H. Burns, J. Bova, K. Story, and K.B. Drennan. 1991. The hypocholesterolemic effects of beta-glucan in oatmeal and oat bran. A dose-controlled study. J. Am. Med. Assoc. 265:1833–1839.

D'Appolonia, B.L., and V.L. Youngs. 1978. Effects of bran high protein concentrate from oat on dough properties and bread quality. Cer. Chem. 55:736–743.

Dawkins, N.L., and I.A. Nnanna. 1993. Oat gum and β-glucan extraction from oat bran and rolled oats: Temperature and pH effects. J. Food Sci. 58:562–566.

Delcour, J.A., S. Vanhamel, and R.C. Hoseney. 1991. Physicochemical and functional properties of rye nonstarch polysaccharides. II. Impact of a fraction containing water-soluble pentosans and proteins on gluten-starch loaf volumes. Cer. Chem. 68:72–76.

Drews, E., and W. Seibel. 1976. Bread-baking and other uses around the world. p. 127–174. *In* W. Bushuk (ed.) Rye: Production, chemistry and technology. Am. Assoc. Cer. Chem., St. Paul, MN.

Fadel, J.G., R.K. Newman, C.W. Newman, and A.E. Barnes. 1987. Hypocholesterolemic effects of beta-glucans in different barley diets fed to broiler chicks. Nutr. Rep. Int. 35:1049–1057.

Fedeniuk, R.W., and C.G. Biliaderis. 1994. Composition and physicochemical properties of linseed (*Linum usitatissimum* L.) mucilage. J. Agric. Food Chem. 42:240–247.

Fincher, G.B. 1975. Morphology and chemical composition of barley endosperm cell walls. J. Inst. Brew. 81:116–122.

Fulcher, R.G., and S.S. Miller. 1993. Structure of oat bran and distribution of dietary fiber components. p.1–24. *In* P.J. Wood (ed.). Oat bran. Am. Assoc. Cer. Chem., Inc., St. Paul, MN.

Gould, M.R., J.W. Anderson, and S. O'Mahony. 1980. Biofunctional properties of oats. p. 447–460. *In* G.E. Inglett and L. Munck (ed.) Cereals for food and beverages. Acad. Press, New York.

Hall, J.M. 1989. A review of total dietary fiber methodology. Cer. Foods World 34:526–528.

Henry, R.J. 1985. Evaluation of methods for the assessment of malting quality in barley breeding. Euphytica 34:135–145.

Henry, R.J. 1986. Genetic and environmental variation in the pentosan and β-glucan contents of barley, and their relation to malting quality. J. Cer. Sci. 4:269:277.

Henry, R.J. 1987. Pentosan and (1→3)(1→4)-β-glucan concentrations in endosperm and whole grain of wheat, barley, oats and rye. J. Cer. Sci. 6:253–258.

Holas, J., J. Prihoda, and J. Kratochvil. 1992. Rye flour and bread. p. 121–127. *In* B. Feillet (ed.) Institut de Recherches Technologiques Agroalimantaires des Céréales, Paris.

Hoseney, R.C. 1984. Functional properties of pentosans in baked foods. Food Technol. 38:114–117.

Jankiewicz, M., and J. Michniewicz. 1987. The effect of soluble pentosans isolated from rye grain on staling of rye bread. Food Chem. 25:241–248.

Jannson, K.W., and L. Lindahl. 1991. Rheological changes in oatmeal suspensions during heat treatment. J. Food Sci. 56:1685–1689.

Jennings, C.D., K. Boleyn, S.R. Bridges, P.J. Wood, and J.W. Anderson. 1988. A comparison of lipid-lowering and intestinal morphological effects of cholestyramine, chitosan, and oat gum in rats. Proc. Soc. Exp. Biol. Med 189:13–20.

Jorgensen, K.G., and S. Aastrup. 1988a. Determination of β-glucan in barley, malt, wort and beer. p. 88–108. *In* H.F. Linskens and J. Jackson (ed.) Modern methods of plant analysis. Vol. 7. Springer-Verlag, New York.

Jorgensen, K.G., and S. Aastrup. 1988b. Quantification of (1→3)(1→4)-β-glucan using Calcofluor complex formation and flow injection analysis. II. Determination of total β-glucan content of barley and malt. Carlsberg Res. Commun. 53:287–296.

Kahlon, T.S., F.I. Chow, B.E. Knuckles, and M.M. Chiu. 1993. Cholesterol-lowering effects in hamsters of beta-glucan-enriched barley fraction, dehulled whole barley, rice bran, and oat bran and their combinations. Cer. Chem. 70:435–440.

Kahlon, T.S., F.I. Chow, J.L. Hoefer, and A.A. Betschart. 1986. Bioavailability of vitamins A and E as influenced by wheat bran and bran particle size. Cer. Chem. 63:490–493.

Kahlon, T.S., R.M. Saunders, F.I. Chow, M.M. Chiu, and A.A. Betschart. 1990. Influence of rice bran, oat bran, and wheat bran on cholesterol and triglycerides in hamsters. Cer. Chem. 67:439–443.

Kasha, K.J., D.E. Falk, and A. Ziauddin. 1993. Potential improvement of barley quality through genetic engineering. p. 419–435. *In* A.W. MacGregor and R.S. Bhatty (ed.) Barley: Chemistry and technology. Am. Assoc. Cer. Chem., Inc., St. Paul, MN.

Kim, S.K., and B.L. D'Appolonia. 1977. Bread staling studies. III. Effect of pentosans on dough, bread, and bread staling rate. Cer. Chem. 54:225–229.

Kinney, A.J., and W.D. Hitz. 2000. Oils. p. 167–180. *In* C.F. Murphy and D.M. Peterson (ed.) Designing crops for added value. Agron. Monogr. 40. ASA, CSSA, and SSSA, Madison, WI.

Kirby, R.W., J.W. Anderson, B. Sieling, E.D. Rees, W.-J.L. Chen, R.E. Miller, and R.M. Kay. 1981. Oat bran intake selectively lowers serum low density lipoprotein cholesterol concentrations of hypercholesterolemic men. Am. J. Clin. Nutr. 34:824–829.

Klopfenstein, C.F., and R.C. Hoseney. 1987. Cholestrol-lowering effect of beta-glucan-enriched bread. Nutr. Rep. Int. 36:1091–1098.

Krishnan, P.G., K.C. Chang, and G. Brown. 1987. Effect of commercial oat bran on the characteristics and composition of bread. Cer. Chem. 64:55–58.

Lehtonen, M., and R. Aikasalo. 1987a. Pentosans in barley varieties. Cer. Chem. 64:133–134.

Lehtonen, M., and R. Aikasalo. 1987b. β-glucan in two- and six-rowed barley. Cer. Chem. 64:191–193.

Lim, H.S., P.J. White, and K.J. Frey. 1992. Genotypic effects on β-glucan content of oat lines grown in two consecutive years. Cer. Chem. 69:262–265.

Mazza, G., and C.G. Biliaderis. 1989. Functional properties of flax seed mucilage. J. Food Sci. 54:1302–1305.

McCleary, B.V., and M. Glennie-Holmes. 1985. Enzymatic quantification of (1→3)(1→4)-β-glucan in barley and malt. J. Inst. Brew. 91:285–295.

Meuser, F., and P. Suckow. 1986. Non-starch polysaccharides. p. 42–61. *In* J.M.V. Blanshard et al. (ed.) Chemistry and physics of baking. R. Soc. Chem., London.

Miller, S.S., and R.G. Fulcher. 1994. Distribution of (1→3),(1→4)-β-glucan in kernels of oats and barley using microspectrofluorometry. Cer. Chem. 71:64–68.

Morgan, A.G., A.A. Gill, and D.B. Smith. 1983. Some barley grain and green malt properties and their influence on malt hot water extract. I. β-glucan, β-glucan solubilase and endo-β-glucanase. J. Inst. Brew. 89:283–291.

Newman, C.W., A.M. El-Negoumy, R.F. Eslick. 1981. Genetic factors affecting the feed quality of barley. p. 299–304. *In* R.N.H. Whitehouse (ed.). Barley genetics. Vol. V. Proc. 4th Int. Barley Genet. Symp. Edinburgh Univ. Press, Edinburgh, UK.

Newman, R.K., A.A. Betschart, C.W. Newman, and P.J. Hofer. 1992. Effect of full-fat or defatted rice bran on serum cholesterol. Plant Foods Hum. Nutr. 42:37–43.

Newman, R.K., S.E. Lewis, C.W. Newman, R.J. Boik, and R.T. Ramage. 1989. Hypocholesterolemic effect of barley foods on healthy men. Nutr. Rep. Int. 39:749–760.

Newman, R.K., and C.W. Newman. 1991. Barley as a food grain. Cer. Foods World 36:800–805.

O'Mullane, J.E., and I.P. Hayter. 1993. Linseed mucilage. Int. Pat. PCT/GB93/00343. Date issued: 3 September.

Oomah, B.D., E.O. Kenaschuk, W. Cui, and G. Mazza. 1995. Variation in the composition of water-soluble polysaccharides in flaxseed. J. Agric. Food Chem. 43:1484–1488.

Parrish, F.W., A.S. Perlin, and E.T. Reese. 1960. Selective enzymolysis of poly-β-D-glucans, and the structure of the polymers. Can. J. Chem. 38:2094–2104.

Peat, S., W.J. Whelan, and J.G. Roberts. 1957. The structure of lichenin. J. Chem. Soc. 3916–3924.

Peterson, D.M. 1991. Genotype and environment effects on oat beta-glucan concentration. Crop Sci. 31:1517–1520.

Pomeranz, Y. 1985. Carbohydrates: Structural polysaccharides, pectins, and gums. p. 91–119. *In* Functional properties of food components. Acad. Press, New York.

Qureshi, A.A., T.D. Crenshaw, N. Abuirmeileh, D.M. Peterson, and C.E. Elson. 1987. Influence of minor plant constituents on porcine hepatic lipid metabolism. Impact on serum lipids. Atherosclerosis 64:109–114.

Ranhotra, G.S., J. Gelroth, and P. Bright. 1987. Effect of the source of fiber in bread-based diets on blood and liver lipids in rats. J. Food Sci. 52:1420–1422.

Ranhotra, G., J.A. Gelroth, K. Astroth, and R.S. Bhatty. 1991. Relative lipidemic responses in rats fed barley and oat meals and their fractions. Cer. Chem. 68:548–551.

Saini, H.S., and R.J. Henry. 1989. Fractionation and evaluation of triticale pentosans: comparison with wheat and rye. Cer. Chem. 66:11–14.

Saastamoinen, M., S. Plaammi, and J. Kumpulainen. 1989. Pentosan and β-glucan content of Finnish winter rye varieties as compared with rye of six other countries. J. Cer. Sci. 10:199–207.

Schuster, V.W., H. Iran-Nejad, and R. Marquard. 1978. Yield performance and some quality characteristics of different linseed varieties (*Linum usitatissimum* L.) in areas with greatly varying environment. Fette Seifen Anstrichmittel 80:133–143.

Scoles, G.J., G.L. Campbell, and J. G. McLeod. 1993. Variability for grain extract viscosity in inbred lines and an F_2 population of rye (*Secale cereale* L.). Can. J. Plant Sci. 73:1–6.

Sharafabadi, S.K. 1987. Chemical and Physical Properties of Mucilage from Canola (*Brassica Campestris* cv. Candle). M.Sc. thesis. Univ. Manitoba, Canada.

Shinnick, F.L., M.J. Longacre, S.L. Ink, and J.A. Marlett. 1988. Oat fiber: Composition versus physiological function in rats. J. Nutr. 118:144–151.

Siddiqui, I.R., S.H. Yiu, J.D. Jones, and M. Kaláb. 1986. Mucilage in yellow mustard. Food Microstruct. 5:157–162.

Smith, D.B, A.G. Morgan, and S. Aastrup. 1980. Variation in the biochemical composition of acid extracts from barley of contrasting malting quality. J. Inst. Brew. 86:277–283.

Southgate, D.A.T., G.A. Spiller, M. White, and R. McPherson. 1993. Glossary of dietary fiber components. p. 29–33. *In* G.A. Spiller (ed.) CRC handbook of dietary fiber in human nutrition. CRC Press, Inc., Boca Raton, FL.

Theander, O., P. Åman, G.E. Miksche, and S. Yasuda. 1977. Carbohydrates, polyphenols and lignin in seed hulls of different colors from turnip rapeseed. J. Agric. Food Chem. 25:270–273.

Theander, O., E. Westerland, and P. Åman. 1993. Structure and components of dietary fiber. Cer. Foods World 38:135–141.

Ullrich, S.E., C.N. Coon and J.M. Sever. 1981. Relationships of nutritional and malting quality traits of barley. p. 225–233. *In* R.N.H. Whitehouse (ed.) Barley genetics. Vol. 4. Proc. 4th Int. Barley Genet. Symp. Edinburgh Univ. Press, Edinburgh, UK.

Vose, J.R. 1974. Chemical and physical studies of mustard and rapeseed coats. Cer. Chem. 51:658–665.

Walker, A.R.P. 1993. Does the dietary fiber hypothesis really "work"? Cer. Foods World 38:128–134.

Wannerberger, K., T. Nylander, and M. Nyman. 1991. Rheological and chemical properties of mucilage in different varieties from linseed (*Linum usitatissimum*). Acta Agric. Scand. 41:311–319.

Weber, F.E., S.A. Taillie, and K.R. Stauffer. 1974. Functional characteristics of mustard mucilage. J. Food Sci. 60:461–466.

Wood, P.J. 1986. Oat β-glucan: structure, location, and properties. p. 121–152 *In* F.H. Webster (ed.) Oats: Chemistry and technology. Am. Assoc. Cer. Chem., Minneapolis, MN.

Wood, P.J. 1993. Physicochemical characteristics and physiological properties of oat (1→3), (1→4)-β-D-glucan. p. 83–112. *In* P.J. Wood (ed.). Oat bran. Am. Assoc. Cer. Chem., St. Paul, MN.

Wood, P.J., J. Weisz, and P. Fedec. 1991. Potential for β-glucan enrichment in brans derived from oat (*Avena sativa* L.) cultivars of different (1→3),(1→4)-β-glucan concentrations. Cer. Chem. 68:48–51.

Wood, P.J., J. Weisz, P. Fedec, and V.D. Burrows. 1989. Large scale preparation and properties of oat fractions enriched in (1→3)(1→4)-β-D-glucan. Cer. Chem. 66:97–103.

Woods, D.L., and R.K Downey. 1980. Mucilage from yellow mustard. Can. J. Plant Sci. 60:1031–1033.

7E Designing Cotton for Added Value

WILLIAM R. MEREDITH, JR.

USDA-ARS
Stoneville, Mississippi

The world textile market is a competitive industry. Cotton from the USA (*Gossypium* spp.) competes with manmade fibers and with cottons produced in other countries, and also with cottons produced in different regions of the same country. To be competitive, much research has been directed toward increasing fiber yield. However, for industrialized nations, such as the USA, cotton must compete through the more efficient use of modern technology that produces not only a greater quantity, but also a better quality product while minimizing damage to the environment. Much effort has been directed to improving cotton fiber through breeding, ginning, quality control, accurate characterization of fiber properties, modernization of the textile industry, and especially the interactions of breeding with other disciplines. Currently, world production is about 19×10^6 t (87×10^6 bales; bale = 218 kg) and is expected to increase to about 22×10^6 t by the year 2000 (Anon., 1995). World production is growing despite no increase in the area planted, and cotton's use is growing faster than population growth. The major cotton-producing countries are China, USA, Pakistan, India, and Uzbeckistan (USDA, 1994).

The USA leads the world in the combined use of improved technology that covers every phase of the cotton industry from production and breeding to textile manufacturing. Research and development efforts are being conducted by both private and public agencies. A review of the changes in technology in the last several decades is followed by a brief review of how cotton fibers develop. The current and future changes taking place with the cotton industry and their effects on the need to develop higher valued cotton cultivars is discussed.

I. INCREASED USE OF COTTON

Through the use of improved technology, cotton's share of the U.S. fiber market has increased from 31 to 40% during the past two decades, and from 1982 to 1994 per capita consumption of cotton has increased from 6.1 to 13.3 kg (Lange, 1995, personal communication). Lange attributed the increase to better quality fiber, accurate fiber characterization, and textile industry modernization. All these factors contribute directly to greater textile mill efficiency and the production of a better quality product.

Other factors that contribute indirectly to the greater demand for cotton are an abundant supply of many fiber-property types, improved marketing procedures, and a rising standard of living.

There are many textile products and textile engineering methods to produce textile products. An abundant supply of fiber-property types gives the textile industry the flexibility to select a fiber type that optimizes their engineering methods to produce the best quality product. Good marketing and promotion of cotton as a natural fiber and it's comfort value have increased the demand for cotton. A rising standard of living also results in greater per capita demands for cotton.

A. Better Quality Fiber

Sasser (1995) summarized fiber property trends of U.S. cottons as reported by the USDA, Agricultural Marketing Service (AMS). Micronaire is a measure of two components, fiber maturity and fineness (diameter) of the fiber. From 1975 to 1995, the yearly average micronaire has been within the desired range of 4.0 to 4.2. Nationally, micronaire has been increasing at a rate of 0.02 units yr^{-1}, probably due to increased maturity of the fiber. From 1975 through 1995, average staple length increased an average of about 1.8 mm yr^{-1}. Fiber uniformity (less short fibers) has been increasing since the first reported averages were given in 1987. The U.S. cotton grades, which are a measure of trash content and color, also have been improving, even though less cleaning operations are being used in ginning the crop. Ginning is the operation that separates the lint fibers from the seed and trash. Less trash results in less textile mill waste and a better product. Improved color aids in the dying efficiency. The greatest improvement in fiber quality has been in the improvement of fiber bunde strength. The U.S. bundle strength has increased at a rate of about 0.34 g tex^{-1} yr^{-1} (where tex = the linear density of fibers, filaments, and yarns expressed as the mass, in milligrams, of a meter of fiber, filaments, or yarn) since 1985. Sasser (1995) attributed these improvements in fiber quality to: (i) improved fiber quality cultivars; (ii) better production practices, primarily in harvesting; and (iii) careful attention to ginning.

B. Accurate Characterization of Fiber

The major change in quality control and marketing has been the development of technology of characterizing the fiber properties of every bale of U.S. cotton that is classed by the USDA, AMS (Chewing, 1992). High Volume Instrument (HVI) testing was first introduced in the 1960s, and after improvements through research, was first used commercially in the late 1970s. Chewing (1992) indicated HVI offered two major benefits: (i) important fiber properties could be determined with accuracy and precision, and (ii) HVI's speed, reliability, and low cost enabled mills to test each bale. It's capacity to accurately test over 131 t (600 bales) in 8 h allowed mills to incorporate quality control procedures to select the bales with desired fiber properties. This resulted in mills operating more efficiently and producing better quality textiles. In 1991, the USDA, AMS began HVI classing of bales. In 1992, it was estimated (Chewing, 1992) that HVI was a factor in selecting 87% of 9.2×10^6 bales consumed by U.S. mills. The HVI system currently measures fiber length, length uniformity, bundle strength, micronaire, color and trash.

C. Modernization of the Textile Industry

A major transition is evolving in the spinning methods used by the U.S. textile industry from 100% ring spinning to rotor (open-end) and air jet spinning (Deussen 1992). Rotor spinning positions accounted for only 1% of the total spinning in 1980. By 1993, they comprised 11% of total spinning (Lange, personal communication). While the total number of spinning positions has declined in the USA in the last decade, the greatly increased efficiency of processing speeds of rotor spinning has resulted in about a 169% increase in the amount of fiber consumed per position. Rotor spinning requires lower labor related costs and is dependent on high-speed and more automated equipment. Cotton's share of all textiles spun by rotor spinning has grown from 58% in 1982 to 76% in 1994 (Lange, personal communication). This evolution of spinning methods will result in a change in fiber property breeding priorities because rotor spinning requires stronger and finer fiber than that required for ring spinning (Deussen, 1992).

An additional factor needed for continued growth in cotton use is to have a large supply of fibers that vary in fiber properties. Different types of textiles require different, optimum combinations of fiber properties. The brief review presented on the increased value of cotton fiber opens the question as to what changes should be considered for future research.

II. NEW INSTRUMENTS TO CHARACTERIZE FIBER PROPERTIES

Each new advance in technology opens new doors for further technological advances. Such is the case with HVI's impact on the cotton industry. Future instruments will require the same fast processing time, lower cost, and acceptable precision and repeatability. Also, HVI furnishes a model to which other useful traits can be added. While the introduction of new instruments in general has as their objective to be added to the HVI system, they also may be used for understanding the basic physiology of fiber development, perfecting better textile processing, and as a tool to be used by breeders. The new instruments are of two types: (i) those that better characterize fiber properties that are already commonly measured, such as length and strength; and (ii) those that measure traits not currently measured, such as neps and sugar content.

One instrument-module combination is the Advanced Fiber Information System (AFIS). Various AFIS modules can measure both the mean and distribution of fiber length, maturity, fineness, trash, dust, and neps. The AFIS measures individual fibers or trash and dust particles and gives their average values and their distributions. The instrument uses an electro-optical system and software models to estimate the various fiber properties.

The AFIS module for length gives the length distribution of individual fibers, which can be compared with the results of the time-consuming Suter Webb array method. Bragg and Shofner (1993) compared the distribution values from AFIS and the Suter Webb analyses of upland cottons and found a linear correlation as $r^2 =$ 0.98, 0.99, and 0.91 for upper quartile length, mean length, and short fiber content, respectively. The short fiber content (fibers < 1.3 cm) is becoming increasingly important as an indicator of fiber breakage caused by harvesting and field stress and over processing at the gin or textile mill. High short fiber content also indicates greater fiber waste during textile processing.

In a similar manner the AFIS module for maturity and fineness gives the mean and distribution for maturity and fineness. The current estimator of these two fiber properties is micronaire, but it reports one value for the combined action of both maturity and fineness. Micronaire also produces values that have no direct relationship with true biological fineness and maturity. The AFIS maturity-fineness module has been shown to be closely correlated to data obtained from image analysis (Chu & Shofner, 1992). Image analysis measures the exact area of a cross-section of fibers to estimate the perimeter, and thus, the fineness of the fiber. It also estimates the proportion of that fiber's lumen that is filled by the secondary cell wall to give an estimate of maturity. Fineness improvement is needed in order to increase spinning efficiency for rotor spinning and production of fine yarns. Fiber maturity is an important component of dying efficiency and yarn quality. Immaturity also is associated with increased levels of neps. Neps are fiber entanglements that resist dying and result in poor yarn appearance. Neps are classified into two types: seed-coat fragments and entanglements of immature fibers (Ghorashi et al., 1994). Another module option of the AFIS can be used to estimate trash and dust particles (Tounes et al. 1992). Foreign matter reduces yarn and fabric appearance and causes reduced manufacturing and processing efficiency.

Fiber strength and elongation of large numbers of individual fibers can be examined by the Mantis (Zellweger Uster, Knoxville, TN). Mantis allows researchers to better understand the fiber components that contribute to bundle and yarn strength. Three components of bundle strength are fiber length, fineness, and individual fiber strength. Meredith (1992) correlated bundle strength with some of its components. The greatest unique contribution to bundle strength was 50% span length that accounted for 22.0% of the total variation among 24 cultivars. Fineness and individual fiber strength accounted for 0.2 and 10.8% of the variability, respectively. However, the combined interactive effect of all three components accounted for an additional 23.8% of the total variation.

III. IMPORTANCE OF FIBER PROPERTIES

Fiber properties determine both how efficiently a textile mill can operate and the quality of the textile produced. For ring spinning, fiber length and length uniformity are the most important criteria (Deussen, 1992). The longer the fiber, the finer the yarn that can be spun. Fiber strength ranks second and since as few as 50 to 60 fibers per cross-section can produce an acceptable yarn structure, fineness is not a major factor in ring spinning. However, according to Deussen (1992) rotor spinning requires good strength and fineness. Strong fibers withstand the many and faster processing steps required to make yarn than that needed for ring spinning. Each processing step (harvesting, seed cotton storage, ginning, cleaning, carding, combing and drafting) results in some loss of strength. Strong fibers also are needed to make strong yarns. The finer the cotton the more fibers there will be in a cross-section of a yarn, allowing faster, more efficient mill operation for rotor spinning and also producing a better quality fabric. While fiber length and length uniformity are important in rotor and air jet spinning, the fiber lengths produced by current cultivars is adequate for efficient spinning of quality fabrics. Fiber maturity is desired as dying quality is directly related to fiber maturity. Color is a traditional marketing criterion and is related to the perceived field weathering, storage and handling of cotton. Bright white cottons imply little weathering or microbial damage. Trash has to be removed before quality yarns can be produced and it also determines the speed with which textile mills can operate.

IV. COTTON FIBER DEVELOPMENT

Cotton is an indeterminate crop with the first fruiting branch (sympodia) being initiated at nodes 5 to 10 depending on cultivar and temperature (Mauney, 1986). The time interval between flowers on successive nodes along individual branches—the horizontal flowering interval—is about 6 d. The time interval between first flowers on successive flowering branches, the vertical flowering interval, is 3 d (Mauney, 1986). Flowering, as with all developmental processes, has great variation due to genetics, environment, and cultural management practices. A flower bud (square) contains three to five locules, with each locule containing 5 to 11 ovules. The number of ovules to reach maturity as seeds varies but is typically about 28 per boll. A cotton fiber develops from a single epidermal cell of the ovule. About 25% of the ovule epidermal cells elongate to form fibers. Fiber initiation begins the day before anthesis, with fibers being first initiated at the chalazal part of the ovule and continuing towards the micropyle. The primary cell wall is not in direct contact with a vascular system or with a nutrient medium. It is not clear how substrate uptake keeps pace with a rapidly growing cell. Benedict et al. (1973) have shown that photosynthesis continues to place labeled C from $^{14}CO_2$ into cotton lint until about 45 d post anthesis or about the same time required for fiber maturation. Maximum fiber diameter is established in about 4 d and it's length is determined in about 20 d. During this time interval the cell is delineated by a primary cell wall and is covered by a waxy layer or cuticle (Triplett, 1992). The secondary cell wall is composed of nearly pure cellulose, a β-D(1-4)-glucan, and is initiated at about day 17. The cellulose microfibrils are deposited inwardly and exhibit daily growth rings that are related to the day and night temperature differentials (Grant et al., 1966). Fiber fineness is apparently determined early in the first 4 d, and fiber length somewhat later in the elongation phase or at about 20 d. Fiber maturity is defined as the wall thickness of the fiber relative to the total cell diameter and is determined during secondary cell wall development, or from day 17 through day 42. Gibson and Joham (1968) found that the rate of cellulose synthesis, and thus maturity, in western Texas, USA, was directly related to night temperature. Any environmental, management or nutritional stress, therefore affects maturity. For example, Pettigrew et al. (1995) reported that the maturity of fiber produced from plants with ample K was 6% higher than that of fiber produced from K-deficient plants. The stage of development that strength is determined is not firmly established. Environmental associations suggest that bundle strength is determined prior to secondary wall development.

Meredith et al. (1968) observed that major changes in fiber bundle strength were associated with cool, cloudy weather when the bolls (developing fruit) were 5- to 14-d postanthesis. However, Ramey (1986) concluded that factors that affect secondary wall development affect fiber strength.

Fiber strength can be measured as bundle strength, individual fiber strength relative to fiber weight, or individual fiber strength per se. Timpa and Ramey (1994) reported that the molecular weight of cotton fiber was positively correlated with HVI bundle strength; molecular weight accounted for 36% of the variability in strength. Benedict et al. (1994) also found a significant association between average cellulose chain length in the crystalline cellulose (measured as the average molecular weight of the crystalline cellulose) and HVI bundle strength. This association accounted for 88% of the strength variability. Thus, the molecular weight of cellulose produced during the secondary wall development is correlated with bundle strength. However, at Stoneville, Mississippi, the correlation of micronaire with the fiber strength of one cultivar Stoneville 213 grown in the same field for 20 yr was not significant ($P < 0.05$) (Meredith, 1985, unpublished data). The developmental stage and environmental factors that influence fiber strength have not been definitely established.

V. GENETIC VARIABILITY OF COTTON FIBER PROPERTIES

Sasser (1995) has indicated that better quality cultivars is one of the reasons for the improved fiber quality of U.S. cotton. To my knowledge all production programs that are known for their high fiber quality have a high quality cultivar as the foundation of the program.

There are 43 species of *Gossypium* (Percival & Kohel, 1990) of which four are domesticated (Lee, 1984). Two of the domesticated species are diploid ($2n = 26$ chromosomes) and are old world cottons. *Gossypium arborem* L. is an important crop in India and *G. herbaceum* L. is grown in Africa and Asia. The New World domesticated cottons are allotetraploids and produce 98% of the total world production (Anon., 1995). *Gossypium barbadense* L. is called extra long-staple, Egyptian, Pima, and Sea Island, and accounts for about 8% of the world production. Upland, *G. hirsutum* L., accounts for about 90% of the world production. In general, *G. barbadense* fiber has longer, stronger, and finer lint but produces less lint per unit area than *G. hirsutum*. However, there has been considerable intercrossing of the two tetraploid species, either by natural means or by plant breeders (Brubaker & Wendel, 1994). In addition, hybridization bridges between *G. hirsutum* and several wild species have occurred resulting in a large genetic variance for fiber properties, such that for the fiber trait distributions of *G. hirsutum* and *G. barbadense*, breeding populations frequently overlap. The wild species offer many potentially useful genetic fiber traits. Genetic engineering offers some additional genetic variability (John & Crow, 1992). Thus far, with each new development in instrumentation to measure fiber properties, researchers have detected ample genetic variability.

To obtain a relative estimate of environmental and genetic contributions to fiber property variability Meredith (1986) estimated the components of variation

Table 7–4. Percentage of variance components associated with U.S. environmental and genetic parameters for fiber properties from 1966 to 1983†.

Variance component	Colorimeter‡ Rd	+b	Micronaire	Length	Yarn strength
Regions (R)	13.5	5.3	8.0	3.5	2.8
Environments/ Regions (E/R)	65.6	77.5	50.8	15.4	14.9
Cultivars (G)	3.6	1.8	25.3	75.1	70.9
G × R	0.1	0.6	2.3	0.3	1.0
(G × E)/R	17.2	14.8	13.6	5.7	10.4

† Meredith (1986).
‡ Rd = brightness, +b = yellowness.

for fine fiber properties. The estimates, presented in Table 7–4, were from regional variety tests that involved about 22 locations from the Carolinas to California for 18 yr. The genetic components of variability, those due to cultivars and their interactions with environments, accounted for most of the length and strength variability. For length and strength variability, 81 and 82% of total variation in these fiber properties, respectively was due to genetic components (Table 7–4). In contrast for fiber color, brightness (Rd) and yellowness (b+), only 21 and 17%, respectively, of the variation was associated with genetic parameters. Micronaire genetic variation was 41% of the total variation. Micronaire is a measure of two components, fiber maturity and fineness (perimeter of the fiber). Later evaluations of these two components (Meredith, 1991) showed that most of the variation in maturity, 82%, was due to environment. Such factors as temperature (Gibson & Joham, 1968) and crop termination with defoliants have major effects on micronaire. Any factors affecting secondary cell wall development affect the maturity of the fiber. Pettigrew et al. (1995) showed that K deficiency resulted in less mature fibers, and Ebelhar et al. (1994) showed that excessive N levels resulted in decreased maturity of the fiber. Variation in perimeter is about equally influenced by genetics and environmental influences. Meredith (1984) summarized data from 15 studies on heterosis of yield, yield components, and fiber properties. He reported that heterosis, defined as $100 \times (F_1 - \text{midparent})(\text{midparent})^{-1}$, was 18% for yield, and 1.5, 13.5, 8.3, and 3.4% for the yield components lint percentage, number of bolls per unit area, boll weight, and seed weight, respectively. In contrast, little or no heterosis for fiber properties was detected. Average heterosis for fiber length, strength, and fineness was 2.0, 0.1, and <0.0%, respectively. This review also showed that the cultivar × environment interactions for fiber properties was much lower than that for yield and its components.

Smoothleaf cultivars have significantly reduced trash content, and consequently they produce higher grades of lint than hirsute cottons (Williford et al., 1984; Novick et al., 1991). In 1982, about 11% of the Mississippi crop was planted to smoothleaf cultivars. Complaints by the textile industry about trash content and discounts on bales with high trash content led to growers shifting from hairy to smoothleaf cultivars. By 1994, 72% of the Mississippi Delta was planted to smoothleaf cultivars (Anon., 1994).

Small increases in grades have also been reported from using the sub okra-leaf trait (L_2^{μ}). Novick et al. (1991) observed a normal leaf grade index of 89.6 for normal leaf and a significantly higher grade index of 91.3 for the sub okra-leaf isoline. Smoothleaf cultivars also have less sticky cotton than hirsute cotton. The major cause of sticky cotton is honeydew from aphids (*Aphis* spp.) and whiteflies (*Bemisia* spp.). Smoothleaf cottons have some tolerance to both aphids (Weathersbee et al., 1994) and whiteflies (Wilson et al., 1992).

The textile industry is placing more emphasis on neps and seedcoat fragments, which reduce processing efficiency in the textile mill and detract from yarn and fabric appearance. No nep or seedcoat fragment genetic studies have yet been published, however, studies involving cultivar differences have consistently shown significant differences for both neps and seedcoat fragments among cultivars (Miravalle et al., 1986). Crop maturity, harvesting methods, and ginning are factors that influence neps and seedcoat fragments (Anthony et al., 1988; Mangialardi & Lalor, 1990; Mangialardi & Meredith, 1990). A general positive relationship between immaturity and fineness is present both for cultivar and crop maturity differences. Seed coat fragments are more dependent on cultivar differences than on crop maturity, harvesting, and ginning methods.

Breeding progress depends on a large genetic variance and programs that detect, evaluate, and utilize that genetic variability. The major source of useful genetic variability for fiber properties has been the wild species and introductions. The USDA, ARS, has been responsible for species maintenance and introductions. The USDA, ARS, genetic enhancement programs also have evaluated genes from wild species and transferred them into more adapted breeding populations. One example of this progress is the program located at Pee Dee, South Carolina. This program has introgressed genes from *G. arboreum*, *G. thurberi* Tod., and *G. barbadense* (Culp et al., 1979). Initial progress was slow because desired fiber properties were rarely found in the segregates. Culp and coworkers reported that in the early part of the program desirable fiber strength segregates occurred only about 0.1% of the time. After much intercrossing among segregates followed by selection the frequency increased to 10%. The problem of low lint percentage of these strains was mostly solved when these researchers introduced ‘C6-5’, a high-lint percentage upland strain that had been obtained by plant introduction. Still the germplasm showed strong negative associations between fiber strength and yield. This obstacle was overcome by several procedures; random intercrossing as reported by Miller and Rawlings (1967) and Meredith and Bridge (1971), frequent crossing of high strength selections with many different cultivar types and backcrossing (Meredith, 1977). The Pee Dee germplasm is currently included in the pedigrees of many successful commercial cultivars.

A second successful high fiber quality program directed by the USDA, ARS, was located in the San Joaquin Valley of California. Introduced upland germplasm and germplasm from the Pee Dee program were used to develop the high-fiber-strength Acala cultivars. This program involved both breeding for fiber strength and a grower organization that specified that released cultivars must meet high fiber strength requirements. The far-sightedness of these growers has resulted in their cottons being in great demand throughout the textile world. This blueprint of having both a high-fiber-strength cultivar and a marketing program that produces excellence in fiber quality has been followed in other areas of the USA. The foundation of high quality programs has usually been a ginning community that gins 4360 to

8720 t (20 000–40 000 bales) of a high fiber strength cultivar. These programs also include harvesting and ginning systems that promote high fiber strength and fiber uniformity. All bales produced in this manner have to meet minimum fiber quality standards as indicated by HVI testing.

Specific cultivar types have evolved for each major regional management system. In the USA, four major cotton production regions have been recognized and one entry from each region is included in all of the National Cotton Variety Tests. The general characteristics of these four cultivar types are indicated in Table 7–5. The information comes from the 1993 and 1994 National Cotton Variety Tests and is from 16 (1993) and 14 (1994) locations (30 environments), which are located in 10 states (Rayburn & Meredith, 1993 and 1994). 'Acala 1517-88' is representative of the general type that predominates the irrigated western USA involving the states of New Mexico, Arizona, and California. Acala 1517-88 is more vegetative in its growth habit than the other cultivars and produces the largest bolls and seed, and strongest and finest lint. 'HS26' is the most widely planted cultivar in the southwestern region of western Texas and Oklahoma (Anon., 1994). Deltapine 50 is the predominant cultivar in the Mississippi River delta. It's lint is the weakest, and its micronaire is the highest. The remaining southeastern USA is represented by Deltapine 90, which produces high fiber strength but coarse fiber. The range of values for the nine characteristics listed in Table 7–5 show that for most characteristics, cultivar differences are much greater than regional differences.

An improved fiber quality program involves all segments of crop culture, ginning, and processing but it is the genetic component that has resulted in the most progress. Areas of the USA, such as the San Joaquin Valley of California, have, as a requirement for production, strict standards on fiber properties, especially fiber strength. This area's reputation of producing high quality cotton has resulted in their receiving premium prices for their cotton. More gin communities are organizing cooperatives that specify specific high-fiber-quality cultivars be grown, and specify specific harvesting and ginning methods for the crop.

The genetic parameters for many fiber characteristics have received only scant attention. For example, there are at least two types of cotton with high fiber bundle strength. The western U.S. Acala types have strong and fine lint while high-fiber-strength mid-South cultivars such as 'MD51ne' and Pee Dee germplasms have high strength but coarse fiber. Since fineness is one of the components of bundle strength, the mid-South high-bundle-strength cultivars must have some other component or intensity of a component in order to produce high strength. The genetics of individual fiber characteristics remains to be studied to determine what additional progress in improving fiber strength can be made. Studies of these individual fiber parameters also can aid in better textile mill processing and yarn and fabric quality. The inheritance of neps and seedcoat fragments also needs investigation.

VI. FUTURE COTTON FIBER CHARACTERISTICS: SUMMARY

Competition for the world textile market is largely responsible for the changes in textile processing methods. These textile engineering changes are designed to both improve textile mill efficiency and also to improve textile quality. Changes

Table 7–5. Average performance of four regional type cultivars grown in 16 (1993) and 14 (1994) National Variety Tests involving four U.S. regions.

Regional cultivar type	Yield	Lint	Boll wt.	Seed wt.	Yarn tenacity	Strength T	2.5% span length	Micronaire	Fineness
	kg ha^{-1}	%	g	mg	—— MN tex^{-1}	——	mm		μg m^{-1}
DPL 90 (East)	1192	38.7	4.97	9.7	134	220	28.6	4.57	48.7
DPL 50 (Delta)	1191	36.4	5.21	10.5	118	190	28.7	4.51	51.5
HS 26 (Plains)	1092	36.6	5.43	11.1	131	219	27.5	4.61	50.7
Ac. 1517-88 (West)	1041	37.9	4.96	10.8	154	240	29.8	4.21	46.1
LSD (0.01)	83	0.9	0.25	0.3	4	7	0.4	0.14	1.0
Regions†									
East	1082b	38.2a	5.22a	10.5a	136a	217a	29.2a	4.50a	51.4a
Delta	1103b	36.6a	5.07a	10.5a	138a	216a	29.1a	4.39a	49.4ab
Plains	906b	37.2a	4.73a	10.2a	133ab	218a	27.6b	4.42a	47.7b
West	1425a	37.7a	5.54a	10.9a	130b	218a	28.7ab	4.57a	48.3b

* Means not followed by the same letter are significantly different at $P = 0.05$ as indicated by the *t*-test using variation among environments within regions as an estimate of error.

also are occurring in the growing, harvesting and storage of seed cotton, and ginning of cotton. The evolution of HVI into the marketing system is resulting in greater quality control throughout the industry. Curently HVI measures fiber color, trash, length, bundle strength, micronaire, and length uniformity. As new instruments are perfected, other fiber properties such as fineness, maturity, sugar content, short fiber content, and neps may be added to the HVI system. Past history of fiber quality successes and a few genetic studies conducted to date indicate that a major component of the future fiber quality programs will be genetic modification. The breeding improvement for these fiber properties will be to use genetic engineering, molecular markers, and conventional breeding. However, at least in the next 5 to 10 yr, most of the breeding improvements will be by way of conventional breeding.

Future cultivars, in addition to being higher in yield, will need to have lint that is stronger, finer, more uniform, have greater maturity, and fewer neps and less trash than current cultivars. The sources of useful genetic variability for fineness, maturity, neps, and short fiber content have not yet been extensively investigated.

REFERENCES

Anonymous. 1994. Cotton varieties planted. 1994 USDA Agric. Market. Serv.-Cotton Div., Memphis, TN.

Anonymous. 1995. Cotton production prospects for the next decade. World Bank Tech. Pap. No. 287. World Bank, Washington, DC.

Anthony, W.S., W.R. Meredith, Jr., J.R. Williford, and G.J. Mangialardi. 1988. Seed-coat fragments in ginned lint: The effects of varieties, harvesting, and ginning practices. Tex. Res. J. 58:111–116.

Benedict, C.R., R.J. Kohel, and G.M. Jividen. 1994. Crystalline cellulose and cotton fiber strength. Crop Sci. 34:147–151.

Benedict, C.R., R.H. Smith, and R.J. Kohel. 1973. Incorporation of ^{14}C-photosynthate into developing cotton bolls, *Gossypium hirsutum* L. Crop Sci. 13:88–91.

Bragg, C.K., and F.M. Shofner. 1993. A rapid direct measurement of short fiber content. Tex. Res. J. 63:171–176.

Brubaker, C.L., and J.F. Wendel. 1994. Reevaluating the origin of domesticated cotton (*Gossypium hirsutum*; Malvaceae) using nuclear restriction fragment length polymorphisms (RFLPs). Am. J. Bot. 81:1309–1326.

Chewing, C.H. 1992. Cotton fiber management using high volume instrument testing and Cotton Incorporated's Engineered Fiber Selection System. p. 29–41. *In* C.R. Benedict and G.M. Jividen (ed.) Cotton fiber cellulose. Proc. Structure Funct. Utilizat. Conf., Savannah, GA. 28–31 October. Natl. Cotton Counc. Am., Memphis, TN.

Chu, Y.T., and F.M. Shofner. 1992. Progress report on fineness and maturity distributions by AFIS. p. 1017–1019. *In* D.J. Herber (ed.) Proc. Beltwide Cotton Prod. Res. Conf., Nashville, TN. 6–10 January. Natl. Cotton Counc. Am., Memphis, TN.

Culp, T.W., D.C. Harrel, and T. Kerr. 1979. Some genetic implications in the transfer of high fiber strength genes to Upland Cotton. Crop Sci. 19:481–484.

Deussund, H. 1992. Improved cotton fiber properties: The textile industry's key to success in global competition. p. 43–63. *In* C.R. Benedict and G.M. Jividen (ed.) Cotton fiber cellulose. Proc. Structure Funct. Utiliz. Conf., Savannah, GA. 28–31 October. Natl. Cotton Counc. Am., Memphis, TN.

Ebelhar, M.W., W.R. Meredith, Jr., and R.A. Welch. 1994. Interaction of nitrogen rates and mepiquat chloride-effects on yield and quality. p. 1577. *In* D.J. Herber (ed.) Beltwide Cotton Prod. Res. Conf., San Diego, CA. 5–8 January. Natl. Cotton Counc. Am., Memphis, TN.

Ghorashi, H.M., J.C. Baldwin, and M.A. Khan. 1994. AFIS advancement in neps and linear measurements. p. 1433–1436. *In* D.J. Herber (ed.) Proc. Beltwide Cotton Prod. Res. Conf., San Diego, CA. 5–8 January. Natl. Cotton Counc. Am., Memphis, TN.

Gibson, J.R., and H.E. Joham. 1968. Influence of night temperature growth and development of cotton (*Gossypium hirsutum* L.) II. Fiber Properties. Agron. J. 60:296–298.

Grant, J.N., R.S. Orr, and R.D. Powell. 1966. Cotton fiber structure an physical properties altered by environment. Tex. Res. J. 36:432–440.

John, M.E., and L.J. Crow. 1992. Gene expression in cotton (*Gossypium hirsutum* L.) fiber: Cloning of the mRNAs. Proc. Natl. Acad. Sci. USA 89:5769–5773.

Lee, J.A. 1984. Cotton as a world crop. p. 1–25. *In* R.J. Kohel and C.F. Lewis (ed.) Cotton. Agron. Monogr. 24. ASA, Madison, WI.

Mangialardi, G.J., Jr., and W.F. Lalor. 1990. Propensity of cotton varieties to neppiness. Trans ASAE 33:1748–1758.

Mangialardi, G.J., Jr., and W.R. Meredith, Jr. 1990. Relationship of fineness, maturity, and strength to neps and seed-coat fragments in ginned lint. Trans. ASAE 33:1075–1082.

Mauney, J.R. 1986. Vegetative growth and development of fruiting sites. p. 11–28. *In* J.R. Mauney and J.M. Stewart (ed.) Cotton physiology. Cotton Found., Memphis, TN.

McCreight, D.J. 1992. Cotton quality requirements for producing world-class textiles. p. 1197–1199. *In* D.J. Herber (ed.) Proc. Beltwide Cotton Prod. Res. Conf., Nashville, TN. 6–10 January. Natl. Cotton Counc. Am., Memphis, TN.

Meredith, W.R., Jr. 1977. Backcross breeding to increase fiber strength of cotton. Crop Sci. 17:172–175.

Meredith, W.R., Jr. 1984. Quantitative genetics. p. 131–150. *In* R.J. Kohel and C.F. Lewis (ed.) Cotton. ASA, Madison, WI.

Meredith, W.R., Jr. 1986. Fiber quality variation among USA cotton growing regions. p. 105–106. *In* J.M. Brown (ed.) Proc. Beltwide Cotton Conf., Las Vegas, NV. 4–9 January. Natl. Cotton Counc. Am., Memphis, TN.

Meredith, W.R., Jr. 1991. Associations of maturity and perimeter with micronaire. p. 569. *In* D.J. Herber (ed.) Proc. Beltwide Cotton Prod. Res. Conf., Nashville, TN. 6–10 January. Natl. Cotton Counc. Am., Memphis, TN.

Meredith, W.R., Jr. 1992. Improving fiber strength through genetics and breeding. p. 289–303. *In* C.R. Benedict and G.M. Jividen (ed.) Cotton fiber cellulose. Proc. Structure Funct. Utilizat. Conf., Savannah, GA. 28–31 October. Natl. Cotton Counc. Am., Memphis, TN.

Meredith, W.R., Jr., and R.R. Bridge. 1971. Breakup of linkage blocks in cotton, *Gossypium hirsutum* L. Crop Sci. 11:695–697.

Meredith, W.R., Jr., R.R. Bridge, and J.F. Chism. 1968. Cotton varieties tested for yield and fiber properties. Cotton Grow. Rev. 45:179–183.

Miller, P.A., and J.O. Rawlings. 1967. Breakup of initial linkage block through intermating in a cotton breeding program. Crop Sci. 7:199–204.

Miravalle, R.J., G.J. Mangialardi, Jr., and W.F. Lalor. 1986. Neps in cotton genotypes grown in six locations across the cotton belt in 1984. p. 426–428. *In* J.M. Brown (ed.) Proc. Beltwide Cotton Prod. Res. Conf., Atlanta, GA. 8–12 January. Natl. Cotton Counc. Am., Memphis, TN.

Novick, R.G., J.E. Jones, W.S. Anthony, W. Aguillard, and J.J. Dickson. 1991. Genetic effects on nonlint trash of cotton. Crop Sci. 31:1029–1034.

Percival, A.E., and R.J. Kohel. 1990. The distribution, collection, and evaluation of *Gossypium*. Adv. Agron. 44:251–284.

Pettigrew, W.T., J.J. Heitholt, and W.R. Meredith, Jr. 1995. Genotypic interactions with potassium and nitrogen in cotton of varied maturity. Agron. J. 88:89–93.

Ramey, H.H. 1986. Stress influence on fiber development. p. 351–359. *In* J.R. Mauney and J.M. Stewart (ed.) Cotton physiology. Cotton Found., Memphis, TN.

Rayburn, S.T.,and W.R. Meredith, Jr. 1993 and 1994. 1993 and 1994 national cotton variety test. USDA, ARS, Stoneville, MS.

Sasser, P.E. 1995. The 1994 crop-quality. p. 3–5. *In* Proc. Beltwide Cotton Prod. Res. Conf., San Antonio, TX. 4–7 January. Natl. Cotton Counc., Memphis, TN.

Schubert, A.M., C.R. Benedict, J.D. Berlin, and R.J. Kohel. 1973. Cotton fiber development—kinetics of cell elongation and secondary wall thickening. Crop Sci. 13:704–709.

Timpa, J.D.,and H.H. Ramey, Jr. 1994. Relationship between cotton fiber strength and cellulose molecular weight distribution: HVI calibration standards. Tex. Res. J. 64:557–562.

Tounes, M.G., J.C. Baldwin, and F.M. Shofner. 1992. A comparison of AFIS T counting and sizing of trash and dust particles with other methods. p. 985–988. *In* D.J. Herber (ed.) Proc. Beltwide Cotton Prod. Res. Conf., Nashville, TN. 6–10 January. Natl. Cotton Counc. Am., Memphis, TN.

Triplett, B.A. 1992. Strategies for improving cotton fiber quality. p. 107–114. *In* C.R. Benedict and G.M. Jividen (ed.) Cotton fiber cellulose. Proc. Structure Funct. Utilizat. Conf., Savannah, GA. 28–31 October. Natl. Cotton Counc. Am., Memphis, TN.

Weathersbee, A.A., III, D.D. Hardee, and W.R. Meredith, Jr. 1994. Effects of cotton genotype on seasonal abundance of cotton aphid (homoptera; Aphididae). J. Agric. Ent. 11:29–37.

Williford, J.R., W.R. Meredith, Jr. and A.C. Griffin, Jr. 1984. Effect of variety, harvest method, and lint cleaners on cotton quality and value in 1983. p. 114–115. *In* J.M. Brown (ed.) Proc. Beltwide Cotton Prod. Res. Conf., Atlanta, GA. 8–12 January. Natl. Cotton Counc. Am., Memphis, TN.

Wilson, F.D., H.M. Flint, W.R. Deaton, D.A. Fischoff, F.J. Perlak, T.A. Armstrong, R.L. Fuchs, S.A. Berberick, N.J. Parks, and B.R. Stapp. 1992. Resistance of cotton lines containing a *Bacillus thuringiensis* toxin to pink bollworm (lepidoptera: Gelechiide) and other insects. J. Econ. Ent. 85:1516–1521.

7F Functional Food Value of Designer Crops

MARY ANN L. SMITH

University of Illinois
Urbana, Illinois

Any food crop is comprised of a host of phytochemicals and other components that, while edible, are not directly related to the food's nutritional value. In addition, a broad range of colorants, preservatives, and other ingredients (both natural and synthetic) are routinely incorporated into modern foods during processing operations and are consumed along with the native food components. The industry is obliged to make generous use of these additives to guarantee the product appearance and quality consumers expect; however, recent controversies have launched a new consumer skepticism, and even apprehension, about potential repercussions from artificial additives. Some of the most urgent research initiatives today concern the identification, characterization, and efficient production of natural food additives and components that are innocuous and have disease-preventative or health-promoting benefits. The term "functional foods" was coined to encompass the known array of foods or parts of foods that provide medicinal and/or health benefits, including the prevention and treatment of disease (DeFelice, 1995). Current breeding efforts center on amplifying the content of beneficial native phytochemicals in food crops to enhance their functional food value. If the product will be processed, research is required to ensure that additives confer no adverse effects.

The catalyst behind one research goal—provision of harmless, preferably natural food products and additives—is the growing base of evidence linking some synthetic additives to negative effects on human health. Many synthetic additives (e.g., colors, flavors, preservatives, and fragrances) have been implicated in cases of hyperactivity, allergies, tumors, brain disorders, heart disease, and a host of other potential health hazards (Timberlake & Henry, 1986; Spears, 1988; Blumenthal, 1991). Adverse publicity has escalated in recent years, and the food industries are pursuing viable alternatives to some commonplace synthetics in current use. For example, researchers are seeking safe pigments of natural origin for use as food colorants, natural flavors for gums, or cosmetics. The most logical and productive sources of innocuous additives are plants that naturally produce the valuable substances that can be extracted and added to other foods during processing (Spears, 1988).

Another research goal is the provision of food products or supplements with enhanced chemoprotective properties. Some key compounds (relatively minor dietary constituents), which are present in small amounts in fruits, grains, foliage, and

 Designing Crops for Added Value, Agronomy Monograph no. 40.

other plant vegetative organs, are naturally capable of disease prevention and therapy (Pszczola, 1992; Wichtl & Bisset, 1994). Dietary supplements, crop-derived foods with superior phytochemical content, herbal products, and processed foods fortified with ingredients specifically to enhance health (e.g., cereals with added fiber, beverages with added Ca) can all be included in the category of functional food products. Functional foods or designer foods also are designations for food products selected or formulated to promote health or treat disease (Haumann, 1993). One significant problem in the whole arena of functional foods research and breeding for food crops with enhanced chemoprotective phytochemical content is that the category lacks a universal definition. An array of pseudonyms, such as nutraceuticals, phytopharmaceuticals, and designer foods have been used, categories that are parallel or overlap significantly (Childs, 1995).

Health professionals have been slow to accept the significance and value of functional foods (Hegenbart, 1993). However, evidence from epidemiological studies, in vitro and in vivo bioactivity tests, and clinical trials supporting functional food benefits is now routinely published in refereed medical journals. Perhaps more significant for development and marketing interests, the general public has embraced this research through the efforts of the media.

In some cases, the functional food is distinguished by reduced levels of a food component that has been identified as a health hazard. Examples include cheese formulated with reduced fat content to reduce cancer risk and low-Na snacks, which acknowledge Na's link to hypertension. Other functional foods are designed to amplify the presence of beneficial food constituents: e.g., fiber-containing grain products or beverages, designed to decrease cancer incidence (Caragay, 1992). Some of the genetic improvements considered in earlier sections of this chapter, such as the development of high-lysine corn (*Zea mays* L.), also are sometimes categorized under the broad blanket of designer foods, or functional foods improvements.

However, perhaps the most promising area for new research is development of functional foods with sharply elevated levels of non-nutrient, naturally occurring, physiologically active phytochemicals, which are present in relatively small amounts in a wide range of crops. The major recognized functional food groups are listed in Table 7–6, including the phytochemicals involved and their indicated health benefits. Candidate crops include, but are not limited to, cruciferous vegetables, grains, some spices, citrus fruits, green tea, garlic, licorice and soybeans (Hasler, 1993). The targeted phytochemicals are active against a diverse range of disease conditions. Phytochemicals have been identified that inhibit cancer, stimulate the immune system, reduce inflammation, prevent blood clots, reduce osteoporosis, and alleviate high blood pressure (Armand et al., 1988; Gutfeld, 1991; Caragay, 1992; Hertog et al., 1992; Kinsella et al., 1993; Schardt, 1994; Bomser et al., 1995; Hasler, 1993). Key phytochemicals include carotenoids, coumarins, flavonoids, glucarates, indoles, isothiocyanates, lignans, mono- and triterpenes, phenolic acids, phthalides, phytates, polyacetylenes, and sulfides; other specific classes such as limonene, allium compounds, or protease inhibitors have been included in the classification (Caragay, 1992; Hasler, 1993). Typically, they act as antioxidants, induce detoxification enzymes, alter hormone metabolism, or otherwise inhibit metabolic damage (Steinmetz & Potter, 1991). The allium compounds (present in garlic and related crops) have been cited for their ability to block the formation of

Table 7–6. Some of the major functional food crops and key phytochemicals under investigation. Functional food benefits for each group are listed, however, each of the phytochemicals may have expanded chemopreventive roles in other metabolic processes.

Phytochemical group	Major crop donors	Functional food benefits
Organosulfur compounds	Garlic (*Allium sativum*), onion (*A. cepa*), leek (*A. ampeloprasum*)	Anticarcinogenic, induces detoxification enzymes
Flavonoids (isoflavones) (catechins)	Fruits and vegetables soybean (*Glycine max*) green tea (*Camellia sinensis*)	Anticarcinogenic/antioxidant, binds estrogen receptors, inhibits cardiovascular disease
Isothiocyanates, dithiolthiones, indoles	Cruciferous vegetables	Blocks/prevents damage from carcinogens, induces detoxification enzymes, affects estrogen metabolism
Lignans	Flaxseed (*Linum usitatissimum*)	Anticarcinogenic/antioxidant
Triterpenes	Fruits and vegetables	Anticarcinogenic/antioxidant
Monoterpenes/limonoids	Citrus	Blocks/prevents damage from carcinogens
Flavonoids, isoflavones, catechins	Fruits, vegetables, wine	Anticarcinogenic/antioxidant, induces detoxification enzymes, prevents cardiovascular disease
Saponins	Soybean, vegetables	Anticarcinogenic
Carotenoids	Orange/yellow and dark leafy green vegetables, fruits	Anticarcinogenic/antioxidant
Coumarins	Citrus, vegetables	Anticarcinogenic, induces detoxification enzymes
Miscellaneous compounds		
Chlorophylline, dithiolthiones, organonitriles	Cruciferous vegetables	Anticarcinogenic, induces detoxification enzymes
Folic acid, Ca, Se	Fruits, vegetables, dairy	Anticarcinogenic
Protease inhibitors, cinnamic acids	Seeds, soybean	Anticarcinogenic/antioxidant

N-nitroso compounds, alter the activity of drug-metabolizing enzymes, and augment cellular repair mechanisms (Milner & Schaffer, 1995). Epidemiologic evidence supporting the anticarcinogenic action of some traditional nutritive compounds (vitamins and minerals) such as folic acid, Vitamin A, and Se, also has accumulated from recent studies. A strict classification scheme for the hundreds of potentially bioactive phytochemicals has not been established, because the compounds can be categorized by chemical name, by primary food source, or even by the mode of chemopreventive action (Napier, 1995). In some cases, bioactive phytochemicals may increase either the concentration or activity of enzymes that act as blocking agents (detoxify or inhibit carcinogens) or as suppressing agents (restraining malignant changes in cells) (Napier, 1995).

These previously underrated, naturally produced phytochemicals with demonstrated efficacy against cancer and other diseases are particularly intriguing research objectives, because once they are identified, they can be exploited in a number of ways. Recommendations for increased dietary intake of certain foods containing the phytochemical or supplements [such as rosemary (*Rosmarinus officinalis*) or garlic extract] can be made, conventional plant breeding or biotechnology can be directed to increase inherent concentrations in a plant, or selective processing can concentrate or incorporate a phytochemical into a new food product (Haumann, 1993). Each of these approaches is a viable option for fostering good health and preventing chronic disease. In many cases, the antioxidant phytochemicals serve a dual role to protect food from discoloration and flavor loss and also to protect the human consumer against degenerative cancers, cardiovascular diseases, cataracts, and diabetes. Antioxidants defend the body by scavenging O_2-containing free radicals that attack and damage lipids, cell membranes, DNA, and protein. Antioxidants provided in functional foods function to bolster the defenses of the body's own immune system.

There is currently a tremendous market-driven demand for new functional food products and for enhanced health/medicinal value in existing food products. The market potential for nutraceuticals or functional foods has recently been gauged at between $27 and $250 billion (DeFelice, 1995). Natural products in general, regardless of functional food value, are in current favor with consumers as alternatives to artificial ingredients or synthetic substances. The medical community is reassessing the role of new natural food products and supplements and genetically engineered designer foods, realizing that they provide a promising new strategy for enhancing the quality of life for an aging population. These products also may dramatically decrease health care costs in the long run (Hasler, 1993). The National Institutes of Health initiated a bionutrition strategic plan in 1993 specifically to expand the scientific knowledge base underpinning this subject area and to increase knowledge of functional food interventions to prevent, cure, mitigate, or interrupt the progression of heart disease, cancer, diabetes, and AIDS.

Many of the value-added features in a functional food crop or crop product can be enhanced by environmental controls. For example, vivid natural anthocyanin pigmentation (valuable as a natural colorant and also because of potent cardioprotective and antioxidant properties) in crops such as grape (*Vitis vinifera*) or cranberry (*Vaccinium macrocarpon*) can be intensified under specific climatic regimes and spectral environments (Mancinelli, 1990; Zhong et al., 1991; Mazza & Miniati, 1993), and the concentrations of aromatic or medicinal oil secondary products in many species are observed to increase when rainfall is ample (Arrebola et al., 1994). However, the goal of this chapter is to explore the potential for genetic improvements in chemoprotective phytochemical content—to breed for crops that are superior dietary sources or can be used as a source of extracts for adding to other foods during processing. This discussion will concentrate on functional foods (edible food crops or dietary supplements from these crops) that specifically contain high native contents of active pharmaceutically important phytochemicals with established efficacy against disease. These constituents of the food crop are alleged to contribute chemopreventive bioactivity (medical or health benefits) beyond subsistence or nutritive value.

In many studies, the key phytochemical has been identified and its biological activity described; in other cases, the effective disease protection or treatment mechanisms have not yet been fully characterized. As scientists isolate new plant substances with bioactivity, they are beginning to discover the routes these substances take to disrupt molecular pathways leading to disease disorders. The challenge is compounded by the complexity of phytochemical composition and potential synergistic effects among chemical compounds (Caragay, 1992). Advances in biomedical research on these plant-derived phytochemicals have occurred in tandem with recent breakthroughs in plant breeding techniques and germplasm manipulation (Pszczola, 1992). So, at the same time as medical and scientific knowledge of the effects of functional foods and identification of key active phytochemicals is increasing, plant breeders and geneticists have new tools at their disposal to incorporate genes to amplify phytochemical concentrations in food crops, or to produce novel compounds.

I. GENETIC ADVANCEMENTS FOR ENHANCED FUNCTIONAL FOOD CONTENT THROUGH CONVENTIONAL BREEDING STRATEGIES

Once an active phytochemical in a plant is identified, and the efficacy against a given disease or benefits towards disease prevention is established, breeding strategies are directed to amplify concentrations of the specific substance in the fruit or edible extract from plant organs other than fruit. Current understanding of genetic and environmental factors controlling nutrient levels, as well as functional food content, of typical vegetable crops is limited by the diversity of crops involved, the number of potential phytochemicals, and lack of characterization research.

Key antioxidant minerals (Se), vitamins (C, E) and other phytochemicals (triterpenes, flavonoids, β-carotene, etc.) associated with some functional foods are believed to help prevent various cancers and cardiovascular disease. Currently, one of the best publicized examples in the category of functional foods is crops with high β-carotene content, such as sweet potato (*Ipomoea batatas*) or carrot (*Daucus carota*). Epidemiological evidence has clearly demonstrated that β-carotene is a strong factor in reduced lung cancer risk for nonsmokers (Mayne et al., 1994). Blue-green algae (*Spirulina* spp.) have recently become a commercial food supplement in many countries due to their high β-carotene content, and the bioavailability of this supplement has been studied in both human and animal models (Kapoor & Mehta, 1993). As the culmination of years of conventional breeding and selection, collaborating research teams from University of Wisconsin-Madison, University of California-Davis, and University of Florida-Gainesville have developed elite carrot lines with up to four times the levels of carotene (Vitamine A precursors) as are found in conventional carrots. Carotenes contribute the majority of Vitamin A in global diets (Simon et al., 1989). Simon et al. (1989) have developed carrots with three- to five-fold greater carotene content compared to conventional carrots. Unlike the Vitamin A found in high amounts in fish oils, the carrots supply pro-Vitamin A in the form of carotene, which is enzymatically broken down to Vitamin A in the human body. However, the carotene will not be metabolized to a toxic level. Commercial production of high-carotene carrots is underway to provide raw material for carotene extraction.

Tomato (*Lycopersicon esculentum*), which contains carotenes in the forms of lycopene (responsible for red pigmentation), β-carotene (provitamin A, responsible for orange pigmentation), and α-carotene (another provitamin A compound), has been one of the best researched crops for increasing key functional food content. Genetic manipulation has resulted in a 10-fold increase in the amount of β-carotene in a tomato line, but the ripe tomato was orange in color and did not meet with consumer approval (Breene, 1994). For this reason, most U.S. breeding efforts concentrate on a high-lycopene, low β-carotene tomato, even though it is less potent in terms of functional food value.

Similarly, antioxidant Vitamins C and E are cited as preventative agents against heart disease and cancer. A diverse range of phytochemicals from crops like soybean, garlic, broccoli [*Brassica oleracea* (Botrytis group)], tomato, carrot, and onion (*Allium cepa*) have been researched to identify the specific compounds responsible for anticancer activity, immunity, and cardiovascular protection.

Several antitumor phytochemicals in soybean and soyfoods, in particular isoflavones and saponins, are linked to reduced cancer risk and decreased incidence of heart disease. Potter et al. (1993) demonstrated that men with elevated blood cholesterol exhibit marked reductions in low-density-lipoprotein (LDL) cholesterol after routine consumption of moderate amounts (50 g d^{-1}) of isolated soy protein incorporated into baked goods, as part of a low-fat, low-cholesterol diet. Even lower amounts (25 g d^{-1} for 30 d) triggered significant reductions in LDL-cholesterol concentrations in subjects with elevated cholesterol. The benefits were attributed to the phytochemical content of the soy protein, and current work is aimed at identifying the mechanisms by which soy phytochemicals affect lipid metabolism. Isoflavonoids in soy have estrogenic effects that decrease cardiovascular risk through decreased LDL cholesterol, and enhanced or unchanged levels of high-density-lipoprotein (HDL) cholesterol; further, they may act as anticarcinogens by blocking estrogen receptors in premenopausal women. Soy saponins: (i) have an inhibitory effect on the infectivity of the AIDS virus and on the activation of Epstein-Barr early antigen, (ii) decrease blood cholesterol, and (iii) serve as adjuvants. Both the saponins and isoflavones are glycosidic compounds present in high levels (2%) in soybean (Potter et al., 1993; Napier, 1995). Traditional breeding and selection techniques are frequently used to intensify native anthocyanin pigments in fruit crops, simply because brighter, deeper red colors are preferred by consumers (Sapers et al., 1983). Only more recently have the cardioprotective and anticarcinogenic properties of anthocyanins (and related phytochemicals in the same biochemical pathway) come to light, including their ability to decrease capillary permeability and fragility, and to act as anti-inflammatory agents (Armand et al., 1988; Khushbaktova et al., 1989; Morazzoni & Magistretti, 1990; Bettini et al., 1991; Hertog et al., 1992). Anthocyanins also function as primary antioxidants and superoxide scavengers. Anthocyanins are present in a wide range of food crops [grape, eggplant (*Solanum melongena* var. *esculentum*), blueberry (*Vaccinium angustifolium*), black currant (*Ribes nigrum*), etc.], which are bred for enhanced anthocyanin-induced color. Comparatively little work has addressed genotypic differences in phytochemical content except when the phytochemicals influence marketability of the crop in ways other than functional food properties. For example, the organic acid compositions of rabbiteye (*Vaccinium ashei*) and highbush (*Vaccinium corymbosum*) blueberries were analyzed because the acid profile differences have bearing on fruit color, decay, and predator susceptibility (Ehlenfeldt et al., 1994).

In a comprehensive screening program funded by the National Cancer Institute, Pezzuto and colleagues (Pezzuto, 1994) identified 70 plant genotypes with chemopreventive activity, and the most promising phytochemical in this group was designated "Brassinin"—a compound with an indole nucleus and an isothiocyanate side chain found in broccoli, brussels sprout [*Brassica oleracea* (Gemmifer group)], cauliflower [*Brassica oleracea* (Botrytis group)], chinese cabbage (*Brassica rapa*) and other cruciferous vegetables. The discovery is particularly significant in light of its high efficacy as a chemopreventive agent against dimethylbenz(a)anthracene (DMBA)-induced preneoplastic lesions in mouse mammary glands. One objective for cruciferous vegetable breeders will now be to enhance levels of the compound in the plant tissues.

Breeding efforts in functional foods have primarily sought to increase total concentrations of crude bioactive phytochemicals in a food crop. Recently, efforts also have turned to the modification of relative levels and chemical profiles of specific phytochemical compounds to enhance efficacy against known disease conditions. In these new research initiatives, not just the crude phytochemical complex, but specifically the most active fractions, are amplified during the breeding process. Toxicological testing of plant extracts using mammalian tissues or other systems aids in precise identification of the most active phytochemical fractions against incipient carcinogenesis, or as treatments for existing cancer (Kapoor & Mehta, 1993; Bomser et al., 1995). Similar work involving blood samples tests plant extracts in cardioprotection assays. Once the precise identities (chemical profile) of active phytochemicals are established, breeders can work to select for these specific substances. This current work is of particular importance, because crude extracts from plants with known nutraceutical/medicinal value may contain contaminants that interfere with efficacy and/or cause unwanted side effects (Wichtl & Bisset, 1994). Research on methods of isolation and purification of the active phytochemicals will enable further research on their mechanism of action and avoid many of the problems inherent in using crude preparations.

Another new research objective is to develop methods to extract these substances and then insert them into other foods, creating designer foods. The National Cancer Institute launched the Designer Foods Research Project to identify active phytochemicals and concentrate them into "superfoods" that can prevent cancer (Gutfeld, 1991). The challenge here is for breeders to work hand in hand with phytochemists who are accumulating toxicological and efficacy data to verify and support functional food claims. Some extracted phytochemicals are hazardous at high concentrations or are adversely affected when diluted with other substances in a food product, such as tannins.

II. IN VITRO SELECTION AND MOLECULAR APPROACHES TO ENHANCED FUNCTIONAL FOOD VALUE

Stable gene transfer systems developed or under development for higher plants have been primarily geared toward reducing dependance on agricultural chemicals: engineering resistance to insects, diseases, or herbicides into a crop. These same approaches also provide a means to increase crop yields and quality, and to improve crops for food processing, specialty chemical, and pharmaceutical industries. These latter industries may, in fact, represent the greatest commercial opportunity for biotechnology (Gasser & Fraley, 1989).

The advent of plant genetic engineering has added an extra dimension to the traditional exploitation of plants for economic content—in particular by targeted modification of plants. The technology now permits plant systems to produce new (bioactive) compounds, or to overproduce, derivatize, or bioconvert compounds that naturally occur in the host plant (Pen et al., 1993). Specific genetic engineering research of direct relevance to food quality includes current projects that seek to incorporate proteins with nutritionally balanced amino acid composition or to intensify quantities of specific fatty acids in oil seed crops. In addition, as the genes involved in pigment biosynthesis are identified, researchers in the food industries are manipulating specific enzymes and genes to enhance natural colorant production, in vivo (in the crop plant) and in vitro (in bioreactor-based cell culture systems). Identification of a gene involved in carotenoid biosynthesis in transformed tomato plants (Bird et al., 1991) has opened up new research towards manipulation of the red pigmentation levels in ripening fruit and the related functional food benefits associated with the carotenoid, lycopene.

In a related branch of health-care research, specific crop modifications using gene technology have been directed towards new production methods for pharmaceuticals and specialty chemicals (Krebbers & Vandekerckhove, 1990). For example, pharmaceutically active foreign proteins from mammalian systems have been synthesized in plant cells. The novel approach permits scale-up production of growth hormones, antibodies, or neuropeptides without the chance of cross-contamination, which may be present when using the original donor tissue as the site of synthesis (Pen et al., 1993). For example, active peptides with medicinal benefits have been produced in the foreign host, oilseed rape (*Brassica napus*).

While technical progress has significantly advanced the development of genetically engineered crops with high bioactive phytochemical content, regulatory approval, proprietary protection, public perception and the ability to make claims about their proposed health benefits have restricted their entry into the marketplace. These restrictions are stumbling blocks at all levels of the process, because no appropriate avenue exists for the review and approval of innovative products that purport to enhance health and prevent disease, or for health claims associated with functional food components (Pszczola, 1992).

REFERENCES

Armand, J., M. DeForni, G. Recondo, L. Cals, E. Cvitkovic, and J. Munck. 1988. Flavonoids: A new class of anticancer agents? Preclinical and clinical data on flavone acetic acid. Prog. Clin. Biol. Res. 280:235–241.

Arrebola, M.L., M.C. Navarro, J. Jiménez, and F. Ocaa. 1994. Variations in yield and composition of the essential oil of *Satureja obovata*. Phytochemistry 35:83–93.

Bettini, V., R. Aragno, M. Bettini, G. Braggion, L. Calor, M. Concoloato, P. Favaro, G. Penada, and S. Montin. 1991. Vasodilator and inhibitory effects of *Vaccinium myrtillus* anthocyanosides on the contractile responses of coronary artery segments to acetylcholine: Role of the prostacyclins and of the endothelium-derived relaxing factor. Fitoterapia 62:15–28.

Bird, C.R., J.A. Ray, J.D. Fletcher, J.M. Boniwell, A.S. Bird, C. Teulieres, I. Blain, P.M. Bramley, and W. Schuch. 1991. Using antisense RNA to study gene function: Inhibition of carotenoid biosynthesis in transgenic tomatoes. Biotechnology 9:635–639.

Blumenthal, D. 1991. Red no. 3 and other colorful controversies. FDA Consumer Magazine. DHHS Publ. 91-2235. FDA Consum. Mag., FDA, Rockville, MD.

Breene, W. 1994. Healthfulness and nutritional quality of fresh vs. processed fruits and vegetables: A review. J. Food Serv. Syst. 8:1–45.

Bomser, J., D.L. Madhavi, K. Singletary, and M.A.L. Smith. 1996. In vitro anticancer activity of fruit extracts from Vaccinium species. Planta Med. 62:212–216.

Caragay, A. 1992. Cancer-preventive foods and ingredients. Food Technology. April 1992, p. 65–68.

Childs, N. 1995. Functional foods and market entry. World of Ingredients. October 1995, p. 36.

Collins, P., and C. Timberlake. 1993. Recent development of natural food colors. IFI (6):32–38.

DeFelice, S. 1995. The nutraceutical revolution: Its impact on food industry research and development. Trends Food Sci. Technol. 6:59–61.

Ehlenfeldt, M., F. Meredith, and J. Ballington. 1994. Unique organic acid profile of rabbiteye vs. highbush blueberries. HortScience 29:321–323.

Gasser, C.S.,and R.T. Fraley. 1989. Genetically engineering plants for crop improvement. Science (Washington, DC) 244:1293–1299.

Gutfeld, G. 1991. Superfoods to fight cancer. Prevention 43:44–47.

Haumann, B.F. 1993. Designing foods. Manipulating foods to promote health. Inform 4:344–373.

Hasler, C. 1993. New weapon on the war against chronic disease. Illinois Res. 35:21–23.

Hegenbart, S. 1993. Nutraceutical reality on the horizon. Food Prod. Design Dec. 1993:19–20.

Hertog, M., P. Hollman, and M. Katan. 1992. Content of potentially anticarcinogenic flavonoids of 28 vegetables and 9 fruits commonly consumed in the Netherlands. J. Agric. Food Chem. 40:2379–2383.

Kapoor, R., and U. Mehta: 1993. Utilization of β-carotene from *Spirulina platensis* by rats. Plant Foods Human Nutr. 43:1–8.

Khushbaktova, Z., V.N. Syrov, and Z. Kuliev. 1989. Study of the hypolipidemic properties of polymer proanthocyanidins from plants used in folk medicine. Khimoko Farmatsevticheskii Zhurnal 23:1111–1115.

Kinsella, J.E., E. Frankel, B. German, and J. Kanner. 1993. Possible mechanisms for the protective roles of antioxidants in wine and plant foods. Food Technol. 47:85–89.

Krebbers, E.,and J. Vandekerckhove. 1990. Production of peptides in plant seeds TIBTECH 8:1–3.

Mancinelli, A. 1990. Interaction between light quality and light quantity in the photoregulation of anthocyanin production. Plant Physiol. 92:1191–1195.

Mayne, S.T., D.T. Janerich, P. Greenwald, S. Chorost, C. Tucci, M.B. Zaman, M.R. Melumed, M. Kiely, and M. McKneally. 1994. Dietary beta carotene and lung cancer risk in U.S. non-smokers. J. Natl. Cancer Inst. 86:33–39.

Mazza, G., and E. Miniati. 1993. Anthocyanins in fruits, vegetables, and grains. CRC Press, Boca Raton, FL.

Milner, J., and E. Schaffer. 1995. Garlic and cancer. Nutrinews. January 1995, p. 4–7.

Morazzoni, P., and M.J. Magistretti. 1990. Activity of myrtocyan, an anthocyanoside complex from *Vaccinium myrtillus* on platelet aggregation and adhesiveness. Fitoterapia 61:13–22.

Napier, K. 1995. Cancer fighting foods: Green Revolution. Harvard Health Lett. 39:9–12.

Pen, J., P.C. Sijmons, A.J. van Ooijen, and A. Hoekema. 1993. Protein production in transgenic crops: Analysis of plant molecular farming. p. 239–251. *In* A. Hiatt (ed.) Transgenic plants. Fundamentals and applications. Marcel Dekker, Inc., New York.

Pezzuto, J. 1994. Faculty feature. Funct. Foods Health News. 2:4–6.

Potter, S., R.M. Bakhit, D. Essex-Soriel, K.E. Weingartner, K.M. Chapman, R.A. Nelson, M. Prabhudesai, W.D. Savage, A.I. Nelson, L.W. Winter, and J.W. Erdman, Jr. 1993. Depression of blood plasma cholesterol in men by consumption of baked products containing soy protein. Am. J. Clin. Nutr. 58:501–506.

Pszczola, D.E. (ed.) 1992. Highlights of the nutraceutical initiative: A proposal for economic and regulatory reform. Food Technology. April, p. 77–79.

Sapers, G.M., J. Phillips, H. Rudolf, and F. DiVito. 1983. Cranberry quality: Selection procedures for breeding programs. J. Am. Soc. Hort. Sci. 108:241–246.

Schardt, D. 1994. Phytochemicals: Plants against cancer. Nutrition Action Health Letter. April 1994, p. 1, 9–11.

Simon, P., X. Wolff, C.E. Peterson, Dale Kammerlohr, V. Rubatzky, J. Strandberg, M. Bassett, and J. White. 1989. High carotene mass carrot population. HortScience 24:174.

Spears, K. 1988. Developments in food colourings: the natural alternatives. TIBTECH 6:283–288.

Steinmetz, K., and J. Potter. 1991. Vegetables, fruit, and cancer. II. Mechanisms, cancer causes control. 2:427–442.

Timberlake, C.F., and B.S. Henry. 1986. Plant pigments as natural food colours. Endeavour 10:31–36.

Wichtl, M., and N.G. Bisset (ed.) 1994. Herbal drugs and phytopharmaceuticals. CRC Press, Inc., Boca Raton, FL.

Zhong, J-J., T. Seki, S-I. Kinoshita, and T. Yoshida. 1991. Effect of light irradiation on anthocyanin production by suspended culture of *Perilla frutescens*. Biotechnol. Bioeng. 38:653–658.

8 Protecting the Value in Value Added Crops: Intellectual Property Rights

P. STEPHEN BAENZIGER and AMIT MITRA

University of Nebraska
Lincoln, Nebraska

IAN B. EDWARDS

Pioneer Hi-Bred International, Inc.
Johnston, Iowa

In understanding intellectual property rights and their use in designing value added crops, it is important to know: what are intellectual property rights, why are they needed, what forms of intellectual property rights are available nationally and internationally, what is covered by them, and how do intellectual property rights and biology interact. Intellectual property rights and commercial biology are dynamic and they co-evolve.

Intellectual property rights can be defined as the legal rights that are given to individuals or institutions that develop new products or uses (the property) by the use of their intellect. The intellectual property rights that are described hereafter are those derived from intellectual property law and court precedent. For reasons of space, this chapter will describe only briefly many aspects of intellectual property rights. Those who are interested in greater detail should consult Bent et al. (1987), Caldwell and Schillinger (1989), or Baenziger et al. (1993).

I. NEED FOR INTELLECTUAL PROPERTY RIGHTS

There is one primary reason for obtaining intellectual property rights: to provide a legal basis to recover the cost of developing a value added crop (Duvick, 1993). Value added crops obviously are developed for economic benefits and these benefits are based on the value added crop being distinguished in the marketplace. There must be market segmentation in one or more of the various markets for crops (seed or propagules, harvested plant parts, or end-use market). Value added crops cannot be treated as commodities and retain their premium value, nor can there be market segmentation unless the costs of market segmentation can be repaid.

Obtaining intellectual property rights and enforcing those rights can be costly. The inventor must determine if these needs and potential economic gain are greater than the cost of obtaining and enforcing intellectual property rights. It is for this reason, lack of economic benefit, that many cultivars are released with no intellectual property rights protection.

Though intellectual property rights refer to those rights conveyed by law or legal precedent, it should be understood that if the purpose for invoking intellectual property rights is to protect the value added crop, then there are ways of supplementing those rights based on law and legal precedence with biology. A discussion of intellectual property rights would be incomplete if it did not include the impact of biology.

II. TYPES OF INTELLECTUAL PROPERTY RIGHTS AND WHAT THEY PROTECT

Intellectual property rights have many forms. Some are specific to plants, such as plant patents and Plant Variety Protection. Others are parts of the more general intellectual property rights such as utility patents, personal property, and trade secrets (Baenziger et al., 1993; Seay, 1993). Initially, living organisms were considered unpatentable because they were products of nature or it was assumed they could not be described as required in patent law. The description of the patented material is important because the inventor provides information to the general public in return for the inventor receiving protection. The invention must be described so that the public could learn from the invention and determine exactly what the invention is. As a result, in 1930 the Plant Patent Act (Seay, 1993; Greengrass, 1993; Weimar, 1993) was passed. This law recognized the contributions of plant breeders and that the products of their work were different than the products of nature. The Plant Patent Act is specifically for the plant and its asexually reproduced clonal progeny. It requires the invention to be distinct (distinguishable from other plants), new (did not exist previously in nature, or was not previously identified in nature), and is limited to the asexual propagation mechanism in plants. It also required the protected plant have a varietal name. The Plant Patent Act allowed the owner to restrict the asexual reproduction of the plant and the selling or using plants that were reproduced asexually. The Plant Patent Act does not cover plant parts, but the whole plant itself. The importance of this coverage will be discussed later. The protection is for 17 yr and the patent is issued in the USA by the Department of Commerce.

Most similar to the plant patent, is the utility patent (Seay, 1993; Weimar, 1993). The utility patent is widely used to protect inventions because the inventor can patent "... any new and useful process, machine, manufacture, or composition of matter ..." (35 U.S.C. 101). It is the breadth of the subject matter area that makes utility patents interesting to biologists. Under utility patents, a segment of DNA, a cell line, an inbred line, or a phenotype defined by composition can be patented. The invention must be new, useful, and nonobvious to someone skilled in the art; and adequately described so that someone skilled in the art could "reproduce" or evaluate the patent (for plants this requirement is often met by describing the material and depositing a sample of material). Each patent application is decided on a case by case basis, hence there is subjectivity in the issuance of patents. Also, patents can be contested and the patent issuance may be determined invalid. The life of a patent has been 17 yr, but effective 8 June1995 amendments to implement U.S. entry into the World Trade Organization have changed the term to 20 yr from the date of filing the application. Up to 10 yr of extension can be obtained under certain circumstances. As with plant patents, in the USA the Department of Commerce issues utility patents.

Utility patents have two other notable aspects. The first is the implicit experimental exception to infringement (often referred to as the research exemption) that allows the individual to use the invention to study it. It does not allow that individual to use the invention to make a new product (e.g., if a genotype is patented, that genotype could not be used to create a new commercial cultivar without the approval of the patent owner during the patent life). The second is the doctrine of equivalents that relates to actions where "an accused produce or process accomplishes substantially the same function or result in substantially the same manner as a patented invention" (Williams & Weber, 1989). The way in which equivalents are determined can vary from country to country but most systems appear to apply the doctrine that pioneering inventions are entitled to a wider scope of equivalents than narrower improvement inventions. The importance of equivalency is that if a phenotype has been patented, equivalent phenotypes developed by different mechanisms may constitute an infringement of the patent. However, the standards for patentability, exhaustion of rights (when the patent is no longer valid), infringement, etc., have not been entirely worked out for the plant area.

Though many intellectual property rights are similar among countries, there were some notable exceptions. For utility patents, in the USA before 8 June1995, the patent was issued to the first to invent the process. Outside the USA, the patent was issued to the first to file the application for patent. Also, intellectual property rights outside the USA do not have different intellectual property rights based on the mode of plant reproduction (asexual vs. sexual).

The U.S. Plant Variety (syn. cultivar) Protection Act was enacted in 1970 (Atchley, 1993; Seay, 1993). The 1970 Act excluded from protection, in addition to fungi and bacteria, first generation hybrids, the seed, plants, or transplants of okra [*Abelmoschus esculentus* (L.) Moench], celery (*Apium graveolens* L.), peppers (*Capsicum spp.*), tomato (*Lycopersicon esculentum* L.), carrot (*Daucus carota* L.), and cucumber (*Cucumis sativus* L.). These vegetables, commonly referred to as the "vegetable group", were originally excluded because it was feared that protection would result in their becoming relatively less available or considerably more costly. After a decade of experience these fears proved to be unsupported, and the 1970 Act was amended in 1980 to remove the vegetable exemption. The Plant Variety Protection Act (Seay, 1993; Greengrass, 1993; Atchley, 1993) has recently been amended to bring it into compliance with the 1991 Act of the International Union for the Protection of New Varieties of Plants (UPOV; Greengrass, 1993) Convention. The relationship between UPOV and Plant Variety Protection Act (Seay, 1993; Weimar, 1993) is that UPOV is an international treaty of which the U.S. is a signatory nation. The 1991 Act of UPOV convention required the signatory nations to make their domestic laws compliant with the act. The 1991 Act sets the minimum standards for protecting varieties and plant breeders rights. Intellectual property rights associated with plant varieties are often referred to as "plant breeders rights". Signatory nations could enact more extensive rights (Greengrass, 1993).

The amended Plant Variety Protection Act, which took effect in 1995, brought the U.S. laws in compliance with the act.

The Plant Variety Protection Act allows protection of "novel varieties" that are distinct (clearly distinguishable from other varieties), uniform, and stable (is true breeding over generations). Each variety must have a unique name. The length of protection is now for 20 yr (25 yr for trees and vines; Seay, 1993) and the Plant Variety Protection Certificate is issued in the USA by the USDA. Plants that are protected under the Plant Variety Protection Act can be used as parents to create new cultivars with certain provisions described below. However, the Plant Variety Protection Act does not require that the protected lines be made available for that use. Lines protected by the Plant Variety Protection Act that are sold are readily accessible; however as described below some lines are Plant Variety Protected and kept as trade secrets, which means they are unavailable for any purpose other than what the inventor/owner permits.

The 1970 Plant Variety Protection Act included a number aspects that were changed in the recently amended Plant Variety Protection Act, and these changes help illustrate how intellectual property rights evolve with experience (see Greengrass, 1993) and biology. First, all sexually or tuber-reproduced plant species will be covered under the new law that removes the uncertainty of the previously exempted species and of new crops. Second, whereas previously the intellectual property rights covered "the production of propagating materials for purposes of commercial marketing" (Greengrass, 1993), the intellectual property rights have been extended to all production of all forms of propagating material. This extension removes the possibility of clonal propagation of plants or cell lines without the owners permission and, in theory, would end "farmer saved seed". At this point, it is important to make the distinction between seed that is used in planting the crop and grain that is harvested and sold for nonseed uses [e.g., in wheat (*Triticum aestivum* L.), the grain is sold to mills to produce flour].

Farmer saved seed is the practice where a farmer purchases seed of a cultivar. The seed is planted and grain is harvested. If the harvested grain has good quality, the farmer could save some of that grain as seed for replanting. Because this seed is not commercially marketed, this use is allowed in many countries. Seed companies have been reticent to enforce their intellectual property rights against their customers who obviously had to buy seed before they could produce farmer saved seed. In some areas, it is common for a farmer to purchase small quantities of new seed every year that is used to produce farmer saved seed for planting the next season. Farmer saved seed is common in many countries and UPOV in 1991 requires statutory authority to allow farmer saved seed to continue in those countries where it is the historical practice as is the situation in the USA. It should be noted that a farmer selling grain that is the expected (implicit) use when seed is purchased, is unaffected by Plant Patent Act (Seay, 1993; Weimar, 1993), Plant Variety Protection Act (Seay, 1993; Greengrass, 1993; Atchley, 1993), and utility patents.

A more serious problem with the original Plant Variety Protection Act was that it allowed a farmer to sell some of the produced grain as seed to other farmers. This provision was unique to the USA. These farmer to farmer seed sales coupled with farmer saved seed were devastating to seed companies in areas where high-quality grain/seed could be routinely produced. In these areas, a new cultivar would be rapidly purchased by farmers who would save some of the seed and sell some of the seed. Very quickly, the market for that cultivar's seed would become saturated and the seed company would have little or no revenue from the sale of its cultivar. In effect, the company's cultivar became the company's competitor. The company would receive significant revenues from its cultivar only for the period from introduction to market saturation despite farmers' planting seed every year. Farmer to farmer seed sales are eliminated in the amended Plant Variety Protection Act, though few companies will enforce their intellectual property rights if a customer due to circumstances beyond their control needs to sell some of their saved seed. For example, if a farmer in a country that allows farmer saved seed produces enough seed to plant half his/her farm, but can only plant one-fourth of the farm due to flooding, the seed company would be encouraged to allow the farmer to sell some of that seed. Though farmer to farmer seed sales have been eliminated in the amended Plant Variety Protection Act, this protection is not retroactive to cultivars protected by the Plant Variety Protection Act of 1970.

Recently a U.S. Supreme Court decision has clarified some of the legal aspects of farmer to farmer seed sales. The U.S. Supreme Court ruled in Asgrow Seed Co. v. Winterboer that the total amount of saved seed a farmer could sell to a neighbor(s) for planting would be the amount of seed that the farmer would plant on his/her farm. For example, if a farmer normally planted 200 units of seed on his/her farm, the total amount of saved seed that could be sold or planted on his/her farm would be 200 units. This clarification of the Plant Variety Protection Act of 1970 will help owners to enforce their intellectual property rights because they may be able monitor farmer to farmer seed sales and determine if a violation has occurred.

The third major change is that intellectual property rights were extended to the harvested material (plants or plant parts). This extension is important because seed could be purchased in a country that is a member of UPOV, taken to a nonmember country and reproduced in an unauthorized manner and used in farming. Previously, seed could not be imported into a UPOV country, but resulting plant products could be imported [e.g., cut flowers, cotton (*Gossypium hirsutum* L.)].

The fourth major change is the concept of essentially derived cultivars. Intellectual property rights have emphasized that cultivars must be distinct, uniform, and stable. In the past, distinctness required that the new cultivar be different for at least one character. The character did not have to be important. As such, new cultivars could be quickly obtained by selecting a variant from the cultivar, by mutation (including somaclonal variation from tissue culture) or backcross breeding, or by genetic engineering. The inventor of the new cultivar owned all of the intellectual property rights for the new cultivar. The amended Plant Variety Protection Act recognized that cultivars developed by the above methods could be considered as essentially derived from the initial cultivar. An essentially derived cultivar is one that is predominantly derived from another cultivar, is distinct from the other cultivar, and except for the distinguishable character has the "essential characteristics" of the original cultivar. The inventor of the essentially derived cultivar must obtain the permission of the inventor of the original cultivar to commercialize the essentially derived cultivar. The purpose of this change was to ensure that plant breeders who spent 7 to 12 yr to release a cultivar, did not lose their intellectual property rights to that cultivar by the addition of a single character that can be quickly added.

The essentially derived concept is a clear example of how the law has been affected by changes in and a better understanding of biology. A full discussion of essentially derived cultivars can be obtained from UPOV and is Appendix 2 in Baenziger et al. (1993).

For plant and utility patents and for plant variety protection, there are strict rules on prior public disclosure. For plant patents, the patent must be applied for within 1 yr of the first public description or availability. For plant variety protection, protection must be applied for within 1 yr of the first public seed sale. For utility patents, one year of prior public disclosure is allowed before filing, otherwise the invention would be public knowledge and not new.

The last two major forms of intellectual property rights protection are trade secrets and contracts (Houser, 1993; Seay, 1993; Greengrass, 1993). Trade secrets are proprietary information or property maintained by the owner through confidentiality or employment arrangements. In hybrid or synthetic crops where the inbreds or clones are not sold, the parent lines can be kept as trade secrets and not made available to the public. The difficulty with the trade secret approach to protecting cultivars is that trade secret protection requires owners to be diligent in protecting their trade secrets. Usually diligence is documented by confidentiality agreements for ideas or by physical barriers for tangible property (locked in safes, behind fences, etc.). In hybrid or synthetic production fields, the trade secret inbred or clone cannot be similarly protected. However, recent legal cases have held that inbred lines can be kept as trade secrets (see Pioneer Hi-Bred International vs. Holden Foundation Seed, as cited by Seay, 1993).

In contracts, the owner may restrict the purchaser in how the invention can be used. For example, an owner could sell seed or plants to someone with the stipulation that all of the seed will be planted and the harvested grain or plant parts sold with no further propagation of the seed (thus eliminating farmer saved seed) or plant parts. If the purchaser decided to propagate the plant material, it would be a breach of contract and the laws protecting contracts could be invoked. Of course, other contractual arrangements could be developed, for example restricting the use of plant gametes to prevent the protected material from being used as parents. For many cultivars that contain patented genes, contract law will be used to restrict their unauthorized use as parental germplasm. Material transfer agreements which are commonly used in plasmid and germplasm exchanges are examples of contracts that control the use of the transferred materials (Houser, 1993).

III. INTELLECTUAL PROPERTY RIGHTS AND VALUE ADDED CROPS

It is clear that numerous possibilities exist for protecting value added crops. How are these rights to be used in practice? An example might be helpful. Suppose that Company A identified a gene for improved crop quality such as one that lengthens postharvest storage capabilities (Keith, 1992) or herbicide resistance (Della-Cioppa et al., 1987; Kishore & Tacob, 1987) and inserts that gene into a cultivar that is owned by Company B. Company A may patent the gene and the plant phenotype with a utility patent. Company B, depending on the crop propagation method, could patent the crop with either a utility patent or plant patent or obtain plant variety protection or obtain a patent and plant variety protection. If the cultivar were patented, Company A could use the transformed cultivar for research purposes under an experimental use exception to infringement, but could not commercialize the transformed cultivar unless it had approval from Company B. After the patent protection is over, Company A could transform the cultivar and commercialize the transformed cultivar without the approval of Company B. A critical issue is whether or not the work must be done after the patent has expired on the cultivar. Transforming the cultivar before the patent has expired with the intent to commercialize the transformed cultivar immediately after the cultivar patent has expired may be considered a patent infringement because the patented cultivar was used with commercial intent (hence the activity may not fall under an experimental use exception to infringement) during the patent life. Cultivars that are patented may be protected by material transfer agreements or sales contracts that will prohibit their use for transformation or as parental germplasm. Using these additional property rights will avoid legal arguments as to what constitutes commercial use or is allowed by the experimental use exemption. If the cultivar were protected by plant variety protection, the transformed cultivar would be considered as an essentially derived cultivar and also could not be commercialized unless approval was obtained from Company B. However, Company A could begin the transformation work during the life of the plant variety protection certificate and commercialize the transformed cultivar immediately after the plant variety protection certificate expired. If Company B had decided that the cultivar did not need patenting or plant variety protection, it would have given up its intellectual property rights and the transformed cultivar would be owned by Company A that could request plant variety protection or a patent.

Though there are many possible ways of protecting value added crops, having protection and being protected are not always the same. As mentioned previously, there were many difficulties with the Plant Variety Protection Act of 1970 (farmer saved seed, farmer to farmer sales, cultivars taken to countries that did not enforce intellectual property rights and plant parts returned to the USA or countries with intellectual property rights). Recent legal changes certainly have reduced the potential for intellectual property rights violations; however, enforcement of an owner's intellectual property rights begins with the owner identifying a violation of the owner's intellectual property rights, which may be difficult (particularly when the violation occurs on a customer's farm), and then seeking legal actions. In many cases, this continues to present the owner with the difficult choice of not pursuing his/her intellectual property rights or suing his/her customer for an inappropriate use of a purchased product. The inappropriate use may have previously been allowed, which further complicates the matter.

The above description of the problems with plant intellectual property rights may be alleviated by value added crop marketing. As mentioned previously, value added crops, in order to retain their value, must be segregated in the marketplace. If the segregation is needed for the end-user, the segregation may require production contracts to ensure that adequate supplies are available for the end-user and that over production is avoided, which could reduce market prices and future production. Production contracts may require and enforce intellectual property rights.

For example, it is common for a contract to stipulate that certified seed be planted (verified by seed purchase receipts), that production fields are visited for crop health, quality, and true to type, and that the harvested product be verified for purity visually or with laboratory assays. In addition, the end-user may be vertically integrated and control the seed source as well as purchase the harvested product. If the value added crop is monitored throughout its production and marketing, there would be little opportunity for intellectual property rights violations by ethical businesses. Crops that are treated as commodities where the identity is lost have greater opportunity for intellectual property rights violations as commonly occurred with seed sales where the seed had clear identity and was protected by intellectual property rights when sold in a segregated market, but the grain was considered a commodity. It should be remembered, however, that if a commodity product contains a patented gene, the gene will not lose its identity and can be easily found, even in commodity products. Hence intellectual property rights violations involving specified DNA sequences will be measurable and identifiable.

On-farm use of protected materials will continue to pose problems for the enforcement of intellectual property rights because there may be few ways of identifying violations. While much of this discussion has used the intellectual property rights of seed as an example, on-farm use of value added crops where the end-product may be protected (such as grain with improved nutritional value for animal feed) also would be difficult to monitor and determine if it is being used within contractual arrangements.

IV. BIOLOGICAL MECHANISMS OF PROTECTING INTELLECTUAL PROPERTY RIGHTS

Without question, the most common form of protecting an owner's intellectual property rights is to put the characteristic of value into a hybrid. If a farmer saved seed or sold the saved seed to another farmer, the resulting crop would have lower yields due to inbreeding depression and may be considerably less uniform, which could cause harvest problems. In addition, if the character of interest is controlled by a single dominant gene for which the hybrid was heterozygous, the saved seed will segregate (3 seed having the character, 1 seed not having the character). Even if the valued trait in the hybrid is controlled by genes that must be homozygous (e.g., a trait controlled by recessive genes), the hybrid would be protected from farmer saved seed by inbreeding depression and segregation in the progeny for other traits. If the character is controlled by multiple dominant epistatic genes, even fewer of the farmer saved seed will contain all of the genes needed for character expression. The biological protection afforded by a hybrid is so great that often the parents are protected only as trade secrets. It should be noted that synthetics do not provide similar protection because once the synthetic is formed, it should be genetically at Hardy-Weinberg equilibrium with the exception of effects due to linkage.

Other biological mechanisms that can be used to bolster intellectual property rights include having genes inserted that are controlled or selected for by proprietary chemicals. In this case, the chemical application gives the added value. While intellectual property rights for chemicals may be easier to enforce than for plants, the cost of registering a new chemical for this use is very expensive, and the protection would only last as long as the chemical was under patent protection. Using a pre-existing, registered chemical would mean that the chemical is available for intellectual property rights violation and would have even less time remaining on the patent life. A second mechanism might be to use known genetic instability where the trait is expressed in the plant but would not be readily transmitted to its progeny. An example of genetic instability is the balanced tertiary trisomic system of Ramage (1965) where the male fertility gene was on a tertiary trisomic chromosome that could not be transmitted through the pollen. It was however transmitted through the female.

A final aspect of biology that affects intellectual property rights is the time required to develop an essentially derived cultivar. Unreleased germplasm is usually shared with some form of material transfer agreement, either a contract or a code of ethics statement, that specifically limits the use of the transferred material. When a cultivar is formally released, it usually is to a select group of producers, such as a crop improvement association or a commercial association, that increase the seed before it is sold widely to the public. Public sale of seed is the first opportunity that someone interested in using the germplasm without negotiating with the owner will have access to the seed. If the publicly available seed (assuming the seed was purchased without a contract that prevented the seed and derived plants from being used as a parent) is then used in a rapid backcross program or for genetic engineering, minimally it will take 1 to 2 yr to develop the essentially derived cultivar. However it also could take another 2 to 4 yr to increase the seed for large-scale commercial sales. While the seed is being increased, the essentially derived line can be field tested to ensure it has the expected level of performance. Hence from the time that a line is publicly available to when an essentially derived line would be available is minimally 3 yr, but more like 5 yr. In these 5 yr, plant breeders should develop new cultivars that are superior to the parental cultivar used to created the essentially derived line. In determining the added value of the essentially derived line with the value added character, the total value (yield × higher purchase price) of the essentially derived line would have to be compared not to the parent line, but to the new cultivars (higher yield × purchase price). One potential benefit of intellectual property rights procedures may be that it should be easier to clearly identify the scope of material transfer agreement as many forms of intellectual property rights require a precise description of the intellectual property. Hence to maximize the added value, inventors who wish to add characters to germplasm developed by others would be wise to negotiate access to unreleased germplasm to avoid the problem of cultivar obsolesce. This access will probably occur through the use of material transfer agreements (Houser, 1993).

V. CONCLUSIONS

Intellectual property rights are becoming increasingly important in crop agriculture and are particularly important in value added crops where the added value provides additional economic incentive to obtain and enforce the intellectual property rights. These rights can be most readily enforced when there are commercial sales of protected materials and violations of the intellectual property rights can be identified. On-farm use will continue to pose problems because it will be more difficult to identify violations. Though intellectual property rights provide legal protection, biological protection is still one of the most effective ways to protect value added crops. Finally as breeding methods change, intellectual property rights law and precedence have evolved to encompass these changes.

REFERENCES

Atchley, A.A. 1993. Status of intellectual property rights protection from the viewpoint of the United States Plant Variety Protection Office. p. 89–92 *In* P.S. Baenziger et al. (ed.) Intellectual property rights: Protection of plant materials. CSSA Spec. Publ. 21. ASHS, ASA, CSSA, and SSSA, Madison, WI.

Baenziger, P.S., R.A. Kleese, and R.F. Barnes (ed.) 1993. Intellectual property rights: Protection of plant materials. CSSA Spec. Publ. 21. ASHS, ASA, CSSA, and SSSA, Madison, WI.

Bent, S.A., R.L. Scwaab, D.G. Conlin, and D.D. Jeffery. 1987. Intellectual property rights in biotechnology worldwide. Stockton Press., New York.

Caldwell, B. E.,and J. A. Schillinger (ed.) 1989. Intellectual property rights associated with plants. ASA Spec.Publ. 52. ASA, CSSA, and SSSA, Madison, WI.

Della-Cioppa, G., S.C. Bayer, M.L. Taylor, D.E. Rocheser, and B.K. Klein. 1987. Targeting a herbicide resistant enzyme from *Escherichia coli* to chloroplasts of higher plants. Bio/Technology 5:579–581.

Duvick, D.N. 1993. Goals and expectations of industry for intellectual property rights for plant materials. p. 21–27. *In* P.S. Baenziger et al. (ed.) Intellectual property rights: Protection of plant materials. CSSA Spec. Publ. 21. ASHS, ASA, CSSA, and SSSA, Madison, WI.

Greengrass, B. 1993. Non-U.S. protection procedures and practices—implications for U.S. innovators? p. 41–59. *In* P.S. Baenziger et al. (ed.) Intellectual property rights: Protection of plant materials. CSSA Spec. Publ. 21. ASHS, ASA, CSSA, and SSSA, Madison, WI.

Houser, D.J. 1993. Exemptions under patents and certificates covering plants and comments on material transfer agreements. p. 107–109. *In* P.S. Baenziger et al. (ed.) Intellectual property rights: Protection of plant materials. CSSA Spec. Publ. 21. ASHS, ASA, CSSA, and SSSA, Madison, WI.

Keith, R. 1992. Safety assessment of genetically engineered fruits and vegetables: A case study of the Flavr Savr tomato. CRC Press, Boca Raton, FL.

Kishore, G.M., and H.S. Tacob. 1987. Degradation of glyphosate by *Pseudomonas* sp. GP2982 via a sarcosime intermediate. J. Biol. Chem. 262:64–71.

Ramage, R.T. 1965. Balanced tertiary trisomics for use in hybrid seed production. Crop Sci. 5:177–178.

Seay, N.J. 1993. Intellectual property rights in plants. p. 61–78. *In* P.S. Baenziger et al. (ed.) Intellectual property rights: Protection of plant materials. CSSA Spec. Publ. 21. ASHS, ASA, CSSA, and SSSA, Madison, WI.

Weimar, E.C. 1993. Update on current status of patent fillings involving plants and other related issues. p. 79–87 *In* P.S. Baenziger et al. (ed.) Intellectual property rights: Protection of plant materials. CSSA Spec. Publ. 21. ASHS, ASA, CSSA, and SSSA, Madison, WI.

Williams, Jr., S.B., and K.A. Weber. 1989. Intellectual property protection and plants. p. 91–107. *In* B.E. Caldwell and J.A. Schillinger (ed.) Intellectual property rights associated with plants. ASA Spec. Publ. 52. ASA, CSSA, and SSSA, Madison, WI.

9 Selling Quality in a Commodity Oriented Marketplace

ROLLIN G. SEARS

AgriPro Wheat
Junction City, Kansas

R. F. BRUNS

AgriPro Wheat
Berthoud, Colorado

Throughout the USA, farm size is increasing, farm numbers are declining, and profit margins are getting smaller. Increasingly, producers are looking for alternative methods to sell their products outside of the traditional commodity system. Although that system serves the immense grain industry in the USA well, producers are challenged with the increasingly higher production costs that are largely decoupled from the global grain pricing system. Second, the movement towards an agricultural sector that is less dependent on government farm subsidies has created an atmosphere of change and futuristic thinking in the farm belt of the USA.

A notable change during the 1990s has been an increase in grower based cooperatives throughout the Midwest. Farmer owned cooperatives are not new, but they offer a number of advantages under this difficult environment. Most importantly, cooperatives offer more local control and are part of the community. They also are traditionally "service oriented" that fits well in the current consumer-driven, service-oriented market. Well-managed cooperatives offer producers a strategy to recapture a part of the downstream segments of the market that are currently being captured by the grain and food processing industries in the USA. Recapturing a share of these markets is not without risk. The landscape is littered with failed organizations and individuals that misread market signals or simply didn't understand the grain industry well enough.

Farmer-owned cooperatives that accept a larger share of the processing by maintaining ownership of the grain longer in the production and food processing stages have an opportunity to increase their share of the consumer's food dollar. Maintaining ownership of the grain through milling has been a strategy of the American White Wheat Producers Association based in Atchison, Kansas. This cooperative is interested in selling a specific type of hard white wheat (*Triticum aestivum* L.) flour to niche markets. The North Dakota Pasta Producers have completed a processing factory in Carrington, North Dakota, and are producing their own durum wheat [*T. turgidum* L. var. *durum*]; milling it on site and extruding the semolina into over 50 pasta products that are marketed for retail, private label, food-service, and ingredient markets. Approximately 5000 spring wheat producers in the four state spring wheat region of the USA have banded together to form the United Spring Wheat Producers Association. Their plans include marketing flour produced from their own mill. Many other examples exist of alternative strategies that producers have utilized to add value to their farm products by retaining ownership of their grain and participating in a portion of the food processing steps.

Identity preservation and/or value-added grains are becoming increasingly important as an alternative to commodity agriculture. Value-added simply means adding value, an example would be protein premiums paid for wheat, or contracts bid for high-oil corn (*Zea mays* L.). No attempts are made to isolate specific cultivars or hybrids. Adding a price premium above the commodity price targets a specific quality attribute. High-oil corn, high-oleic sunflowers (*Helianthus anuus* L.), low-linolenic soybeans (*Glycine max* L.) are just a few examples of value-added crops. Value-added high-oil corn has become so common that many now consider it a commodity in its own right. Identity-preserved products require an additional step of isolation, keeping the agricultural product separate and thus unique. Proponents of identity-preserved or value-added production are optimistic that genetic manipulation of quality traits through biotechnology will expand the opportunities in the 21st century. The American Angus Association set a precedent some 20 yr ago by introducing its Certified Angus Beef Program in the USA. Before most other programs, they anticipated the growing consumer demand for food quality and reliability. Today's consumers worldwide are demanding to be more involved in the food production process. They are asking for greater choices, more convenience, and better assurances of quality. New to the formula, and of interest to all members of the food industry, is their willingness to pay for that quality as well.

Contractual production has been, and will increasingly become, a component of either value-added grain or identity-preserved production. Contractual vertical integration does allow for efficient use of resources and provides a more consistent supply of raw material at a negotiated price. There are numerous programs available to corn and soybean growers to supply contracted grains to the feed industry.

Extensive vertically integrated contract production is employed to service and expand the large animal based processing industry in the Midwest. The southern Great Plains has been the center of cattle feeding, utilizing huge amounts of corn (*Zea mays* L.), sorghum (*Sorghum bicolor* L. Moench), and wheat (*Triticum aestivum* L.). This vast area is now rapidly becoming a large hog-feeding region as well. Expanding markets by efficiently feeding animals and thus exporting meat and not grain is a proven way to add value. The People's Republic of China has increased its pork consumption some 60% from 1990 to1995 (Brown, 1998) and is expected to increase consumption of pork, beef, and chicken significantly in the coming decades. An ample supply of high quality feed grains produced close enough to the feeding industry can reduce transportation costs and increase value. Producing feed grains with genetically added traits to improve consumption, or rates of gain or with higher oil or protein concentrations (high oil or large germ corn) developed specifically for hogs or cattle will provide an attractive alternative to simply selling the product to a commodity marketplace. While most of these current programs are perceived to create value at the grower level, some of the most highly integrated programs in the poultry and hog industries have contracts with marginal profits without assurances of future contracts and have been viewed suspiciously by grain producers.

When high-cost transgenic plants with genes expressing specific products are developed, contractual production with the owner of the genetics is expected to be the norm and not the exception. If plants become small factories for the production of specific value-added food or feed products, industrial applications, or pharmaceutical products for either animals or humans, then specific contract production to control both the genetics and the amounts of product produced through identity-preserved, vertical integration appears to be a good choice.

Burchett, (1994) suggested that the nation's certified seed industry is the ideal model for operation of this type of specific, contractual, identity-preserved system. Certified seed growers already are identity-preserved producers of cultivars or inbred lines. They have the experience and capability to produce either small or large quantities of identity-preserved seed. For any of these vertically integrated systems to work effectively, all components in the system must receive a fair profit and have reasonable expectations for the future. Producers, handlers, processors, and consumers all need to share honest information about cost to ensure equal profit incentives for all participants.

Selling quality to the marketplace requires not only valuable and proven intrinsic compositional qualities or processing traits but also a keen understanding of the entire industry. Plant breeding, seed production, product production, milling, product fabrication, and marketing are all complex industry components that view "value" and "quality" differently. In a cooperative setting or an identity-preserved setting the key to success will be communication and sharing of information within all segments. All components of the system must be willing to share both production and economic information to ensure that everyone shares in the profits (Bruns, 1997).

In the commodity-oriented wheat marketplace the definition for quality can be considered as "whatever quality attribute the industry doesn't have at the time". With this definition in place, it has been exceedingly difficult to identify quality targets for which the baking industry would be willing to pay premiums on a contractual basis. In 1 yr protein concentrations may be low, and premiums are paid for higher protein grain. The next year may result in an environment that produces higher protein concentration resulting in premiums that are nonexistent or even inverse. Under these conditions, it is difficult for producers to plan ahead and manage for a specific protein target (Hucl et al., 1998). When the first limiting ingredient in a complex processing step defines quality, and that ingredient may change from year to year, it is exceedingly difficult for producers to target and supply that market with hopes of a dependable price reward. In these cases, vertical integration, where all participants share a portion of the risk and reward should be considered.

Plant breeding programs also are handicapped because the small-scale methods utilized to evaluate quality are often inadequate to estimate the subtleties needed to enhance value. Wheat products are exposed to a range of commercial processes resulting in products such as white pan bread, buns, tortillas, noodles, and crackers. Most of the small-scale methods were designed to predict general quality parameters necessary for the commodity market. Specific end-use tests, which could provide targeted estimates of use and value, are needed for an array of specific processes. In the oilseed and other refining and processing industries, many utilization and processing methods are considered proprietorial and not generally made available to plant breeders.

The various participants also view quality very differently in the marketing system. Grain producers view quality largely as their return on investment, i.e., dollars returned per hectare of grain harvested. Yield and sound grain are the important determinants. A number of other tangibles such as high protein, high oil, altered amino acid contents, or improved brewing or baking qualities can be added to the model, but most importantly, the cultivar or hybrid has to be a competitive performer in the field. Small increases in price can't offset the penalty of lower grain yields.

Grain handlers base their view of quality upon efficiency and special handling criteria. Grain handling is a volume business more than a price business. Value-added crops must provide extra value in efficiency of handling and transportation.

Millers value quality in terms of efficiency and uniqueness. Grains with easier throughput rates in the mill, higher extraction percentages, and unique traits that provide increased marketing value are all valuable considerations. They also are interested in uniformity, which can describe many different criteria in a grain crop. However, to a miller, uniformity principally means uniform throughput and production of a product that is predictable and consistent for the next stage user, whether it is a baker, exporter, or a feedlot operator.

The end-users as a group are oriented towards consistency and uniqueness with less emphasis on efficiencies. If cultivars or hybrids could be developed that were uniform for important quality traits across all environments the end-user industry would be ecstatic.

One universal definition of value-added quality combining all the above views could be: "whatever quality attributes the industry needs and doesn't have at the time". The major risk for the supplier is accurately estimating how long the "needs" part of the definition will remain of interest and economically viable in the marketplace. For the wheat industry, the "needs" component has been variable and inconsistent. Trying to target a specific-value added trait or group of traits that ultimately will pay for the development costs has been frustrating. Equally frustrating for the developer of technology has been the rapid change in consumer preference and the changing processes required for matching those demands by the consumer. Plant breeding is a long-range process that works best with well described, stable quality targets that over years plant breeders can select for and gradually improve. Rapid shifts in food processing steps and changes in product properties are all extremely difficult to deal with in a plant breeding program.

Whatever the industry doesn't have and needs also is a definition that applies to the commodity mentality we have for grains in the USA, but doesn't fit well in the developing identity-preserved arena. Because of the lack of testing equipment discussed above and the time frame involved in development, there is an inherent risk in developing plants that produce products outside of the qualities defined by the traditional markets. The development of waxy wheat is an example of this predicament. Product value and potential applications are relatively unknown. Research and development is required in both areas to establish value. Returns on investment then must cover not only the development of waxy wheat, but also the new potential applications. In this case, value of the waxy grain must be sufficient to cover both costs. This same dilemma exists for new concepts like increased lysine, increased methionine, decreased P (phytic acid), improved starch digestibility, and reduced oligosaccharides.

To be successful, a "quality" oriented project or a value-added project must create enough value somewhere in the chain of industries to overcome the inherent added cost incurred in the development phases. The following are examples of costs associated with quality: technology costs, identity-preserved production costs, transportation and storage costs, special handling costs, market development costs and nongrade disposition costs. Historically, the value of quality developed from the technology phase has been overemphasized. The cost of implementation and the return on investment often have been underestimated. In addition, how the market measures or estimates quality often has been obscure and difficult to determine. Aside from direct measurements of protein, oil or carbohydrates, assessing baking quality is difficult without baking the flour into bread. Assessing brewing quality of malting barley (*Hordeum vulgare* L.) is difficult without making the beer. Feed quality models don't always reflect performance in the feedlot.

A wheat miller or baker's most frequent question when asked about paying a premium for a specific value-added grain almost always is: " How do I know I'm getting the quality I'm paying extra for?" With wheat as well as other grains, their concern is justified; most of the variability in bread baking quality is caused by environmental effects (Peterson et al.,1992; Baenziger et al., 1985). Year to year variation in quality far exceeds the differences among cultivars or locations within a particular year (Graybosch et al., 1996). While grain buyers can rely on identity-preserved systems to provide a margin of improvement through better management practices, sounder grain, and cultivar or hybrid specifications, the role of environment always will be a major factor to contend with. Additional methods of verification will increase the willingness of grain buyers to pay a premium for quality. Adoption of near infrared (NIR) technology as that critical missing link to justify identity-preserved production is getting closer. Rapid assessment of end use quality at the initial receiving point for many important traits like protein, specific protein qualities, oil content, oil composition, fiber characteristics, mineral spectrums, etc., will help to establish immediate value above the baseline commodity price.

If the wheat grain buyers have access to data on predicted flour milling performance and dough mixing properties, as well as basic information on dockage, kernel size, moisture and protein concentration, they can make more accurate estimations of value. Correlation coefficients have been improving every year as technical improvements are made in NIR devices. A wider range of near infrared light spectrum can now be visualized, and a range of quality estimates determined on whole grain samples in less than 60 s. Recent work in Canada (Pawlinsky & Williams, 1998) has established that farinograph absorption, development and stability times, and mixing tolerance on whole grain samples can be estimated, apart from the influence of protein concentration, by NIR instruments. Use of NIR to estimate both protein quantity and quality in combination with identity-preserved production and improved management could provide enough critical economic information to establish value. At that point, a premium for better quality grains is much more likely.

The Single Kernel Characterization System (SKCS) has been commercially available since 1994. Originally developed to assist in the classification of hard and soft wheat and mixtures of each class, the instrument is now finding acceptance in providing information on wheat milling quality, tempering processes and milling flour yields (Psotka, 1997). Applications with other grains are proving to be useful. These types of equipment, capable of measuring potential end use quality at the initial point where grain enters the market will provide identity-preserved grain a substantial advantage in the future. Many of these technologies will have applications across grains for specific needs.

Ways to identify the true value of quality can be grouped into three categories: (i) improved efficiencies in the system, (ii) improvements to existing technologies, and (iii) new technologies. While often overlooked, efficiency in a system can create more value than many of the new technologies currently being developed. An example of improved efficiency might be higher production yields, improved conditional quality (grain soundness), lowered transportation costs and lower storage and segregation costs. Any inefficiency in a system can consume the value of a new technology very quickly. For example, lower yields are extremely difficult to overcome in the grain industry, as are increased costs associated with handling.

Improvements in existing technologies usually are associated with incremental improvements to milling, compositional traits, ingredient replacement and consistency/uniformity traits. This arena has received, and will continue to receive, considerable attention. The American White Wheat Producers Association grows hard white wheat that has a slight advantage in milling yields, especially when whole-wheat flour is being made. High-oil corn or high protein soybean or wheat cultivars create value by incrementally improving existing systems. Overcoming environmental variation in these types of compositional traits often is more difficult than initially estimated. In wheat for example, 80% of the variation in bread-baking quality can be attributed to environmental effects and only 20% to genetic effects (Peterson et al., 1997). Grower groups can address this type of variation more easily than plant breeding. Proper management with a specific end product in mind, such as, protein concentration or large uniform kernels can be addressed at the production level. Linking growers together, with a focused and targeted management plan, to produce for a specific industry, has been the successful equation for several farm-based cooperatives.

Ingredient replacement and consistency/uniformity aspects have been difficult to address from a technology or plant-breeding standpoint. To date, variability in compositional traits caused by environmental fluctuation has limited the impact these two components have had on the quality-value-added equation. Certainly in the future, replacing wax esters with modified soybean oil, coconut (*Cocos nucifera* L.) oil replacements through the use of modified lauric acids, and gelatin replacements through modified corn starch systems will become realities in the marketplace.

New technologies offer the industry unique opportunities in efficiencies and improved products. Specific "tailor-made" genetics created by emerging techniques in molecular biology offer hope that specific value-added projects could be addressed. Many of these types of opportunities are discussed in the preceeding chapters. Successful implementation will rely, in part, on how well plant breeders and processors can share information in the early phases of product development. Plant breeders and molecular biologists need insight and information from the food industry to match existing or new genetic diversity with new value added products. If plant breeders continue to evaluate experimental material utilizing conventional small-scale tests developed to predict "commodity quality" before searching for value-added applications they'll continue to find frustration.

The most successful future commercialization of improved quality will involve projects that combine multiple points of added value in the system, while at the same time minimizing the inefficiencies created. Accomplishing this will require a keen understanding and awareness of the entire industry. The lack of such comprehensive understanding has been a major reason why successful ventures thus far have been limited.

REFERENCES

Baenziger, P.S., R.L. Clements, M.S. McIntosh, Y.T. Yamazaki, T.M. Starling, D.J. Sammons, and J.W. Johnson. 1986. Effect of cultivar, environment, and their interaction and stability analyses on milling and baking quality of soft red winter wheat. Crop Sci. 22:5–8.

Brown, L.R., and J. Mitchell. 1998. Building a new economy. p. 168–187. *In* L.Starke (ed.) State of the world 1998. W.W. Norton & Company, Inc., New York.

Bruns, R.F. 1997. Taking quality to the market place. p. 213–215. *In* J.L. Steele and O.K. Chung (ed.) Proc. Int. Wheat Qual. Conf. Grain Indust. Alliance, Manhattan, KS.

Burchett, L. 1994. Why is it so difficult to launch IP programs? Seed & Crops Industry. October, p. 4–11.

Hucl, P, K. Preston, P. Williams, and S. McColl. 1998. Spring wheat cultivar variation for protein concentration and grade in Saskatchewan . p. 100–109. *In* D.B. Fowler et al. (ed.) Wheat protein production and marketing. Univ. Ext. Press, Univ. Saskatchewan, Saskatoon, Canada.

Graybosch, R.A., C.J. Peterson, D.R. Shelton, and P.S. Baenziger. 1996. Genotypic and environmental modification of wheat flour protein composition in relation to end-use quality. Crop Sci. 36:296–300.

Pawlinsky, T., and P. Williams. 1998. Use of near-infrared reflectance spectroscopy to predict wheat bread-baking functionality in whole kernels. p. 201–206. *In* D.B. Fowler et al. (ed.) Wheat protein production and marketing. Univ. Ext. Press, Univ. Saskatchewan, Saskatoon, Canada.

Peterson, C.J., R.A. Graybosch, P.S. Baenziger, and A.W. Grombacher. 1992. Genotype and environment effects on quality characteristics of hard red winter wheat. Crop Sci. 32:98–103.

Peterson, C.J., and R.G. Sears. 1997. Breeding for quality. p 175–184. *In* J.L. Steele and O.K. Chung (ed.) Proc. Int. Wheat Qual. Conf. Grain Indust. Alliance, Manhattan, KS.

Psotka, J. 1997. Single kernel characterization system. p. 129–140. *In* J.L. Steele and O.K. Chung (ed.) Proc. Int. Wheat Qual. Conf. Grain Indust. Alliance, Manhattan, KS.

SUBJECT INDEX